SWEDEN

SK	Skåne	VR	Värmland
BL	Blekinge		Dalarna
HA	Halland		Gästrikland
SM	Småland		Hälsingland
ÖL	Öland		Medelpad
GO	Gotland		Härjedalen
GS	Gotska Sandön	JÄ	Jämtland
ÖG	Östergötland	ÅN	Ångermanland
VG	Västergötland	VB	Västerbotten
BO	Bohuslän	NB	Norrbotten
DS	Dalsland	ÅS	Åsele Lappmark
NÄ	Närke	LY	Lycksele Lappmark
SÖ	Södermanland	PI	Pite Lappmark
UP	Uppland	LU	Lule Lappmark
VS	Västmanland	TO	Torne Lappmark

NORWAY

Ø	Østfold	HO	Hordaland
AK	Akershus	SF	Sogn og Fjordane
HE	Hedmark	MR	Møre og Romsdal
O	Oppland	ST	Sør-Trøndelag
B	Buskerud	NT	Nord-Trøndelag
VE	Vestfold	Ns	southern Nordland
TE	Telemark	Nn	northern Nordland
AA	Aust-Agder	TR	Troms
VA	Vest-Agder	F	Finnmark
R	Rogaland		

n northern s southern ø eastern v western y outer i inner

FINLAND

Al	Alandia	KB	Karelia borealis
Ab	Regio aboensis	Om	Ostrobottnia media
N	Nylandia	Ok	Ostrobottnia kajanensis
Ka	Karelia australis	ObS	Ostrobottnia borealis, S part
St	Satakunta	ObN	Ostrobottnia borealis, N part
Ta	Tavastia australis	Ks	Kuusamo
Sa	Savonia australis	LkW	Lapponia kemensis, W part
Oa	Ostrobottnia australis	LkE	Lapponia kemensis, E part
Tb	Tavastia borealis	Li	Lapponia inarensis
Sb	Savonia borealis	Le	Lapponia enontekiensis

RUSSIA

Vib Regio Viburgensis   Kr Karelia rossica   Lr Lapponia rossica

The Sepsidae (Diptera) of Europe

FAUNA ENTOMOLOGICA SCANDINAVICA
Volume 37 2002

The Sepsidae (Diptera) of Europe

by

Adrian C. Pont and Rudolf Meier

Brill
Leiden · Boston · Köln

This book is printed on acid-free paper.

PRINTED IN THE NETHERLANDS

Editor-in-chief: N.P. Kristensen
Desk editor: V. Michelsen

ISBN 90 04 12477 2
ISSN 0106-8377

Library of Congress Cataloging-in-Publication Data

The Library of Congress Cataloging-in-Publication Data is also available.

Die Deutsche Bibliothek - CIP-Einheitsaufname

Pont, Adrian C.; Meier, R.:
The Sepsidae (Diptera) of Europe / by Adrian C. Pont and R. Meier. –
Leiden ; New York ; Köln : Brill, 2002
(Fauna entomologica Scandinavica ; Vol. 37)
ISBN 90–04–12477–2

Cover illustration:
Sepsis punctum (Fabricius, 1794)

Authors' addresses:

Adrian C. Pont
Oxford University Museum of Natural History
Parks Road
Oxford OX1 3PW
UK

Rudolf Meier
Zoological Museum
University of Copenhagen
Universitetsparken 15
DK-2100 Copenhagen Ø
Denmark

Contents

Abstract

This volume deals with the European species of the family Sepsidae, a small family of acalyptrate flies. The taxonomy, biology and faunistics of all the European species are revised with emphasis on the Fennoscandian species, and the detailed distribution of the species known from Fennoscandia, the adjacent areas of Russia, Germany and Great Britain is tabulated in a catalogue. Keys are given to generic level for eggs, larvae and adults, and to species for the adults. Descriptions of the adults are provided for genera and species, together with diagnostic notes on the immature stages. The distribution and biology of the European species is summarized, and the results of extensive type-studies are presented. Illustrations are given of the male fore legs and genitalia of all the European species, and also of other characters of diagnostic importance for the egg, larval and adult stages.

Nine genera and 44 species are dealt with, and of these 8 genera and 34 species are known from Fennoscandia. 23 species are listed from Denmark, 26 from Norway, 32 from Sweden and 31 from Finland. One new species is described, *Themira ringdahli* Pont, sp. n., from North Sweden (Abisko). New life-histories and immature stages are described for 3 species from 2 genera. 23 lectotypes are designated. The following 17 new synonyms are established: *Themira* Robineau-Desvoidy, 1830 (syn: *Annamira* Ozerov, 1999; *Nadezhdamira* Ozerov, 1999); *Themira annulipes* (Meigen, 1826) (syn: *brunicosa* Robineau-Desvoidy, 1830; *marginata* Hoffmeister, 1844); *Themira lucida* (Stæger, 1844) (syn: *roseni* Becker, 1915); *Saltella nigripes* Robineau-Desvoidy, 1830 (syn: *gagatea* Robineau-Desvoidy, 1830); *Saltella sphondylii* (Schrank, 1803) (syn: *melanocephala* Drapiez, 1820; *longipes* Robineau-Desvoidy, 1830; *nana* Robineau-Desvoidy, 1830); *Sepsis flavimana* Meigen, 1826 (syn: *pygmaea* Robineau-Desvoidy, 1830); *Sepsis fulgens* Meigen, 1826 (syn: *vibrans* Harris, 1780 [preocc.]; *minimus* Harris, 1780 [preocc.]; *fulgida* Robineau-Desvoidy, 1830; *nitida* Robineau-Desvoidy, 1830; *atripes* Robineau-Desvoidy, 1830); *Sepsis punctum* (Fabricius, 1794) (syn: *cornuta* Meigen, 1826; *rufocincta* Hoffmeister, 1844).

Introduction

The Sepsidae is a relatively small, morphologically and ecologically uniform family of the Sciomyzoidea in the acalyptrate series of Diptera. Most species are rather slender, ant-like flies (plate 1: *Sepsis punctum*), 2-6 mm in length, with spherical head, mainly dark bodies, petiolate abdomen, and relatively few setae or hairs. They are active flies, with a characteristic rapid gait accompanied by wing-waving. *Orygma luctuosum* Meigen, a seashore species, is an exception to this and is flattened, robust, richly setose (plate 2), rather secretive in habits and disinclined to fly.

There is no vernacular English name for this family. The American 'black scavenger flies' is hardly likely to catch on in Europe. In Swedish they are 'svängflugor'. In German they were called 'Heckenfliegen' by Wiedemann (1830: 467), but are now usually known as 'Schwingfliegen' (e.g. Püchel, 1992: 417).

Family SEPSIDAE Walker, 1833

Diagnosis: Adult Sepsidae can be distinguished from other families of the acalyptrate Diptera by the presence of one or more setae on the hind margin of the posterior spiracle (Fig. 3). The palpi are minute, except in *Orygma* (Fig. 12a).

The family-group synonyms have been listed by Sabrosky (1999: 280-281). To these we can add Sepseidae Naswaty (1960: 75), an emendation based on the declension of the Greek noun 'sepsis' (= 'putrefaction'), genitive 'sepseos' and thus with the stem 'sepse-'.

The Sepsidae is a well-founded monophyletic family, and its sister-group is probably the Neotropical family Ropalomeridae. At present it comprises some 283 described species in 30 genera (A. L. Ozerov, pers. comm.). Many new species have recently been described from the hitherto poorly known Afrotropical and Oriental faunas (Iwasa, 1982a, b, 1984a, 1986, 1987, 1989, 1994, 1996; Iwasa & Jayasekera, 1994; Iwasa & Tewari, 1990, 1994; Iwasa & Thapa, 1994; Iwasa et al., 1991; Ozerov, 1990, 1991b, 1992a, c, 1993b, 1994b, 1996, 1997b, 1998a, c, 1999b, c, f, 2000a, b, d; Zuska, 1968). Nearctic *Sepsis* have been revised by Ozerov (2000e). The Neotropical fauna has recently been revised (Ozerov, 1992b, 1993a, 1994a, 1998c; Silva, 1990, 1992a, b, 1993, 1995). Our knowledge of sepsid alpha taxonomy is thus well above average for a family of the acalyptrate series. However, the largest genus of Sepsidae, *Sepsis,* remains to be revised and there is a need for an up-to-date species catalogue and/or for a new world monograph replacing Duda's monograph of 1926 (Duda, 1926a, b) and Hennig's (1949) treatment of the Palaearctic fauna for 'Die Fliegen der Palaearktischen Region'. The key to British species by Pont (1979) will enable the common European species to be identified. Additional regional treatments are available for Russia (Ozerov, 1999e), Japan (Iwasa, 1980, 1981, 1984b, 1985, 1995), and Korea (Suh & Kwon, 1994; Iwasa et al., 1994). The catalogue of Palaearctic species by Zuska & Pont (1984) is rapidly being superseded by the work of Ozerov (e.g. 1986b, 1993d, 1997a).

The present work is based on the literature published before January 2001 and we are dealing with the European fauna, which comprises 9 genera with 44 species. Of these, 8 genera and 34 species are known from Fennoscandia. Europe as defined here extends from the North Cape in the north, to the countries bordering the Mediterranean Sea (Spain, Italy, Greece) in the south; and the British Isles in the west, to approximately longitude 40°E in the east, though we have also included the genus *Susanomira* from the Caucasus Mts. Iceland and the Faroes are included; the Macaronesian islands, North Africa, the Middle East and Russia east of the Ukraine are excluded. A few species from peripheral areas are mentioned at the conclusion of this work. Many Sepsidae are expanding their ranges, and it is likely that there will be further species that are able to establish themselves on mainland Europe, from the south and from the east.

Sources of material and acknowledgements

The following abbreviations are used for the location of types and of other material specifically mentioned in the text:

ASSP – Institute of Zoology, Academy of Sciences, St Petersburg, Russia.

BAA – Naturhistorisches Museum der Benediktiner-Abtei, Admont, Austria.

BMNH – The Natural History Museum (formerly British Museum (Natural History)), London, U.K.

DEI – Deutsches Entomologisches Institut, Eberswalde, Germany.

IRSNB – Institut Royal des Sciences Naturelles de Belgique, Brussels, Belgium.

LSL – Linnean Society of London, U.K.

MGAB – Muzeul de Istoria Naturala 'Grigore Antipa', Bucharest, Romania.

MHNL – Muséum d'Histoire Naturelle, Lille, France.

MHNN – Muséum d'Histoire Naturelle, Neuchâtel, Switzerland.

MNHNP – Muséum National d'Histoire Naturelle, Paris, France.

MNM – Magyar Természettudomanyi Muzeum Allattara, Budapest, Hungary.

MZLS – Museo Zoologica 'La Specola', Florence, Italy.

NMI – National Museum of Ireland, Dublin, Republic of Ireland.

NMV – National Museum of Victoria, Abbotsford, Australia.

NMW – Naturhistorisches Museum, Vienna, Austria.

NRS – Naturhistoriska Riksmuseum, Stockholm, Sweden.

NSMT – National Science Museum, Tokyo, Honshu, Japan.

OUAVM – Obihiro University of Agriculture and Veterinary Medicine, Obihiro, Hokkaido, Japan.

SMN – Staatliches Museum für Naturkunde, Stuttgart, Germany.

TAU – Tel-Aviv University, Tel-Aviv, Israel.

UMO – Oxford University Museum of Natural History, Oxford, U.K.

UQC – Private collection of Ullmar Qvick, Eskilstuna, Sweden.

USNM – National Museum of Natural History (formerly United States National Museum), Washington, U.S.A.

UZMH – University Zoological Museum, Helsinki, Finland.

VMUT – Vitenskapsmuseet, Universitetet i Trondheim, Trondheim, Norway.

ZATM – Zoologisk Avdeling, Tromsø Museum, Tromsø, Norway.

ZMAN – Zoologisch Museum, Universiteit van Amsterdam, Amsterdam, The Netherlands.

ZML – Museum of Zoology, University of Lund, Lund, Sweden.

ZMHU – Zoologisches Museum der Humboldt-Universität, Berlin, Germany.

ZMUB – Zoological Museum, University of Bergen, Bergen, Norway.

ZMUC – Zoological Museum, University of Copenhagen, Copenhagen, Denmark.

ZMUM – Zoological Museum, Moscow State University, Moscow, Russia.

ZMUO – Zoologisk Museum, Universitetet i Oslo, Oslo, Norway.

For the loan of material and/or for information, or for otherwise making types and other material available, the authors thank:

Dr Paul Ardö, Dr M. Barták, Professor Rudolf Bährmann, Dr Irina Brake, John Chainey, Dr Ruth Contreras-Lichtenberg, Dr Richard Dahl, Dr Roy Danielsson, Dr R. Defrétin, Dr Michael Fitton, Dr Patrick Grootaert, Dr Lita Greve Jensen, Terje Jonassen, Dr Herman de Jong, Dr Karin Kiontke, Dr Theowald van Leeuwen, Dr Bernhard Lindeberg, Dr Leif Lyneborg, the late Dr Loïc Matile, Dr George C. McGavin, Dr Verner Michelsen, the late Dr Günter Morge, Dr Arturs Neboiss, Dr Tore R. Nielsen, Dr James O'Connor, Dr Thomas Pape, the late Per Inge Persson, Frank Püchel, Ullmar Qvick, Dr J. E. Raastad, Dr Knut Rognes, Dr Hubert Schumann, Dr J. O. Solem, Dr G. E. E. Søli, the late Dr Árpad Soós, Dr Bo Svensson,

Dr Aurelia Ursu, Dr Pekka Vilkamaa, Dr Richard Vockeroth, Dr Hella Wendt, Dr Joachim Ziegler, and Dr Alexey Zinoviev.

ACP's studies and fieldwork on Sepsidae, including study-visits to the museums of Paris, Lille, Brussels, Vienna, Copenhagen, Lund, Stockholm, and extensive fieldwork in Scandinavia and western Europe, were supported from 1964 to 1989 during his employment at the then British Museum (Natural History), London. A visit to Helsinki in 1988 was supported by the Scandinavian Science Press and to Berlin in 1999 by the British Council. A NERC Grant for Taxonomic Publications, administered by the Linnean Society of London, supported the preparation of new illustrations for this work. Fieldwork over many years at Abisko, North Sweden, has been facilitated by the Royal Swedish Academy of Sciences. RM acknowledges help from Monika Varain and financial support from the Zoological Museum Copenhagen and the Danish Natural Science Research Council (grant Nr. 28808). The habitus illustrations of *Sepsis punctum* and *Orygma luctuosum* were prepared by Francis L. Fawcett. Finally, a generous grant from the Carlsberg Foundation (Ans. 990474/20-856) has enabled this work to be completed.

Dr Leif Lyneborg provided extensive information on Danish localities, and Dr Knut Rognes supplied a modernised list of Zetterstedt and Siebke Norwegian localities. An earlier version of the keys was tested by Dr Jean-Paul Haenni, who provided several constructive comments.

At the outset of this project, much helpful advice was given by Dr Jan Zuska, whose pioneering studies on Sepsidae are here gratefully acknowledged. Dr Mitsuhiro Iwasa has advised on a number of problems. Dr Lorenzo Munari has been a constant source of ideas, and his work on the *Sepsis punctum*-group has been particularly appreciated. Finally, Dr Andrew Ozerov has at all times generously shared his unrivalled knowledge of world Sepsidae, has provided reprints and much invaluable material for study, and has critically read this manuscript.

To all these institutions and individuals we offer our deep thanks and appreciation.

Morphology of adults

(European species only)

The morphological terms used in this volume follow McAlpine (1981: 9-63), with few modifications.

Head (Fig. 1) roundish or long-oval, dichoptic in both sexes. Eyes large or very small (*Orygma*, Fig. 12). Fronto-orbital plates and frontal vitta distinct; ocellar tubercle and frontal triangle not developed. Setae few in number (in pairs): 1 outcurved postocellar (absent: *Ortalischema*), 1 incurved inner vertical, 1 outcurved outer verti-

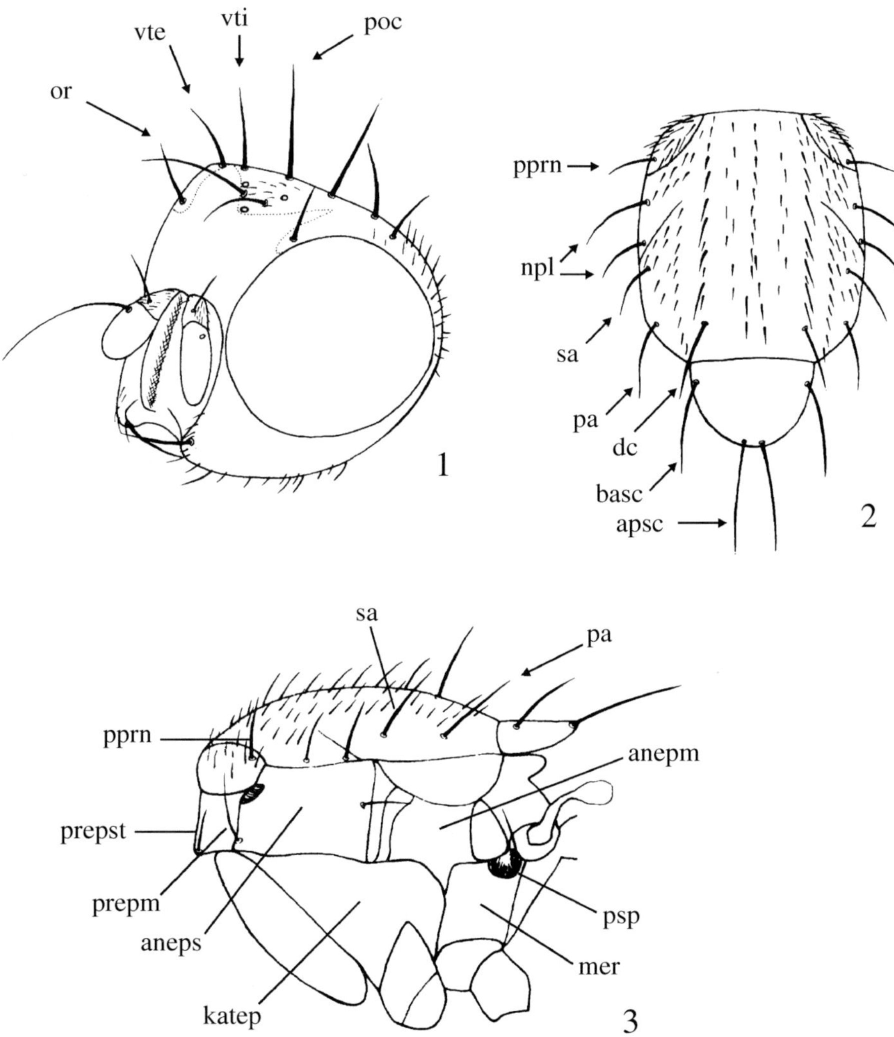

Figs 1-3. *Saltella sphondylii* (Schrank). 1: head, oblique lateral view; 2: scutum and scutellum; 3: pleura. (anepm – anepimeron, aneps – anepisternum, apsc – apical scutellar seta, basc – basal scutellar seta, dc – dorsocentral seta, katep – katepisternum, mer – meron, npl – notopleural setae, or – orbital seta, pa – postalar seta, poc – postocellar seta, pprn – postpronotal seta, prepm – proepimeral seta, prepst – proepisternal seta, psp – posterior spiracle, sa – supraalar seta, vte – outer vertical seta, vti – inner vertical seta.)

cal (vestigial or absent: *Themira*, *Ortalischema*, *Nemopoda speiseri*), 1 strong ocellar, 1 orbital (weak or absent: *Nemopoda*, *Ortalischema*, *Sepsis*), 0 frontal (2: *Orygma*). Antennal flagellomere long-oval or round; arista bare. Parafacial linear and gena narrow (except *Orygma*). Vibrissal seta weak to absent. Proboscis of the usual lapping type. Palpus minute (except *Orygma*, Fig. 12a).

Thorax with postpronotal lobes distinct. Scutum (Fig. 2) matt or shining, without a dusted pattern (except weakly in *Orygma*). Setae (in pairs): acrostichals absent; 1-2 prescutellar dorsocentrals (3 well spaced pairs in *Orygma*), rarely preceded by some additional short setae; 0-1 postpronotals, 2 notopleurals, 1 supraalar (absent: *Themira*), 0 intraalar, 1 postalar (minute: *Sepsis*; 2: *Orygma*). Pleura (Fig. 3) wholly to partly dusted, to mainly to wholly shining. Prosternum setulose or bare. Anepimeron bare (except *Orygma*); katepisternum without setae (1: *Orygma*); meron bare. Postcoxal bridge absent. Scutellum short and wide, with usually only the apical pair of setae strong, but in *Saltella* very enlarged, flattened, with strong basal and apical pairs of setae (Fig. 2); basal setae well-developed also in *Orygma*, *Ortalischema* (Fig. 15) and, moderately so, *Zuskamira*.

Legs rather long and slender, with few setae (short and robust, with dense setulae and stout setae in *Orygma*, plate 1). ♂ fore legs unmodified in *Orygma*, but in all other genera with stout setulae, spinules, spines, pegs, teeth, tubercles, protuberances or excisions (e.g. Figs 74, 140, 404). ♂ mid and hind legs also sometimes with secondary ornamentation. ♂ hind tibia in several genera (especially *Themira* and *Meroplius*) with a large dorsal field-like scent organ ('osmeterium').

Wings clear, tip of vein R_{2+3} with a smoky area (*Nemopoda*) or a dark oval spot (*Sepsis*, Fig. 14). Basal section of vein R (stem-vein) setulose ventrally, other veins bare except costa. No costal breaks. Basal medial and basal radial cells distinct (fused in *Saltella*: Fig. 13, and *Sepsis niveipennis*). Vein A_1 (anal vein) long (*Orygma*) or short, not reaching wing-margin; vein A_2 absent. Upper calypter with a fringe of long hairs; lower calypter absent.

Abdomen shining to subshining, dusted only in *Orygma*. Consisting of four visible segments (Figs 54, 79); segment 1+2 narrow and constricted in *Sepsis*; tergite 7+8 external in ♂ (Fig. 79). Tergites with few ground-setulae; with marginal setae on at least tergites 4 and 5. Sternites 1-3 in ♂, 1-5 in ♀, narrow. ♂ sternite 4 absent in *Sepsis*. ♂ sternite 4, or sternite 5 (*Sepsis*), often highly modified and enlarged, with tufts or brushes of setae (e.g. Fig. 81). ♂ tergite 6 absent (present: *Orygma*, *Ortalischema*); sternite 6 absent. ♂ tergite 7+8 present[1], with 2 spiracles in lower corner (tergite 6 and spiracle 6 in McAlpine, 1989: 1454), sometimes a partial suture visible, indicating the original two tergites, in which case the spiracles are in the anterior (tergite 7) element; sternite 7+8 present (absent: *Zuskamira*; present as the original 2 sternites: *Orygma*) (sternite 7 in McAlpine, 1989: 1454). ♂ hypopygium (e.g. Figs 57-58) with a pair of surstyli, completely or incompletely fused to epandrium, sometimes asymmetrical; cercal plate small, usually poorly developed; aedeagal complex with hypandrium elongated and band-like, sperm pump (ejaculatory apodeme) very large, tapering in *Orygma* and *Ortalischema*, otherwise shaped like a dumb-bell, gonopod (praegonite) and paramere (postgonite) present, phallus long with an expanded, asymmetrical, heavily sclerotised, spinulose ventral plate, and distiphallus terminating in a brush-like group of setae. ♀ ovipositor short (Figs 4-7); segments 6 and 7 with complete tergites and sternites, spiracles 6 and 7 in the tergites; segment 8, when not invaginated, produced ventrally and posteriorly, displacing the hypoproct and cerci dorsad, the tergal element usually forming a pair of narrow bars that do not quite touch apically, the sternal element absent or also forming a pair of narrow plates that do not touch apically; cerci small, round, with 1-2 strong setae; epiproct present or absent; hypoproct not always distinguishable, bare or setulose. Two spermathecae, spherical, strongly sclerotised, one larger than the other.

Variation. – Some species exhibit considerable variation, which has led to a proliferation of specific and infra-specific names. Colour variation is

[1] Rather than engage in the controversies surrounding the homology and nomenclature of the male postabdominal segments, we have adopted a neutral, topographical approach, which we believe will be more easily followed by the users of this volume. Thus, tergite 5 is followed by tergite 6 (when present) and then by tergite 7+8 (which has been variously called synsternite 7+8, syntergite 6+8, syntergosternite 7+8); and sternite 5 is followed not by sternite 6 (which is absent in Sepsidae) but by sternite 7+8 (usually represented by a narrow band).

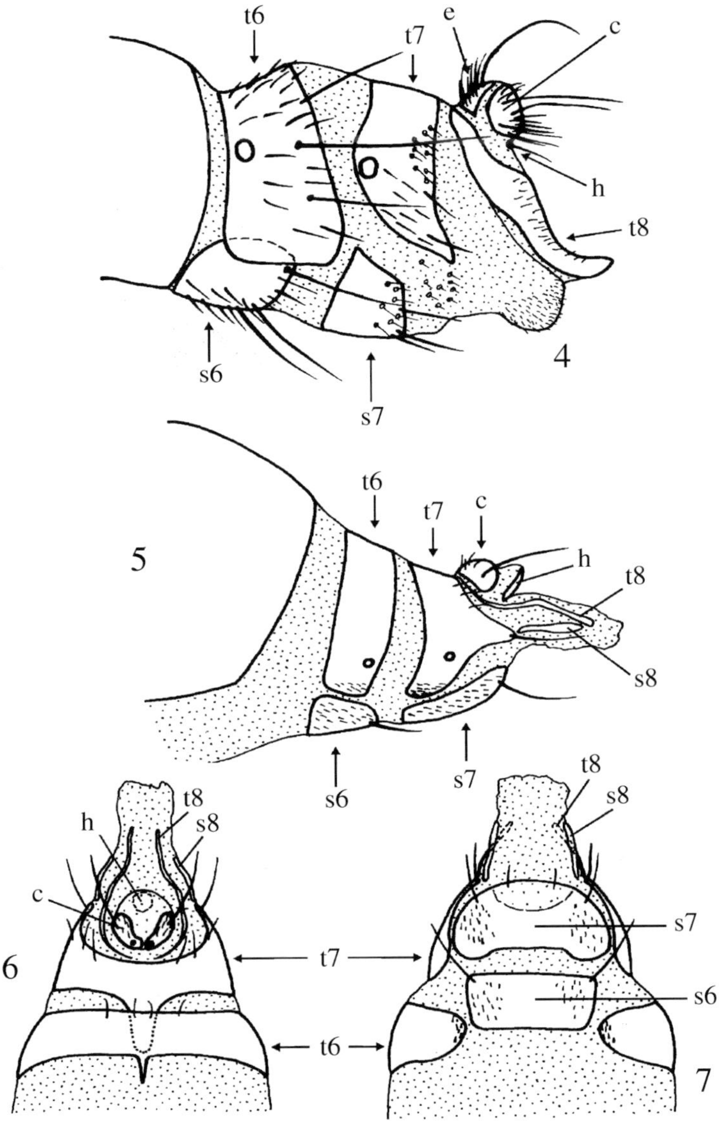

Figs 4-7. ♀ ovipositor. 4: *Orygma luctuosum* Meigen, lateral view; 5-7: *Sepsis punctum* (Fabricius); 5: lateral view; 6: dorsal view; 7: ventral view. (c – cercus, e – epiproct, h – hypoproct, s – sternite, t – tergite.)

usually sex-specific, with the males exhibiting yellow areas on the head (parafacial, face, gena), the pleura, the base of the abdomen, the genitalia, and of course the legs. In Europe, *Sepsis punctum*, *Sepsis thoracica*, *Sepsis neocynipsea*, *Sepsis biflexuosa* and *Saltella sphondylii* provide the most striking examples, and such pale individuals are also larger, more robust, and with more strongly developed setae and other ornamentation. They occur alongside males that are smaller, darker and less robust and that resemble the females in these respects. In at least some *Sepsis* species this variation has a significant genetic component (Reusch & Blanckenhorn, 1998: 114), but body size is also strongly correlated to the availability of food for the larvae. Zerbe (1993: 25) was able to breed *Sepsis punctum* siblings whose dry weight varied by a factor of 10.

Morphology of immature stages

(European species only)

Eggs

The description of the egg stage is based on material of all the European genera. Lists of species for which the eggs are known are included under each genus in the descriptions below.

Egg long-oval, dorsally flattened (Figs 8-10), excluding egg filament 0.51-1.0 mm long (contra Hinton, 1960, for *Saltella*); *Ortalischema* without respiratory filaments (Fig. 8); *Orygma* with two that have distinctly clubbed ends (Fig. 9); remaining species with a single filament 0.35-3.75 mm in length, thus ranging from little more than half to more than four times the length of the egg, but normally 2.5-3.5 times as long (Fig. 10); filament normally with a single central canal (Figs 21, 23-25; except for *Orygma*: Fig. 22) and in most species plastron in contact with surrounding air through aeropyles of various sizes (Figs 26-30); main body of egg in at least some genera with a polygonal pattern of ridges; chorion then porous within polygons; egg without a dorsal ridge or operculum (except for *Ortalischema*: Fig. 8); micropyle cup-shaped only in *Orygma* and *Ortalischema* (Fig. 28), in all other genera micropyle at least partially fused to dorsal side of filament base (Fig. 29); hatching line normally indicated by a weak fold of varying length around micropyle (Figs 9-10, 28-29); fold only conspicuous in *Ortalischema* (Fig. 8).

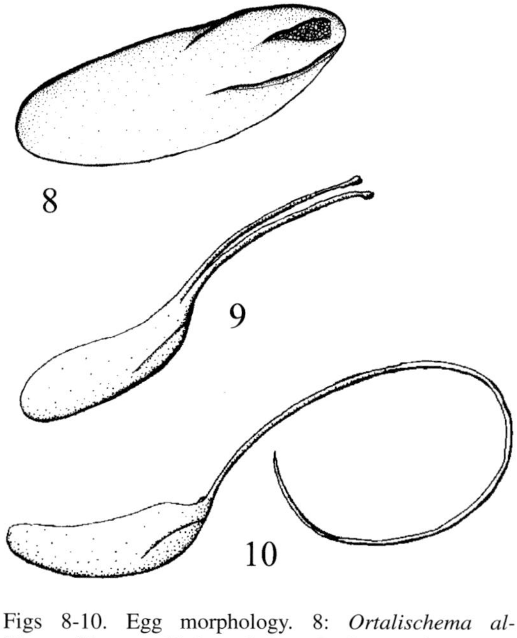

Figs 8-10. Egg morphology. 8: *Ortalischema albitarse* (Zetterstedt), lateral view; 9: *Orygma luctuosum* Meigen, lateral view; 10: *Sepsis* sp., lateral view.

Larvae

The following description applies to third instars only and is restricted to characters visible using light-microscopy, in the course of which it is necessary to make slide preparations in some cases. The description is based on material of all the European genera. Lists of species for which the larvae are known are included under each genus in the descriptions below.

Body elongate, cylindrical and of typical cyclorrhaphan shape, but posterior end in preserved specimens often conspicuously bulbous and usually bearing numerous, sometimes dark-pigmented spines (Figs 45-46, 60, 189); cuticle of most segments white to yellow-white, shining and normally smooth (matt and hirsute in *Nemopoda*; Figs 43, 231-232); cephalic lobes normally with cirri (Figs 59, 72, 86, 164, 186, 199, 228, 304) and facial mask usually with oral ridges, rarely only with combs or cirri (Fig. 199: e.g. some *Themira*); segmentation: number of body compartments is 11, 3 thoracic, 7 abdominal segments and the anal division, a composite compartment consisting of the four last abdominal segments which are fused (Figs 34, 39, 43-44, 51-53); thoracic segments and the first five abdominal segments on anteroventral side with several (6-12) rows of small spinules; abdominal segments 1-7 (Fig. 11; *Orygma*, *Ortalischema*, *Themira leachi*) or 1-6 (Fig. 168; remaining sepsids) with creeping welts on anteroventral side of segments, usually consisting of two (Figs 45, 189), rarely three (Figs 46, 60) long rows of reclinate spines spanning the entire width of segment; most species with a few additional median spines on segment preceding the one carrying the creeping welt (Fig 45); anal division ending in stigmatophores and in most genera with a pair of

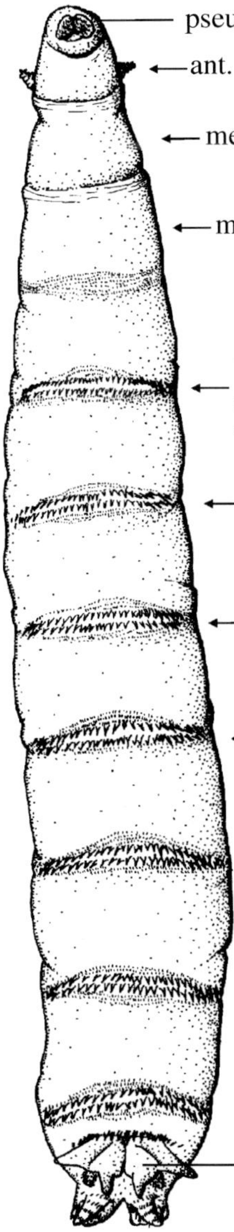

Fig. 11. Larval habitus of *Themira leachi* (Meigen). (abd. – abdominal, ant. – anterior.)

subanal, lateral (absent in *Orygma*: Fig. 38, *Ortalischema*: Fig. 32) and ventral tubercles (absent in *Ortalischema*: Fig. 32; full complement of tubercles in Fig. 53). *Saltella* with bifurcate ventral tubercles (Fig. 34); stigmatophores dorsally also with a tubercle (Fig. 38; absent in *Ortalischema*: Fig. 32); surface of the anal division usually covered with numerous spines and/or spinules (Figs 34, 43-44, 51-53); anal opening surrounded by a usually large anal pad consisting of 1-2 pairs of lobes (Figs 34, 51-53, except for *Orygma*); anal division ventrally normally with small to large preanal tubercles and the unpaired anal tuft, shape of anal division varying from wider than long (Fig. 52) to several times as long as wide (Fig. 396); dorsal side of anal division often with three double rows of spines aligned along three grooves (Figs 43-44, 357-358). Cephaloskeleton dark brown to black; mouthhooks well developed, paired, separate, normally invariable within genera; dental sclerites present or absent, when present then paired and either separate or fused to mandibular sclerites; labial sclerites present or absent (Figs 35-36, 47-48); if present then paired or fused into a V-shaped sclerite (Figs 35-36); epistomal sclerite present, usually longer than wide, rarely vice versa, with or without posterior bars (Figs 136-138); intermediate sclerite H-shaped with a bridge at middle (Fig. 138) or anterior third (Fig. 136), anteriorly with (Fig. 138) or without a flat plate (Fig. 136); parastomal bars of variable length, from not reaching bridge (Figs 47, 137) to extending beyond bridge of intermediate sclerite (Fig. 48); posterior ends fused to basal sclerite; basal sclerite with dorsal cornua shorter and normally narrower than ventral cornua; basal sclerite longer than high.

Anterior spiracles consisting of 3-40 lobes, spiracle situated laterally or dorsolaterally on prothorax; spiracular openings at the end of finger-like projections; lobes in a fan- (Figs 37, 73), tree- (Figs 41-42, 307, 308), or bush-like arrangement (Fig. 40).

Posterior spiracles situated on stigmatophores which are separated at bases and distally often dark-pigmented (Fig. 53); spiracular openings usually almost equal in length and shape (Figs 50, 61, 190, 301), except in some *Sepsis* where two are long and the third short (Figs 49, 398-399, 402); openings usually situated on a single oval, sclerotized spiracular plate, except *Sepsis* where one opening leaves the plate (see Figs 49, 398-399, 402), often being situated on a small separate extension; arrangement of openings on plate either circular around the rim (Fig. 61), triangular (Fig. 50), whorl-like (Figs 49, 398-399, 402), or two spiracular openings parallel with the third one perpendicular to the other two (Figs 301, 167, 190).

Classification and phylogeny

The phylogenetic relationships of the family within the Sciomyzoidea have been discussed most recently by Griffiths (1972: 184-185) and McAlpine (1989: 1453-1454, fig. 116.4). The sister-group of the Sepsidae is probably the Ropalomeridae, a largely Neotropical American family of some 8 genera and 30 species. McAlpine (l.c.) listed 8 synapomorphies for the Ropalomeridae + Sepsidae in terms of the ground-plan of the Sciomyzoidea, and these were discussed in some detail by Meier (1995b: 100-101) who pointed out that all but two of them also occur in other Sciomyzoidea and/or more distantly related acalyptrates. The most important support for the sister-group relationship between the Ropalomeridae and Sepsidae lies in the presence of a group of setae on the margin of the posterior spiracle and, possibly, the complete fusion of the ♂ aedeagal apodeme (phallapodeme) to the hypandrium.

The monophyly of the Sepsidae has never seriously been in doubt. Griffiths (1972: 185) listed six apomorphous character-states that characterise the Sepsidae in terms of the ground-plan of the Ropalomeridae + Sepsidae. One of these (♂ hypandrium and aedeagal apodeme completely fused) was mentioned above, and was listed by McAlpine (1989: 1453) as being part of the ground-plan of the Ropalomeridae + Sepsidae, and to the remaining five McAlpine (1989: 1454) added a further two characters. These have been carefully evaluated by Meier (1995b: 100), who listed three as being undisputed:

(1) ♂ spiracle 6 situated in tergite 6 (our spiracle 7 in tergite 7+8), a condition unknown in any other Sciomyzoidea.

(2) ♂ sternite 7 not delimited, either lost or fused with sternite 8 (the resulting sclerite termed by us sternite 7+8), such fusion in other Sciomyzoidea, when it occurs, being indicated by a suture.

(3) Vein A_1 shortened and not reaching the wing-margin (but secondarily complete in some *Paratoxopoda*).

The other apomorphies discussed by Griffiths and McAlpine are less convincing, as subsequent work has shown that they are prone to homoplasy within the Sciomyzoidea or more widely in the acalyptrates. However, one could add to the list a peculiarity of the female postabdomen which was first noted by Hennig (1937: 35). Segment 8, when not invaginated, is produced ventrally and posteriorly, thus displacing the hypoproct and cerci dorsad. This constellation is apparently unknown from any other sciomyzoid family.

More problematic is the inclusion of the genus *Protorygma* Hennig in the Sepsidae, which was described from Baltic amber (Eocene) (Hennig, 1965: 77, figs 78-88). Hennig assigned *Protorygma* to the family Sepsidae and considered it to be the sister-group of the remainder of the family. We have not seen this specimen, but McAlpine (l.c.) discussed its characters and concluded that 'on the basis of these considerations its placement in the Sepsidae should probably be reassessed'. We believe that, on the evidence given by McAlpine, a better case can be made for regarding *Protorygma* as part of the stem-group of the Ropalomeridae or even of the Ropalomeridae + Sepsidae. The apomorphous precoxal bridge certainly excludes it from the Sepsidae (precoxal bridge absent plesiomorphously), but is shared with the Ropalomeridae. The plesiomorphously setulose vein R_1 and ♀ abdominal spiracle 6 in the membrane also exclude it from at least the Sepsidae-crowngroup (apomorphously R_1 bare and ♀ spiracles 6 and 7 in tergites 6 and 7 respectively).

There have been few attempts at reconstructing the phylogenetic relationships within the family. Hennig (1949) is a fruitful source of ideas, but the only available cladograms are by Meier (1995b: fig. 76; 1996: figs 471-472) which are based wholly (1995b) or mainly (1996) on larval characters (58 out of 85 characters). The sister-group relationship and recognition of the monotypic Orygmatinae and the Sepsinae is uncontroversial, and is well-supported by larval and adult characters (Meier, 1996: 139). Evidence has also been accumulating for resolving the basal relationships within the Sepsinae. The position of *Ortalischema* as sister-group of all the remaining Sepsinae is now well-supported by egg, larval and adult characters (Meier, 1996: 139; see also below, p. 41). Thereafter there is some conflict between larval and adult data. Considering just the European genera, Meier's (1995b: fig. 76) cladogram shows *Themira* following *Ortalischema*, but with the species placed by some authors in the subgenus (or genus) *Enicita* as outside and more derived than the remaining *Themira*; then *Susanomira* + *Saltella* as a single clade; *Nemopoda* (including *Pseudonemopoda*); *Meroplius* (including *Xenosepsis*); and *Sepsis* (including *Sepsidimorpha* and *Australosepsis*).

We also carried out a cladistic analysis of the adults of European Sepsidae, fully aware that this is not a representative cross-section of world Sepsidae, in order to investigate the patterns based solely on adult data. 58 characters were used, from external morphology and ♂♀ postabdomina and genitalia; and 16 European species representing all 9 genera, to which were added *Willistoniella pleuropunctata* (Wiedemann) (Ropalomeridae, outgroup), *Toxopoda* sp. (Thailand), *Platytoxopoda* sp. (Ivory Coast) and *Paratoxopoda amonane* Vanschuytbroek (Zimbabwe). A consensus tree based on these data showed *Orygma* as the basal clade (Orygmatinae and Sepsinae as sister-groups) and, within the Sepsinae, the clade *Toxopoda* + *Platytoxopoda* + *Paratoxopoda* as sister-group of the other Sepsinae. The tree for these remaining Sepsinae was poorly resolved and consisted of a polytomy of nine clades: *Ortalischema*, *Nemopoda nitidula*, *Nemopoda speiseri*, *Saltella*, *Meroplius fukuharai*. *Meroplius minutus*, *Themira*, *Zuskamira* + *Susanomira*, and *Sepsis*. This tree only served to confirm the monophyly of *Themira* as defined in this paper. Individual parsimonious trees based on adult characters alone diverged considerably from trees based on characters of the immature stages.

Combining adult, egg, larval and biological data gave a tree with few divergences from that of Meier (1996: figs 472-473): the new tree showed *Nemopoda* and *Sepsis* as sister-groups, and *Saltella* and *Meroplius* as sister-groups, as follows: *Ortalischema* + (*Toxopoda*-group + (*Themira* + {(*Zuskamira* + *Susanomira*) + [(*Saltella* + *Meroplius*) + (*Nemopoda* + *Sepsis*)]})).

These results are preliminary in the extreme, and serve only to underline the difficulties in arriving at a well-founded classification of sepsid genera. More species and more characters need to be incorporated into the analysis, and more detailed morphological studies need to be made of the ♂ genitalia, especially the aedeagal complex, and of the ♀ ovipositor.

Biology

The Sepsidae are among the most common acalyptrate flies in Europe, and after years of neglect the recent period has seen a flurry of new studies on the natural history of many species. Initially, the mating behaviour was at the centre of interest and the family proved interesting in many respects. Elaborate courtship behaviour before mounting seems rare in Sepsidae, although there are some exceptions: Hammer (1941: 83) described a 'dance' for *Sepsis duplicata* and *Saltella sphondylii*, Mangan (1976: 1026) mentioned courtship involving leg displays in *Themira annulipes, T. putris,* and *T. minor,* and the authors of this volume have seen a courtship similar to *Themira* in *Paratoxopoda amonane*, *Toxopoda* sp. and *Ortalischema albitarse*. However, as a rule males of most species will skip tedious pre-mounting courtship and will attempt to mount any sepsid fly within jumping reach, in the course of which the males of many species defend small temporary territories on the breeding substrate (e.g. Mangan, 1976: 1026; Zerbe, 1993: 26-27; 1995: 121). If the potential partner turns out to be of the wrong species, sex, or an unreceptive female, the male will dismount more or less quickly and wait for the next opportunity. Opportunities are usually plentiful, since the two sexes of most species cluster around the patchily distributed breeding substrates. Even when the female is receptive, she will still often attempt to dislodge the male, possibly to test the male for fitness (Allen & Simmons, 1996: 737, 740).

Receptivity in females is apparently related to ovarian status. In almost all the species studied, females will only copulate when the ovaries are devoid of mature oocytes (Parker, 1972a: 194; Eberhard, 1999: 129-130; Schulz, 1999: 42, 59-60, 67, table 2.2). This might be related to an apparently early origin within the Sepsidae of a large spermatophore which would interfere with oviposition if copulation were immediately to precede oviposition (Schulz, 1999: 39-40). Such a spermatophore is known for a relatively basal sepsid like *Paratoxopoda depilis* (Kotrba, 1996: 307) as well as for the derived *Archisepsis*, and *Sepsis* (Eberhard & Huber, 1998: 237-238; Schulz, 1999: 71). However, there are apparently also species which copulate when gravid. This has been observed repeatedly in, for example, *Saltella sphondylii* (Schulz, 1999: 65, 68-69). As spermatophore production requires long copulation times, its short copulation duration suggests that a spermatophore might be lacking in *Saltella* (Schulz, 1999: 72). *Saltella sphondylii* is also the only known sepsid with a postcopulatory guarding phase, i.e. the pair first copulates and then the male remains mounted on the female during oviposition, which takes place immediately afterwards (Foster, 1967: 1035-1036; Schulz, 1999: 65, 68-69, table 2.2).

The precopulatory guarding of many derived sepsids in the *Sepsis*-group (Hennig, 1949: 10) is much better studied than this postcopulatory guarding of *Saltella sphondylii*. Parker (1972a; 1972b) was the first to describe the bewildering sequence of events for two *Sepsis* species, *S. cynipsea* and *S. punctum*. The male will mount the female on the breeding substrate. She will then proceed with oviposition before the pair leaves the substrate in order to copulate in the surrounding vegetation. In *Sepsis cynipsea* 60% of the precopula pairs will eventually copulate (Parker, 1972a: 203), whereas in *S. punctum* the rate is considerably higher (Zerbe, 1993: 38; 1995: 124). This kind of precopulatory guarding is exceedingly rare in Insecta (Schulz, 1999: 88), but widespread in Sepsidae. It is now known to occur within the following sepsid genera: *Meroplius*, *Archisepsis*, *Palaeosepsis*, *Microsepsis*, *Parapalaeosepsis*, *Dicranosepsis*, *Sepsis* (Schulz, 1999: 94; RM, pers. obs.). However, not all the species in these genera exhibit it, and there are even species in which some populations guard and others do not (e.g. *Sepsis punctum*, *S. biflexuosa*, *S. neocynipsea*; Schulz, 1999: 95, 115). This population-level variability allowed Schulz to study the genetics of precopulatory guarding in *Sepsis punctum*, although the viability of hybrids generated from crosses involving European and North American flies was somewhat reduced (Schulz, 1999: 92, 147). She found that the behaviour has a genetic basis, is polygenic, and is inherited from the mother, whereby at least some of the genes are located on the x-chromosome (Schulz, 1999: 154-156; at least 4 chromosomes were seen in slide preparations: Schulz, 1999: 45). The evolutionary lability of precopulatory guarding is best expressed by its intraspecific variability in several species. It makes it difficult to answer the question of whether guarding evolved multiple times, or only once and was then reduced in some species and/or populations (see Schulz, 1999: 100, fig. 3.2).

But what is the adaptive function of precopulatory guarding? From a female perspective, the precopulatory guarding phase might have the advantage of allowing the screening of males prior to copulation (Allen & Simmons, 1996: 737, 740; Schulz, 1999: 97, 114). Moreover, females would probably rarely be allowed to oviposit without being constantly attacked by males (Zerbe, 1993: 38-39; 1995: 128). From a male perspective, the behaviour is thought to be a response to male-male competition due to a strongly male-biased sex ratio at the breeding sites (Parker, 1972b; Zerbe, 1993: 38; 1995: 128; Schulz, 1999: 112, 129). The male-male competition hypothesis is indirectly supported by the observation that only the males of *Sepsis punctum* populations with precopulatory guarding are significantly larger than the females (Schulz, 1999: 113, 121).

The precopulatory guarding of different species differs with regard to the success rate of male 'take-overs' of females that are already mounted by other males. In a species like *S. punctum*, take-over attempts are often successful, can be very violent, and are usually won by the large male (Zerbe, 1993: 30; 1995: 121, 124). Small males thus have almost no chance of copulating with large females at attractive rendezvous sites. In other species such as *S. cynipsea*, take-overs are rarely successful and the attacking males are not very insistent. However, Ward (1983: 75) nevertheless found that large males have an advantage over small males.

The copulation posture of most sepsids also deserves some discussion. It is of great importance for understanding the male fore leg morphology. The often strongly modified fore femora and fore tibiae pinch the base of the female wings, thereby attaching the male securely to the female prior to genital contact. Once copulation starts, the wing-hold is released in some but not all species (e.g. Draeger, 1997; Schulz, 1999: 102). This copulation posture is derived within the Sepsidae. It is absent in the most basal sepsid, *Orygma luctuosum*, as well as in all the outgroup families of the Sciomyzoidea for which the copulation posture has been described. The importance of male fore leg design for mating has recently been confirmed by Allen & Simmons (1996: 740) who have shown that the more symmetrical the legs of *Sepsis cynipsea* males are the higher is their mating success. During copulation, at least *Sepsis punctum* males actively interfere, most likely by removal, with the sperm that might have previously been deposited in their mates (Schulz, 1999: 35). This explains the overwhelming last-male sperm precedence in this species (Schulz, 1999: 31).

The normally large male claspers are also subject to sexual selection (Eberhard & Pereira, 1996: 43). After copulation, the male and female often have difficulty in separating and it may require a 180° torsion of the male or female to free the complex aedeagus from the female reproductive system (Parker, 1972a: 194-195; Zerbe, 1993: 30; 1995: 124; Eberhard & Pereira, 1996: 56). Other features of Sepsidae are almost certainly also related to mating behaviour, although their functions are very poorly known. Males of many species in different genera have so-called osmeteria on their hind tibia. These consist of a more or less extensive elongate patch of hollow hairs that are connected to gland cells beneath. The function of the osmeterium and its effect on female behaviour remains unknown. Similarly common and mysterious, but not necessarily found in the species with osmeteria, are the sometimes elaborate hair brushes on sternites 4 or 5 of males (see Hennig, 1949: plates I & II; and the illustrations in the present volume). Their function during mating remains largely unstudied (but see Eberhard in prep.). Lastly, an even more conspicuous behaviour, the well known wing-waving of Sepsidae, is probably not related to mating. Both sexes engage in this activity (mostly in derived genera) and it is also observed when flies of only one gender are present. Its function remains one of the biggest mysteries of sepsid biology. Laboulbène (1896: 111-112) pointed out that adults also wave the halteres as well as the wings, though these two actions are not synchronised.

Sepsidae are the only known family of Diptera with a gland found in both sexes that is presumably used for defence. This 'Dufour gland' (Dufour, 1851) is responsible for the sweet odour that has been noted repeatedly in the literature (e.g. Mik, 1886: 279; see also Pont, 1987a: 310-311). One or two gland reservoirs are attached to the rectum and contain a mixture of solvents and defensive chemicals (Zenk, 1990: 26-31). The gland is known to be absent in *Orygma*, *Ortalischema, Platytoxopoda* and *Toxopoda*. Its defensive function has been shown in feeding experiments involving ants (Zenk, 1990: 9-11) and orb-weaving spiders (Bristowe, 1979: 17-19). It is also suggested by the observation that predators such as *Scathophaga* and *Vespula* dislike feeding on Sepsidae and often abandon attacks (Parker, 1972b: 248-249). Sepsids nevertheless do occasionally fall prey to other insects (Tuomikoski, 1952: 171; Parker, 1972b: 248-249). The scent from the gland is often also associated with large swarms of mostly *Sepsis fulgens*, which may contain up to an estimated 50,000-100,000 individuals. The swarms can be stationary for up to three months (Pont, 1987a: 307) and may re-occur at the same sites in consecutive years (Pont, 1987a: 307-308). Swarming is thought to be related to hibernation (Pont, 1987a: 313-314).

Adult sepsids are dependent on the availability of two kinds of foods. The females of at least most species are incapable of producing eggs without first feeding on decaying organic material for several days. What is much less known is that sepsids provided only with dung will still die rather rapidly. The second kind of food required is carbohydrates (e.g. Zerbe, 1995: 117). They are mostly obtained from flowers and there are numerous reports of flower feeding for a large number of species (Table 1). Other known sources of carbohydrates are rotting fruits, tree sap, plant secretions, aphid honeydew, and the secretions of dripping ergots growing on wild grasses (Minder, 1963: 388; Draeger & Scheurer, 1999: 69-71, table 1).

The breeding substrates of Sepsidae are very varied. The only common theme is that they have to be decaying organic material, and, based on the presence of a filter apparatus in the cephalopharyngeal skeleton, bacteria are the most likely food item. Breeding substrates range from a wide variety of bird and mammal droppings, through vertebrate carrion or decaying vegetation, to such odd materials as slime moulds and dead insects or snails (see Table 1). However, it is safe to say that cow dung is the one substrate that is suitable for the largest number of species (see Table 1). Waterfowl dung is also very important for a number of *Themira* species and horse dung for genera like *Ortalischema* and *Zuskamira*. The most basal sepsid, *Orygma luctuosum*, breeds exclusively on decaying brown algae. When the breeding substrates are mapped on to a phylogenetic tree for the Sepsidae (Meier, 1996: 136), it is noticeable that the 'basal' taxa utilize a variety of resources

Table 1. Natural history of European Sepsidae: substrates for breeding and adult feeding and list of life history stages known to hibernate. a = adult collecting record, l = larvae found in substrate, r = reared from, c = culture maintained in the laboratory.

Species	cow dung	horse dung	pig dung	sheep dung	human faeces	waterfowl dung	other dung	sewage	vertebrate carrion	other substrates	hibernating stage	carbo-hydrates
Orygma luctuosum										brown algae: a,l,r,c		
Ortalischema albitarse		a,c									p (diapause)	flowers
Zuskamira inexpectata		a,c									p (diapause)	
Themira annulipes	a,c					a,l,c		a				fruits
Themira arctica									a			
Themira biloba	c					a						
Themira gracilis	a	a,c							a		p (diapause)	fruits
Themira leachi	a				l					decaying vegetation: l	l	
Themira lucida						a,c			a			fruits
Themira minor	a,r,c	a	a,r					a,r		garbage: l	l	flowers
Themira nigricornis	a,l				r		chicken dung: l		a	fungi: a; birch sap: a	l	
Themira pusilla							muscrat: a			algae: a		

Table 1. (Continued.)

Species	cow dung	horse dung	pig dung	sheep dung	human faeces	waterfowl dung	other dung	sewage	vertebrate carrion	other substrates	hibernating stage	carbo-hydrates
Themira putris	a,l,c		a		a			a,l,r	a	garbage: a; decaying vegetation: l,r; bird nest: r	l,p	flowers, honey dew
Themira simplicipes										decaying vegetation: l		
Themira superba						a,c						
Susanomira caucasica	a,l	a										
Saltella nigripes	a,l											
Saltella sphondylii	a,l,r,c	r	r				yak: a				p	flowers
Nemopoda nitidula	r,c				r				r,c	fungus: r; slime mould: r; rotting log: r; dead snail: r; invertebrate carrion: l	l	
Nemopoda pectinulata	l				a					bird nest: r		
Nemopoda speiseri									l			flowers

Table 1. (Continued.)

Species	cow dung	horse dung	pig dung	sheep dung	human faeces	waterfowl dung	other dung	sewage	vertebrate carrion	other substrates	hibernating stage	carbo-hydrates
Meroplius fukuharai			a				chicken dung: a, c		a,l	garbage: a		
Meroplius minutus	a,c		a,l		a,l		rabbit: l		a,l	garbage: a; rotting vegetation: a; rotting fungi		flowers
Sepsis barbata	a						buffalo: a					
Sepsis biflexuosa	a,l,r,c	a	a				buffalo: a,r; small mammal: a	a,l				flowers
Sepsis cynipsea	a,l,r,c	a		a,r							p,a	flowers
Sepsis duplicata	a,l,r,c						red deer: a				a	
Sepsis flavimana	a,l,r,c	a			l		red deer: a					flowers
Sepsis fulgens	a,l,r,c	a,l	a,r	l			chicken: l	a	a	compost: a; silage: a	l,a	flowers
Sepsis lateralis	r,c	a	a				camel: a; chicken: r; ostrich: c; dog: r			dead locust: r		
Sepsis luteipes	r					a						flowers

Table 1. (Continued.)

Species	cow dung	horse dung	pig dung	sheep dung	human faeces	waterfowl dung	other dung	sewage	vertebrate carrion	other substrates	hibernating stage	carbo-hydrates
Sepsis neocynipsea	a,r,c	a,l	a,l	a	a,l		small mammal: a; mink and fox: a, yak: a		a			plant secretion
Sepsis nigripes	r					a						
Sepsis niveipennis	a,l,r,c											
Sepsis orthocnemis	a,l,r,c	a		a,l	a				a		a	flowers
Sepsis punctum	a,l,r,c	a	l	l	a		small mammal: a; fox: a; red deer: a; rabbit dung: l; dog: l; brown bear: l	l	l	rotting vegetation: a, l	l,a	flowers
Sepsis thoracica	a;l,c	a	a	a,r	a		donkey: a; yak: a; buffalo: r				l	
Sepsis violacea	a,l,r,c		l		l,r		rabbit: l; bear: l	a,l		rotting vegetation: r	l,p,a	

while the species in the very speciose *Sepsis*-group, comprising the vast majority of species, are largely specialists on ruminant dung. Accordingly, the sepsid fauna is richest in Africa and rather small in South America. It is also certain that the advent of cattle farming by humans has had a positive influence on at least population sizes and has modified species distributions.

There are few species of Sepsidae that are capable of breeding in more or less the full range of substrates (e.g. *Nemopoda nitidula, Sepsis lateralis, S. fulgens*). Most species have rather specialized breeding requirements, which explains the large number of sympatric species in some genera (Minder, 1963: 385; Zuska, 1970: 605; Bährmann, 1993a: 105-106). There are different kinds of specialization. Some species will only use a very narrow range of substrates (Table 1). Others will only use dung of a certain age (see comments under the various *Sepsis* species). This specialization for dung of a certain age may explain the otherwise puzzling coexistence of as many 8-9 species of Sepsidae on cow dung in Central Europe. Yet other species can breed in a variety of substrates, but they have to be found in a particular habitat. For example, *Nemopoda nitidula* and *Sepsis punctum* are generalists with regard to substrate choice, but both species are only found in rather damp and cool places (Zerbe, 1995: 118). Humidity, temperature and wind velocity are in general important factors determining sepsid activity (Modassir, 1993b: 243-246). Extreme conditions are avoided and the same species may be active all day during a cold season, while activity is restricted to the morning during summer (Modassir, 1993b: 244-245). Flies generally prefer to settle on darkly coloured substrates rather than white or yellow surfaces (Modassir, 1993a: 83).

Sepsid eggs are invariably laid into the substrate with the main body of the egg submerged and, if present, the respiratory filament(s) remaining on the substrate surface. The filament is usually considerably longer than the egg itself (Meier, 1995b: 427, but see *Themira annulipes*) and is only absent in *Ortalischema*. In most other Sepsidae, there is only one such filament, the only known exception being *Orygma* with two. The filament is thought to channel air to the embryo and to prevent drowning in cases of substrate inundation (Hinton, 1961: 145). In many species eggs are laid singly (e.g. Zerbe, 1995: 123). However, in a number of species such as *Saltella sphondylii*, groups of 10-20 eggs are oviposited with the filaments emerging from a single point in the dung and being arranged rosette-like around the oviposition site. It is thought that the filaments are thereby able to trap an air bubble when the substrate is flooded (Ozerov, 1986c: 46). At least *Sepsis duplicata, S. flavimana* and *Saltella sphondylii* are known to enter the tunnels of dung beetles for oviposition (Hammer, 1941: 113; Minder, 1963: 387). Egg development times are largely dependent on temperature and species (e.g. Modassir, 1995: 281), and range from 6 hours in *Sepsis lateralis* (Hafez, 1939: 320) to 2 days in *Orygma luctuosum* (Egglishaw, 1960: 122). In several species, a single egg may be incubated within the female and the larva can hatch immediately upon oviposition (Kotrba & Meier, in prep.).

The larvae of most Sepsidae are rather slow moving within the substrate. Exceptions include the larvae of *Ortalischema albitarse*, which move very rapidly in horse dung. Most larvae in dense substrates like cow dung remain in contact with the air via their large hind spiracles, which are withdrawn into the substrate when the larvae are disturbed. During feeding on dung, the mouthhooks of the larvae can often be seen scraping substrate from the surface of the dung into the mouth opening, i.e. the anterior and posterior body of the larva are exposed while the mid-section is buried in the substrate. However, the position of the larvae in dung varies from species to species (see Bishop et al., 1996: 211). Pupariation can be initiated by a larva having reached its maximum size. Alternatively, the substrate may be used up, which is frequently the case for fly larvae living in ephemeral substrates. Most sepsids can thus vary tremendously in body size. For example, *Sepsis punctum* can vary by a factor of 10 in dry weight (Zerbe, 1993: 25). This strongly influences the morphology (e.g. Zerbe, 1995: 118-120) and is partially to blame for the large number of synonyms in existence for some species. Maggots ready to pupariate are characterized by a usually bright-yellow fat body which appears one to several days before pupariation. The larvae stop feeding, empty the digestive system, and wander to the pupariation site. Depending on the species, pupariation takes place within the substrate or in the soil or sand surrounding the substrate. If the species pupariates within a dense substrate, the anterior end and thus the anterior spiracles of the puparium usually penetrate the substrate surface. In at least two species of Sepsidae, *Orygma luctuosum* and *Ortalischema albitarse*, a rapidly-hardening white substance is spread over the surface of the puparium. Male and female larvae pupariate at different times in at least *Sepsis cynipsea* (Blanckenhorn, 1997: 344). Different hatching times for males and females from the same batch of eggs are typical for all sepsids reared to date, and the larger sex is are not necessarily the one to emerge last.

Development times from egg to adult vary dramatically with species, temperature and substrate (e.g. Modassir, 1995: 281). The fastest and slowest times reported in the literature for rearing under field conditions, from oviposition to adult emergence, are 6-7 days for *Sepsis lateralis* (Hafez, 1939: 325) and 26-39 days for the much smaller *Sepsis duplicata* (much longer in overwintering larvae and species with diapause). As a rule of thumb, species specialized for breeding in old dung are slow, whereas the development of specialist species ovipositing only in fresh bird droppings or ruminant dung is fast. The correlation between development times and temperature is best documented for *Sepsis cynipsea*. Laboratory cultures of the same population require approximately 8 days at 30°C and 26 days at 15°C, from oviposition to adult emergence (Blanckenhorn, 1997: 348). Some authors have claimed that sepsids produce only 2-4 generations per year in temperate regions (Hammer, 1941: 186-190; Minder, 1963: 386; Randall et al., 1981: 188; Draeger & Scheurer, 1999: 67), but this is certainly a low estimate for many countries. Most species appear to reach population peaks in late summer (Hammer, 1941: 187-188; Minder, 1963: 385; Bährmann, 1993a: 109). A few species such as *Themira nigricornis* seem to be active only in early spring to early summer.

Which stage in the sepsid life cycle hibernates seems to vary between and within species, the latter not being surprising considering the wide distribution of many species. The only stage not known to hibernate is the egg. Hibernation by larvae, puparia and adults is commonly reported (Table 1). However, it would be wrong to conclude that all stages of all species can hibernate. Blanckenhorn (1998: 6) demonstrated in a simulated winter experiment that Swiss populations of *Sepsis cynipsea* are unlikely to hibernate as puparia. In these populations hibernation was mainly initiated by temperature, and the adult stage fared best under the simulated winter conditions. It is also notable that so far the only adults to have been caught during the winter are species of *Sepsis*. There is no evidence of hibernating adults in the other genera, and the three species for which a winter diapause appears likely hibernate as puparia. If adult flies hibernate, it has been suggested that hibernation is preceded by migration to dry habitats (Bährmann, 1993a: 111), which could explain why *Sepsis fulgens* swarms are often found far from potential breeding sites (Van der Goot, 1986b: 33; Pont, 1987a: 309).

Economic importance. – Sepsidae undoubtedly play an important role in the decomposition of cattle dung and can be considered mildly beneficial insects in this respect (Snowball, 1944: 234; Papp, 1971a: 105). They are particularly important members of the alpine cow pat fauna (Papp, 1976: 136). However, sepsids may be considered as potentially harmful to humans, because they regularly carry pathogens. Their synanthropy and its implications have been discussed by Lobanov (1962b: 125-128), Zuska & Laštovka (1969: 201-221), and Řehůřková (1988: 231, table 5), who found especially *Nemopoda nitidula* and *Sepsis punctum* to be of a high dominance grade of synanthropy. Recently different pesticides have been tested in an effort to control Sepsidae in India (Modassir, 1993a: 82-83), where they are the dominant flies visiting dung (Modassir, 1993b: 241). But since Sepsidae generally avoid the inside of buildings (e.g. Minder, 1963: 385-386; Van der Goot, 1987a: 82), they have a low potential as transmitters. Only a few species are regularly found in slaughterhouses (Zuska & Laštovka, 1969: 205-207). The large outbreaks of *Themira putris* in sludge fields and around the sewage tanks of large-scale pig farms might be a greater concern for public health. Papp (1974b: 139) has estimated that in the Hungarian summer up to 300,000 flies can emerge every three days from a 270 m^3 storage facility for pig dung.

Conservation. – Most species of Sepsidae have benefitted from the transformation of natural forests into cattle farms and are probably more common nowadays than ever before. However, modern industrial agriculture is a mixed blessing for sepsids. The reluctance of most species to enter buildings effectively removes species from modern cattle rearing facilities and the only species to profit are those that regularly go indoors and/or breed in the sewage sprayed on to fields (Van der Goot, 1987a: 82). For example, although common earlier this century, *Meroplius minutus* has undoubtedly become rare in Central and Western

Europe, and this has been attributed to improved hygienic conditions (Van der Goot, 1987b: 82). However, this species is also known to breed in decaying rodent carcasses (Ozerov, 1991a: 54), so that this explanation is not entirely convincing.

Synecology. – Sepsidae are known to be the carriers for phoretic nematode species (Kiontke, 1996: 356; Chizhov, 1996: 185). One of these nematodes, *Diplogaster coprophila* Sudhaus & Rehfeld, is unable to complete development without going through a phoretic phase on a limited range of dung insects (Kiontke, 1996: 355). But species of *Sepsis* are also host for a parasitic tylenchid nematode (Chizhov & Sturhan, 1998: 5). Predation on the adult flies is presumably reduced by the presence of the defensive gland (Parker, 1972b: 248-249; Bristowe, 1979: 17-19; Zenk, 1990: 9-11). However, the larvae have been seen to be eaten by larval and adult staphylinids, larval *Sphaeridium* (Sowig, 1995: 477), as well as by ants. Puparia are known to be hosts for various hymenopterous parasitoids. Adults are also carriers for a range of phoretic and parasitic mites.

The following parasitic Hymenoptera have been reared in Europe from Sepsidae immatures (P = reared from puparium):

Parasitoid	Host
BRACONIDAE	
Alysia manducator Panzer	*Nemopoda nitidula*
Aphaereta minuta var. *cephalotes* Haliday	*Orygma luctuosum*
Gen. & sp. indet.	*Orygma luctuosum*, P
DIAPRIIDAE	
Coptera depressus Kieffer	*Sepsis fulgens*, P
Platymischus dilatatus Westwood	*Orygma luctuosum*, P
EUCOILIDAE	
Cothonaspis pusilla Giraud	*Saltella nigripes*, P
Cothonaspis gracilis Hartig	*Saltella nigripes*, P
	Sepsis cynipsea, P
	Sepsis thoracica, P
Cothonaspis nigricornis Kieffer	*Sepsis* sp., P
Cothonaspis (Erisphagia) sp.	*Sepsis cynipsea*
Kleidotoma brevicornis Thomson	*Sepsis thoracica*, P
Kleidotoma certa Belizin	*Sepsis* sp., P
Kleidotoma geniculata Hartig	*Sepsis fulgens*, P
	Sepsis thoracica, P
Kleidotoma scutellaris Thomson	*Sepsis thoracica*, P
Kleidotoma subaptera Walker	*Orygma luctuosum*, P
Kleidotoma truncata Cameron	*Themira putris*, P
EULOPHIDAE	
Melittobia acasta Walker	*Nemopoda nitidula*
PTEROMALIDAE	
Isocyrtus sp.	*Orygma luctuosum*, P
Muscidifurax raptor Girault & Sanders	*Sepsis* sp.
	Themira putris, P
Spalangia endius Walker	*Sepsis* sp.
	Themira putris, P
Spalangia subpunctata Forster	*Themira putris*, P
Urolepis maritima Walker	*Themira putris*, P

The following species of mites have been found on adult Sepsidae in Europe:

MESOSTIGMATA:	
PARASITIDAE	
Parasitus coleoptratorum (Linnaeus)	Sepsidae indet
Parasitidae indet., 2 spp.	*Sepsis biflexuosa*
	Sepsis cynipsea
	Sepsis fulgens
	Sepsis niveipennis
	Themira lucida
MACROCHELIDAE	
Macrocheles insignitus Berlese	*Sepsis fulgens*
Macrocheles plumiventris Hull	*Sepsis violacea*
Macrocheles subbadius Berlese	*Meroplius minutus*
VEIGAIIDAE	
Veigaia sp.	*Sepsis fulgens*
EVIPHIDIDAE	
Thinoseius fucicola (Halbert)	*Orygma luctuosum*
Parasitiformes indet.	*Sepsis violacea*
PROSTIGMATA:	
TROMBIDIIDAE	
Trombidium sp., larva	Sepsidae indet.
ASTIGMATA:	
HISTIOSTOMATIDAE (= ANOETIDAE)	
Anoetus sapromyzarum (Dufour)	*Nemopoda nitidula*
Bonomoia sphaerocerae Vitzthum	*Sepsis barbata*
	Sepsis cynipsea
	Sepsis fulgens
	Sepsis thoracica
	Sepsis violacea
Myianoetus dionychus (Oudemans)	*Sepsis* sp.

Distribution and zoogeography

The distribution and zoogeography of Sepsidae in Fennoscandia can be briefly summarised as follows. Most species (27 species, 79%) have entered from the south, along routes to the west and the east of the Baltic Sea. To varying degrees they have extended their ranges northwards, to 62°N or 66°N or to the farthest north. If one of the effects of global warming is for milder winters in Fennoscandia, then it may be expected that ranges will continue to expand northwards and species such as *Saltella sphondylii*, *Themira minor*, *Sepsis biflexuosa*, *Sepsis thoracica*, which are at the limit of their distribution in southern Fennoscandia, will regularly be found north of their present limits.

Of the remaining seven species (21%), three are endemic to this area: *Zuskamira inexpectata*, from central Sweden, *Themira ringdahli*, from subarctic

Sweden, and *Themira paludosa*, from the eastern rim of the Baltic Sea (Estonia, Russian Karelia, Finland). Two species, both Holarctic and both tundra species, have entered Fennoscandia from the north-east: *Themira arctica* and *Themira malformans* have both spread south to 65°N, and *malformans* also to montane Oppland in southern Norway. A third species, *Ortalischema albitarse*, appears to have followed this route, but is probably a Mongolian element and has reached further south into Småland in Sweden; the absence of this species from Denmark, Germany, France, where favourable conditions appear to exist, is certainly puzzling. Finally, *Orygma luctuosum* is an amphi-Atlantic seashore species, confined to the wrack beds in which its larvae live.

Of the 34 Fennoscandian species, 16 are Holarctic in distribution whilst 18 are confined to the Palaearctic region (or extend into the immediately adjacent areas of the Oriental region).

Many species of Sepsidae, especially those breeding in mammal dung, are common and widespread throughout Europe or even the whole Palaearctic region and beyond. Of the remaining European, non-Fennoscandian species, this is certainly the case with *Sepsis barbata*, *S. fissa*, *S. lateralis* and *S. niveipennis*. *Sepsis pseudomonostigma* is confined to south and south-east Europe, whilst *S. neocynipsea*, a Holarctic species and an upland species in Europe, is known from the British Isles and the middle belt of Europe but is absent from Fennoscandia and European Russia. The only European *Themira* species not found in Fennoscandia is *T. simplicipes*, which is known from Hungary, Romania and North Ossetia. *Saltella nigripes* is a species of Central and Southern Europe. *Susanomira caucasica* is restricted to the Caucasus Mts (North Ossetia and Georgia). Finally, *Meroplius fukuharai* is known from a number of localities in the middle belt of Europe, from Great Britain to Moscow and North Ossetia.

Many Sepsidae disperse easily, not only through the movement of domestic livestock but also through human commercial activities, e.g. the Oriental *Decachaetophora aeneipes* (Meijere) to west coast North America; the Australian *Lasionemopoda hirsuta* (Meijere) to New Zealand; and several Holarctic species of *Sepsis* (*biflexuosa*, *lateralis*, *thoracica*) to the Hawaiian Islands. In Europe, the most striking newcomer of recent years is *Meroplius fukuharai*, first described from East Asia in 1984 but subsequently discovered at various localities in west, central and eastern Europe (Moscow district and North Ossetia, the Czech and Slovak Republics, Germany, France and Great Britain). It seems highly likely that this species, like *Themira biloba* and possibly *Sepsis pseudomonostigma*, is expanding its range and has not simply been overlooked by collectors.

Review of Fennoscandian and other faunistic literature

The first Sepsidae from Scandinavia were described by the early pioneers of European entomology: C. Linnaeus (1758: *cynipsea*, *putris*), J. C. Fabricius (1794: *cylindrica*), C. F. Fallén (1810: *Sepsis*; 1820a: *scutellaris*; 1820b: *lutaria*, *nitidula*), and J. W. Zetterstedt (1838: *coxarum*; 1847: *albitarse*, *gracilis*, *pusilla*; also 1849, 1852, 1855, 1860). Zetterstedt's work spanned the whole of the Fennoscandian area, and he dealt with a total of 21 species. Subsequent regional, primarily faunistic, investigations can be summarised as follows:

Denmark

The first review of Danish species was by Stæger (1844), who recorded 14 species including three new ones: *ciliata* (now = *Themira superba* (Haliday)), *falleni* (now = *Themira nigricornis* (Meigen)), and *lucida* (now *Themira lucida*).

Little further work was done on the family until relatively recent times. Hammer (1941), in an extensive and pioneering biological study of flies associated with pasturing cattle, dealt with 9 species of Sepsidae. Hammer's material no longer exists (L. Lyneborg, pers. comm.). Otherwise, the only recent faunal list is one dealing with the Hansted nature reserve (N W Jutland), published by Lyneborg (1965, 9 species).

When misidentifications and synonyms have been removed from the list, the number of validly-recorded species from Denmark now stands at 24.

Norway

Work on Norwegian Sepsidae is sparse, and there is no up-to-date list of Norwegian species. Old investigations by Siebke (1853, 1866a, 1866b, 1870, 1873) culminated in an annotated catalogue (Siebke, 1877) in which 14 species were listed. More modern records are contained in papers by Duffield (1937, 2 species), Ringdahl (1954a, 3 species), and Dahl (1968, 3 species).

When misidentifications and synonyms have been removed from the list, the number of validly-recorded species from Norway now stands at 26.

Sweden

After Zetterstedt, who dealt with very extensive collections made in different parts of Sweden by himself, C. H. Boheman, N. J. Andersson, A. E. Holmgren, P. F. Wahlberg, etc, the next sepsid specialist was E. Wahlgren. In the series *Svensk Insektfauna*, Wahlgren (1919) gave the first review of Swedish species: 20 species were recorded, in addition to *Orygma luctuosum* (Wahlgren, 1917). Work was continued by O. Ringdahl, who published an important review (Ringdahl, 1934) which included a key to genera, a key to species of *Themira*, and notes on a number of species including several additions to the Swedish list. A supplement to this paper appeared later (Ringdahl, 1948). A new species (*Themira biloba*) was described by Andersson (1975), and a new genus and species (*Zuskamira inexpectata*) by Pont (1987c).

Other significant faunistic papers are as follows: species from the marine shore dune ecosystem (Ardö, 1957, 7 species), Kullaberg in Skåne (Ringdahl, 1952, 1960, 12 species), Råå in Skåne (Ringdahl, 1954b, 15 species), the sublittoral of N W Skåne (Ringdahl, 1959b, 10 species), Hallands Väderö (Ringdahl, 1941, 1959a, 1960, 13 species), Gotska Sandön Island (Jansson, 1925, 1 species), Öland Island (Wahlgren, 1915, Ringdahl, 1932, 1947, Persson, 1983; 6 species), Sarek National Park, Lule Lappmark (Poppius, Lundström & Frey, 1916, 4 species), Abisko National Park, Torne Lappmark (Ringdahl, 1931, 6 species), species from the Swedish mountains (Ringdahl, 1951, 19 species), and species from the regio arctica (Ringdahl, 1939, 1 species).

When misidentifications and synonyms have been removed from the list, the number of validly-recorded species from Sweden now stands at 33. Of the 34 Fennoscandian species of Sepsidae, only *Themira paludosa* is not known from Sweden.

Finland

The first list was contained in the pioneering work of Frey (1908), who wrote the first 'modern' work on European sepsid taxonomy. He recorded 20 species from Finland. Revised notes and keys were given in a later paper (Frey, 1925), which is marred by a somewhat cavalier approach to nomenclature. Records of Sepsidae visiting flowers are given by Poppius (1903) and Silén (1906, 1909), but the identifications in these old works are certainly not reliable.

Several other faunistic papers have been published: species of the shifting sand systems (Krogerus, 1932, 9 species), Tana älv at Utsjoki (Frey, 1950, 8 species), N Savolaks (Frey, 1917, 2 species), Pisavaara National Park (Lindberg & Saris, 1952, 5 species), S W Häme province (Hedström, 1964, Brander, 1965, 12 species), the subarctic Mt Ailigas (Hedström & Nuorteva, 1971, 2 species), the Finnish moors (Krogerus, 1960, 1 species), culminating in a new checklist (Hackman, 1980) in which 24 species were listed. After a study of the UZMH collection, Pont (1990) added 8 species, confirmed one (*nigripes*) and deleted one (*luteipes*).

When misidentifications and synonyms have been removed from the list, the number of validly-recorded species from Finland now stands at 31. Of the 34 Fennoscandian species of Sepsidae, only *Orygma luctuosum*, *Zuskamira inexpectata* and *Themira ringdahli* are not known from Finland.

NW European Russia

Very few faunistic papers have been published from this area. There is an old list of species from the Arkhangelsk area (Frey, 1918, 7 species). A list of species from the Leningrad district was published by Shtakel'berg (1958, 22 species) and this was followed by keys to the Sepsidae of the European part of the Russia (Shtakel'berg, 1970). Kandybina (1984) recorded 10 species from the Vyborg reserve. An important study of the Sepsidae of the European Russia was published by Minder (1963), who recorded 20 species from the

Moscow district and dealt with their taxonomy, biology and ecology. A number of new records have recently been published by Ozerov (1986a, 1986d, 1999d, 1999e), whose work on the Sepsidae of Russia is ongoing.

Baltic Countries

A few records were given by Dampf (1924, 1927) from the Estonian moors, and Ahlqvist (1935) listed 4 species from Estonia. Elberg (1963) described *Themira paludosa* from Estonia, and recorded *Sepsis biflexuosa* from Estonia (Elberg, 1966). Karps (1986) listed 19 species from Latvia, and Pakalniškis (1989) recorded 10 species from Lithuania.

Arctic

All pre-1910 records from the arctic area (Greenland, Iceland, Faroes, and islands of the Arctic Ocean) were summarised by de Meijere (1910). No Sepsidae, nor *Orygma*, were recorded. No records were given in Wahlgren's (1920) review of Diptera from Jan Mayen, Bear Island, Spitsbergen, Franz Josef Land and Novaya Zemlaya.

Four species are now known from the Faroes (Becker, 1915, Lyneborg, 1968): *Orygma luctuosum*, *Themira lucida* (syn: *roseni*), *T. arctica* (syn: *dampfi*), *T. putris*.

Four species are known from Iceland (Nielsen et al., 1954, Andersson, 1967, Ólafsson, 1991): *Orygma luctuosum*, *Themira arctica* (syn: *dampfi*), *T. leachi*, *T. pusilla*. *Themira arctica* and *T. pusilla* were also found on the new island of Surtsey and *O. luctuosum* on Heimaey but not on Surtsey (Lindroth et al., 1973).

Two species are now known from Greenland (Vibe, 1950): *Orygma luctuosum* and *Themira arctica*, though *luctuosum* was not found by Madsen (1936) in his work on the Greenland littoral fauna.

General European works

The monographs of Duda (1926a) and Hennig (1949) have been fundamental for the study of European Sepsidae. However, these papers contain few records from the Fennoscandian area. Duda saw very little Fennoscandian material, except for Stæger's collection and some of Lundbeck's collection from ZMUC, and some Norwegian material located in ZMHU. (For information on Duda's own collecting localities, see Wheeler, 1969.) He discussed the work of Zetterstedt and Frey from their publications and not from their material. Hennig, with his customary thoroughness and care, included many published records in his revision but saw little new Fennoscandian material.

Faunistic records up to 1982 were contained in Volume 9 of the *Catalogue of Palaearctic Diptera* (Zuska & Pont, 1984).

Since Hennig's monograph of 50 years ago, the principal taxonomic advances have been in papers by Soós (1959), Zuska (passim; especially 1960, 1965, 1968, 1970b, 1971, 1977a), Gregor (1966), Pont (1979, 1986a, 1987c), Munari (1985b) and Munari & Vienna (1982). Extensive faunistic records have been published for the British Isles (Pont, 1987b), the Netherlands (Van der Goot, 1985a, b, 1986a, b, 1987a, b, 1989), Germany (Püchel, 1992), Switzerland (Haenni, 1997), NE Alps of Austria (Franz, 1989), Romania (Ursu, 1965, 1969a, 1973, 1984), Canary Islands (Báez, 1982), Spain (Carles-Tolrá, 1993), mainly Italy (Munari, 1984, 1985a, 1986, 1987a, b, c, 1991a, b, 1993, 1997, 1998; Munari & Vienna, 1977, 1979a, b, c; Munari & Roháček, 1990), and Malta (Schembri et al., 1991). Surveys of synanthropic species in eastern Europe were conducted by Gregor & Povolný (1959a, b, 1960a, b, 1961, 1964). In the last decade, checklists have been published for Poland (Nowakowski, 1991), Belgium (Grootaert, 1991), Germany (Bährmann, 1993b; Meier, 1999), Czech and Slovak Republics (Zuska & Barták, 1997), Italy (Munari & Rivosecchi, 1997), Switzerland (Haenni, 1998), and the British Isles (Chandler, 1998).

A paper by Meier & Pont (2000) reviews the Palaearctic fauna. Recent papers by Iwasa (Oriental, Malagasy), Silva (Neotropical) and Ozerov (all regions) are dealing with the non-Palaearctic fauna – see the bibliography.

Collecting, preserving and identification

Adult Sepsidae may be found in many different habitats – grasslands, meadows, riverbanks, lake shores, sand dunes, even the seashore (*Orygma*), deciduous woodlands, from sea-level to high altitudes/latitudes, from broadleaved forests to montane and arctic tundra. Most species are attracted to dung, and will be found at isolated droppings, dungheaps, manure heaps, middens, animal stalls in general. Cow dung is particularly attractive: *Saltella* spp. and *Sepsis duplicata* are found on older cowpats, whilst *Sepsis cynipsea* and *S. flavimana* are found on freshly deposited dung. The dung of other mammals (horse, sheep, pig, rabbit, etc) will also attract Sepsidae. For example, *Ortalischema* and *Zuskamira* are found only on horse droppings. The droppings of waterfowl, whether on riverbanks, lake margins or around small ponds, are particularly good for species of *Themira*.

Adult Sepsidae are best collected by sweeping or by hand-netting over substrates where adults are gathering for feeding, mating or oviposition. Faeces-baited traps and Malaise traps are useful adjuncts to hand-collecting.

Adults are best preserved by micro-pinning. They should be pinned whilst still fresh through the side (pleura), with the pin entering and leaving by a different pleural plate so that the same character is not destroyed on both sides. The legs should be drawn away from the body, so that the bristling and ornamentation can be seen, and, in the case of males, the genitalia should be hooked back to reveal the shape of the surstyli. Specimens preserved dry can be gummed to points, but this method is only satisfactory if extreme care is taken not to break off setae during the mounting process: as the ground-colour of most sepsids is black and shining, it is often difficult to determine whether a seta is absent (no scar) or was once present (with scar). If it is necessary to see the armature of specimens with bunched legs, the femoro-tibial joint can be temporarily softened with a drop of volatile liquid (e.g. xylene) and then the tibia gently teased away from the femur with a fine pin. Specimens in alcohol present no problems if they are males: the ornamentation of the fore legs (and often of other legs too) and the genitalia are highly species-specific and are easily seen in alcohol material. Females in alcohol are more difficult, as much reliance has to be placed on pleural pruinosity which is difficult to see when wet. If specimens, particularly females, are needed for genitalia studies, they can be kept live and fed on sugarwater, which causes the abdomen to bloat and the genitalia (and especially the ovipositor) to extrude, at which point they can be killed and preserved in alcohol.

Identification of adults must be made with a light microscope giving high magnification and good light quality in order to detect some of the subtleties of bristling and pruinosity against the black background of the sepsid integument. When dissection of the ♂ genitalia or ♀ ovipositor is necessary, it can be done following the usual procedures: soften and clear in KOH, wash in water, study and draw in alcohol or glycerine, preserve either in a microvial of glycerine or, after passing through glacial acetic acid and beechwood creosote or clove oil, mount in Canada balsam or euparal on a glass slide or celluloid mount.

Larval Sepsidae are generally abundant in a large variety of decaying organic materials. However, the identification of the larvae of many very common species in *Sepsis* and *Themira* is difficult and often impossible with light or even scanning electron microscopy. An identification of larvae via an association with adults on the substrate in which the larvae were found is not recommended. Most popular sepsid breeding substrates support multiple, often congeneric species and the adults visit the substrate during a very short time window, while larvae can be present for many days. For example, larvae of *Sepsis cynipsea* will be present in cow dung for several days, while the adults will only visit the dung for up to 6 hours and will later be replaced by other *Sepsis* species.

The safest path to obtaining correctly identified immature material is therefore through rearing. For many species it is neither very difficult nor time-consuming, since the generation times are usually short. Females are collected at the breeding substrate. They need to be provided with some sugarwater during transportation, because sepsids die very quickly without a source of carbohydrate. In the laboratory the females can be sorted to species while they are still alive, or single females can be used to start the cultures. Identification of the living flies is easiest when they are anaesthetized by carbon dioxide, which can either be bought commercially or produced by adding a strong acid to limestone. Alternatively, one female at a time can be placed in a glass vial and some cottonwool can be used to immobilize it against the wall of the vial so that it can be identified under a microscope. In species where copulation takes place regularly, one can also identify the female through its mate.

For breeding, the female(s) are placed in a container that is provided with (1) a source of moisture (e.g. moist cottonwool), (2) a source of carbohydrate (e.g. sugarwater), and (3) the breeding substrate, ususally in a petri dish. The breeding substrate has to be 'virgin', i.e. it must be free of viable eggs or larvae. The first choice, in the case of dung, is material whose production has been witnessed. Alternatively, the substrate can be frozen before use, but substrates prepared in this manner will often fail to induce oviposition by females. If females have been caught away from any potential breeding substrate, cow dung or waterfowl dung have the highest chance of being suitable. The most common causes of death under laboratory conditions are too much water (e.g. condensation in badly aerated containers), not enough water, and exposure of the culture container to sunlight. Any open water will invariably trap and kill the flies, and a drought will prove similarly fatal within a short period of time. Whether eggs have been laid in the substrate is usually easy to ascertain based on the presence of the conspicuous egg filaments which will be exposed on the surface of the breeding substrate. Larval development takes several days at least and usually one can see conspicuous holes in the substrate when it is time to harvest the third instars. One should not collect all larvae as some immatures need to be bred to the adult stage for identification.

The three different larval instars can be distinguished by the number of lobes on the anterior spiracles. In Sepsidae, the first instar larva has no externally visible lobes, the second instar has one, and the third instar has at least three but normally many more. If multiple instars are present at the same time, they can also be distinguished by the size and degree of sclerotization of the cephaloskeleton. The first instar is the only stage with only a partially sclerotized cephaloskeleton. Second and third instar larvae can be separated by the discontinuous size variation of the cephaloskeleton.

Before eggs or larvae are preserved, they need to be carefully cleaned. This step is essential for producing good immature material, and since sepsid immatures breed in very filthy substrates cleaning is a daunting task. It has to precede preservation of the larvae in alcohol because the 'dirt' tends to cake on to the specimens once they are exposed to ethanol. All immature stages are best cleaned in lukewarm water to which a few droplets of dishwashing liquid have been added. The eggs and larvae can then be sucked in and out of a pipette until inspection under a binocular microscope shows that they are clean. Eggs and third instar larvae can then be transferred to 70% alcohol. However, all larvae first have to be killed in near-boiling water. If they are placed directly in alcohol, the pseudocephalon and the anterior spiracles will often be withdrawn. After killing, the first two instars need to go through a graded alcohol series (10% steps; 30 minutes each), as they shrivel if placed immediately in 70% alcohol. If an alcohol series is not available, one can mimic the effect of such a series by placing the larvae in low-concentration alcohol and regularly removing some and replacing it with high-concentration ethanol until a 70% concentration is reached, which is the best storage medium for eggs and larvae.

All immature stages are best studied using scanning electron microscopy after critical-point-drying and sputter-coating. The larvae should be placed on to the SEM stubs so that at least some can be studied from a dorsal, lateral and ventral view. The latter is by far the most informative view. When studying larvae with light-microscopy, high magnification and a good source of lighting are necessary as many characters are otherwise difficult to see. The main problem is the general lack of pigmentation of integumental features, which are creamy-white on a creamy-white background. The preparation of slide mounts of the integument can help here. The cephaloskeleton is also best studied as a slide mount after it has been extracted by soaking the anterior section of a larva in KOH for some time and then carefully removing the integument and tissue remains. Detailed instructions for making larval slide mounts can be found in Ferrar (1987: 10-11). If eggs cannot be studied with SEM, then slide mounts should be prepared which allow focusing through the filament in order to deduce its three-dimensional structure. However, the surface structure of the filament will be difficult to see, and viewing a broken-off filament by SEM is an easier way to study the plastron of the filament.

Checklist of European Sepsidae

Full references, details of type-localities and type-specimens, and distribution will be found in the main body of the text.

ORYGMATINAE

Orygma Meigen, 1830
  Psalidomyia Doumerc, 1833
luctuosum Meigen, 1830
  fucicola Doumerc, 1833

SEPSINAE

Ortalischema Frey, 1925
  Protothemira Duda, 1926
albitarse (Zetterstedt, 1847)
  rossica Frey, 1918

Zuskamira Pont, 1987
inexpectata Pont, 1987

Themira Robineau-Desvoidy, 1830
  Enicopus Walker, 1833, preocc.
  Cheligaster Macquart, 1835
  Enicita Westwood, 1840
  Halidaya Rondani, 1856
  Enicomira Duda, 1926
  Cheligastrula Strand, 1928
  Annamira Ozerov, 1999, **syn. n.**
  Nadezhdamira Ozerov, 1999, **syn. n.**
annulipes (Meigen, 1826)
  brunicosa Robineau-Desvoidy, 1830, **syn. n.**
  varipes Meigen, 1838
  marginata Hoffmeister, 1844, **syn. n.**
  crassiseta Duda, 1926
  nigra Duda, 1926, unavailable
arctica (Becker *in* Becker et al., 1915)
  dampfi Becker, 1915
biloba Andersson, 1975
germanica Duda, 1926
gracilis (Zetterstedt, 1847)
  plurisetosa Duda, 1926
leachi (Meigen, 1826)
  phantasma Robineau-Desvoidy, 1830
  tarsalis Walker, 1833
  fumipennis Walker, 1833
  coxarum Zetterstedt, 1837, nomen nudum
  coxarum Zetterstedt, 1838
  curvipes Van der Wulp, 1864
  umbripennis Van der Wulp, 1871
lucida (Stæger *in* Schiødte, 1844)
  roseni Becker, 1915, **syn. n.**
malformans Melander & Spuler, 1917
minor (Haliday, 1833)
  halidayi Curtis, 1831, nomen nudum
  halidayi Curtis, 1837, nomen nudum
  dentimana Van der Wulp, 1864
  consobrina Van der Wulp, 1871
nigricornis (Meigen, 1826)
  falleni Stæger *in* Schiødte, 1844
paludosa Elberg, 1963
pusilla (Zetterstedt, 1847)
  spinosa Verrall, 1886, nomen nudum
  spinosa Verrall, 1886
putris (Linnaeus, 1758)
  fimeti Linnaeus, 1761
  conssencis Harris, 1780
  pilosa Robineau-Desvoidy, 1830
  setosa, error
ringdahli Pont, **sp. n.**
simplicipes (Duda, 1926)
superba (Haliday, 1833)
  ciliata Stæger *in* Schiødte, 1844
  nigra Duda, 1926, unavailable
  fasciculata Duda, 1926, unavailable

Susanomira Pont, 1987
caucasica Pont, 1987

Saltella Robineau-Desvoidy, 1830
  Brachygaster Meigen, 1826, preocc.
  Pandora Haliday, 1833, preocc.
  Anisophysa Macquart, 1835
  Pseudopandora Rapp, 1946
nigripes Robineau-Desvoidy, 1830
  ferruginea Robineau-Desvoidy, 1830
  gagatea Robineau-Desvoidy, 1830, **syn. n.**
sphondylii (Schrank, 1803)
  scutellaris Fallén, 1820
  melanocephala Drapiez, 1820, **syn. n.**
  analis Meigen, 1826
  varia Meigen, 1826
  hemorrhoidalis Meigen, 1826, unavailable
  longipes Robineau-Desvoidy, 1830, **syn. n.**
  nana Robineau-Desvoidy, 1830, **syn. n.**
  basalis Haliday, 1833
  scutellata Macquart, 1835
  ruficoxa Macquart, 1835
  albipennis Macquart, 1835
  spondylii Schrank *in* Gistel, 1837
  sellata Curtis, 1837, nomen nudum
  sellata Haliday, 1838
  distincta Meigen, 1838

analis Roser, 1840, preocc.
humeralis Roser, 1840
parmensis Rondani, 1874
nigerrima Rondani, 1874
nigripes Duda, 1926, unavailable
fumipennis Duda, 1926, unavailable
floreus Duda, 1926, unavailable
chaerophylli, error or emendation

Nemopoda Robineau-Desvoidy, 1830
Pseudonemopoda Duda, 1926
nitidula (Fallén, 1820)
cylindrica Fabricius, 1794, preocc.
putris Robineau-Desvoidy, 1830
viridis Macquart, 1835
pectinulata Loew, 1873
speiseri (Duda, 1926)

Meroplius Rondani, 1874
Xenosepsis Malloch, 1925
Pseudomeroplius Duda, 1926
fukuharai (Iwasa, 1984)
minutus (Wiedemann, 1830)
lutaria Fallén, 1820, unavailable
stercoraria Robineau-Desvoidy, 1830
nigrilatera Macquart, 1835
rufipes Meigen, 1838
varipes Walker, 1871, preocc.
polita Duda, 1926, unavailable

Sepsis Fallén, 1810
Threx Gistel, 1848
Acrometopia Lioy, 1864
Beggiatia Lioy, 1864
Sepsidimorpha Frey, 1908
Australosepsis Malloch, 1925
Saltelliseps Duda, 1926
Lasiosepsis Duda, 1926
barbata Becker, 1907
chopardi Séguy, 1932
pamirensis Enderlein, 1934
biflexuosa Strobl, 1893
flavipes Goetghebuer & Bastin, 1925
desultor Séguy, 1932
cynipsea (Linnaeus, 1758)
elongata Müller, 1764
hilaris Meigen, 1826
barbipes Meigen, 1826
vivida Robineau-Desvoidy, 1830
incisa Strobl, 1894
duplicata Haliday, 1838
pilipes Van der Wulp, 1871
pilipes Loew, 1873, preocc.
loewi Hendel, 1902
fissa Becker, 1903
flavimana Meigen, 1826
ruficornis Meigen, 1826
pygmaea Robineau-Desvoidy, 1830, **syn. n.**
maculipes Walker, 1833
simplex Goetghebuer & Bastin, 1925
borealis Frey, 1925
melanopoda Duda, 1926
kerteszi Duda, 1926
meijerei Duda, 1926
fulgens Meigen, 1826
vibrans Harris, 1780, preocc., **syn. n.**
minimus Harris, 1780, preocc., **syn. n.**
fulgida Robineau-Desvoidy, 1830, **syn. n.**
fulvida, error
nitida Robineau-Desvoidy, 1830, **syn. n.**
atripes Robineau-Desvoidy, 1830, **syn. n.**
concinna Walker, 1833
communis Frey, 1925
tonsa Duda, 1926
lateralis Wiedemann, 1830
complicata Wiedemann, 1830
inpunctata Macquart *in* Webb & Berthelot, 1839
algira Macquart, 1843
hyalipennis Macquart, 1851
rufa Macquart, 1851
melitensis Rondani, 1874
schembrii Rondani, 1874
fragilis Becker, 1903
luteipes Melander & Spuler, 1917
lamellifera Frey, 1917, nomen nudum
lamellifera Frey, 1925
helvetica Munari, 1985
neocynipsea Melander & Spuler, 1917
nigripes Meigen, 1826,
minima Strobl, 1893, preocc.
niveipennis Becker, 1903
orthocnemis Frey, 1908
pseudomonostigma Ursu, 1969
punctum (Fabricius, 1794)
stigma Panzer, 1798
cornuta Meigen, 1826, **syn. n.**
ornata Meigen, 1826
pectoralis Macquart, 1835
rufocincta Hoffmeister, 1844, **syn. n.**
zernyi Duda, 1926
quadrisetosa Duda, 1926
icaria Séguy, 1932
meridionalis Séguy, 1932
setulosa (Duda, 1926)
thoracica (Robineau-Desvoidy, 1830)
tridens Becker, 1903

consanguinea Villeneuve, 1920
goetghebueri Frey, 1925
idmais Séguy, 1932
ino Séguy, 1932

violacea Meigen, 1826
fuscipes Roser, 1840
uncta Becker, 1912
ciliforceps Duda, 1926
schineri Duda, 1926

List of Fennoscandian Sepsidae

1 *Orygma luctuosum* Meigen, 1830
2 *Ortalischema albitarse* (Zetterstedt, 1847)
3 *Zuskamira inexpectata* Pont, 1987
4 *Themira annulipes* (Meigen, 1826)
5 *Themira arctica* (Becker, 1915)
6 *Themira biloba* Andersson, 1975
7 *Themira germanica* Duda, 1926
8 *Themira gracilis* (Zetterstedt, 1847)
9 *Themira leachi* (Meigen, 1826)
10 *Themira lucida* (Stæger *in* Schiødte, 1844)
11 *Themira malformans* Melander & Spuler, 1917
12 *Themira minor* (Haliday, 1833)
13 *Themira nigricornis* (Meigen, 1826)
14 *Themira paludosa* Elberg, 1963
15 *Themira pusilla* (Zetterstedt, 1847)
16 *Themira putris* (Linnaeus, 1758)
17 *Themira ringdahli* Pont, **sp. n.**
18 *Themira superba* (Haliday, 1833)
19 *Saltella sphondylii* (Schrank, 1803)
20 *Nemopoda nitidula* (Fallén, 1820)
21 *Nemopoda pectinulata* Loew, 1873
22 *Nemopoda speiseri* (Duda, 1926)
23 *Meroplius minutus* (Wiedemann, 1830)
24 *Sepsis biflexuosa* Strobl, 1893
25 *Sepsis cynipsea* (Linnaeus, 1758)
26 *Sepsis duplicata* Haliday, 1838
27 *Sepis flavimana* Meigen, 1826
28 *Sepsis fulgens* Meigen, 1826
29 *Sepsis luteipes* Melander & Spuler, 1917
30 *Sepsis nigripes* Meigen, 1826
31 *Sepsis orthocnemis* Frey, 1908
32 *Sepsis punctum* (Fabricius, 1794)
33 *Sepsis thoracica* (Robineau-Desvoidy, 1830)
34 *Sepsis violacea* Meigen, 1826

Key to the European genera of Sepsidae

Adults

1 Eye small and gena very deep, the depth below eye equal to about twice an eye-width, the face strongly receding (Fig. 12). 1 orbital and 2 frontal setae (Fig. 12). Anepisternum, anepimeron, katepisternum and disc of scutellum densely setulose. 3 strong dorsocentral and 2 postalar setae. Large bristly species with robust legs and short tarsi .... subfamily ORYGMATINAE *Orygma* Meigen (p. 36)

– Eye large and gena narrow, the depth below eye at most half an eye-depth, often considerably less, the face not receding (Fig. 16). 0-1 orbital and 0 frontal seta (Fig. 17). Anepisternum and katepisternum with only a few setulae, anepimeron and disc of scutellum bare. 1-2 dorsocentral setae, or if with 4-5 (*Susanomira*, *Zuskamira*) then 1 postalar seta. Smaller species with more delicate legs and longer tarsi .... subfamily SEPSINAE 2

2(1) Basal radial and basal medial wing-cells fused (Fig. 13). Wing-tip clear, unmarked. Scutellum strikingly flattened and large, as long as broad, with a strong pair of basal and apical setae (Fig. 2). 1 weak proepisternal and 1 weak (*nigripes*) or strong (*sphondylii*) proepimeral setae. Fore femur with a complete row of posterodorsal setae. Face with a conspicuous carina between the antennae (as in Fig. 20), and (in *sphondylii*) lower margin projecting forwards .... *Saltella* Robineau-Desvoidy (p. 107)

– Basal radial and basal medial wing-cells distinct (Fig. 14), or if fused (*Sepsis niveipen-*

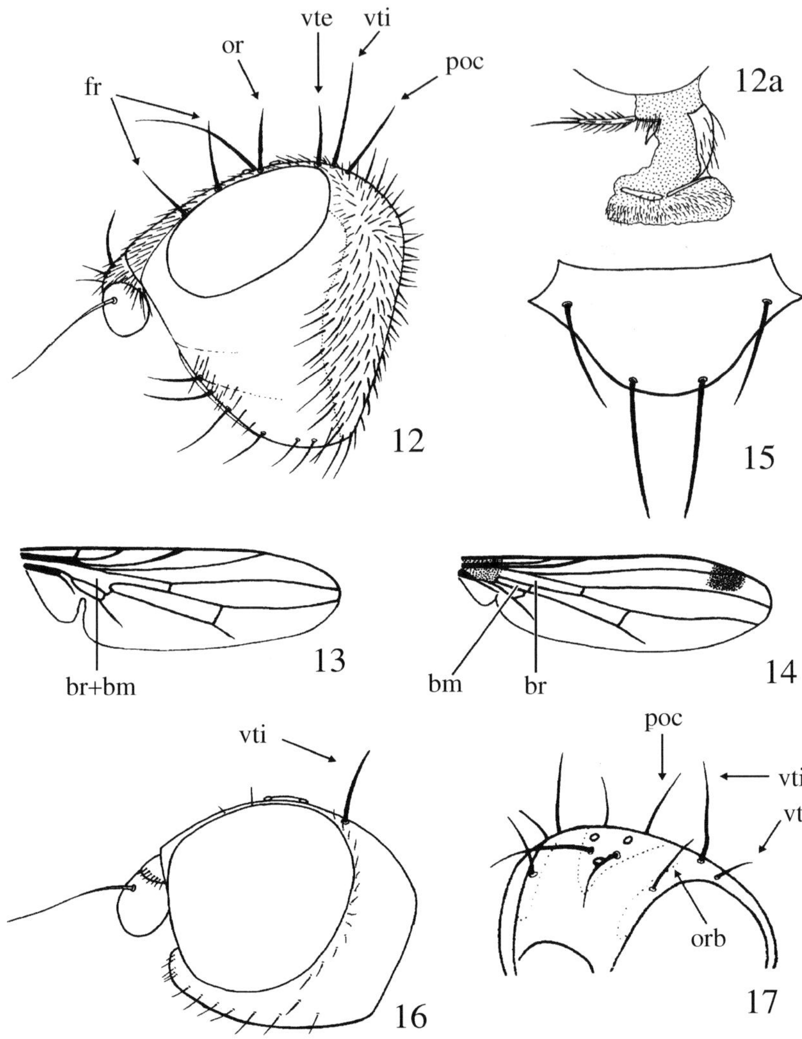

Figs 12-17. 12: *Orygma luctuosum* Meigen, head, lateral view; 12a: proboscis and palpus. - 13: *Saltella sphondylii* (Schrank), wing. 14: *Sepsis fulgens* Meigen, wing. 15: *Ortalischema albitarse* (Zetterstedt), scutellum. 16: *Ortalischema albitarse* (Zetterstedt), head, lateral view. 17: *Meroplius minutus* (Wiedemann), frons, oblique view. (bm – basal medial cell, br – basal radial cell, fr – frontal seta, or – orbital seta, – poc – postocellar seta, vte – outer vertical seta, vti – inner vertical seta.)

nis) then wing with a dark spot near tip. Scutellum normal, convex and very short, half as long as broad, with the basal seta reduced or absent (strong in *Ortalischema* and *Zuskamira*). Proepisternal and proepimeral setae represented only by tiny hairs (but a short proepisternal present in *Susanomira* and *Meroplius*). Fore femur at most with a few posterodorsal setae near tip (some *Themira*), otherwise bare on this surface. Face without a conspicuous carina and lower margin not projecting forwards (except in *Susanomira*, Fig. 19) 3

3(2) Postocellar seta absent (Fig. 16) 4

– Postocellar seta present (Fig. 17) 5

4(3) Outer vertical and orbital setae absent (Fig. 16). Postpronotal seta absent. Anepisternal seta very short. Basal scutellar seta strong (Fig. 15). Prosternum bare. ♂: katepisternum with a stout seta at lower anterior angle, and hind coxa with a ventral projection covered with setulae (Fig. 67); hind tibia without an osmeterium...... *Ortalischema* Frey (p. 40)

– Outer vertical and orbital setae present (Fig. 17). Postpronotal seta present. Anepisternal seta long and strong. Basal scutellar seta short and weak. Prosternum setulose. ♂: katepisternum and hind coxa not modified as above; hind tibia with an osmeterium...... .... *Meroplius* Rondani [in part] (p. 128)

5(3) Supraalar seta absent. Outer vertical seta absent; orbital seta present (as in Fig. 17). Anepisternal seta vestigial or absent....... .... *Themira* Robineau-Desvoidy (p. 49)

– Supraalar seta present. Outer vertical seta present, but if absent (*Nemopoda speiseri*) then orbital seta also absent. Anepisternal seta present..6

6(5) Wing with a dark spot near tip (Fig. 14) (except in the European *lateralis*). Postalar seta very short. Orbital seta minute or absent. Postpronotal seta present. Either mid or hind tibia with an anterodorsal or posterodorsal seta. ♂: hind tibia without an osmeterium .... *Sepsis* Fallén (p. 136)

– Wing without a dark spot. Postalar seta strong. Orbital seta present even if short, but if absent (*Nemopoda speiseri*) then postpronotal seta absent. Neither mid nor hind tibia with any dorsal setae. ♂: hind tibia often with an osmeterium..7

7(6) Only 1 dorsocentral seta. Katepisternum largely glossy with white pruinosity confined to upper posterior corner. Prosternum setulose. Gena linear, its width below lowest eye-margin less than half width of antennal flagellomere...8

– Dorsocentral setae 1+3-4. Katepisternum entirely pruinose, white or brownish-grey. Prosternum bare. Gena broader, its width below lowest eye-margin at least half width of antennal flagellomere (*Zuskamira*, Fig. 18) or one-third of eye-depth (*Susanomira*, Fig. 19) 9

8(7) Orbital seta strong (Fig. 17). Basal medial cell and basal part of discal medial cell without clothing-hairs, absolutely hyaline. Katepisternum without a seta, white pruinose only in upper posterior corner; postpronotal seta present. Fore femur in male with 2 stout peg-like anteroventral setae (Figs 359, 364), and in female with only 1-2 anteroventral spinules at apical third. ♂: hind tibia with a well-developed field-like osmeterium................................. .... *Meroplius* Rondani [in part] (p. 128)

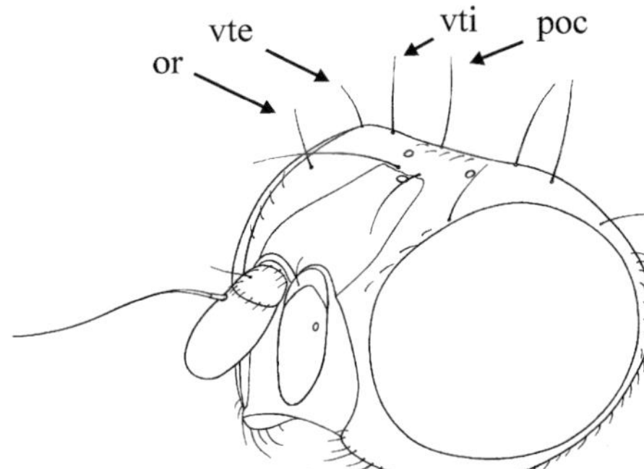

18

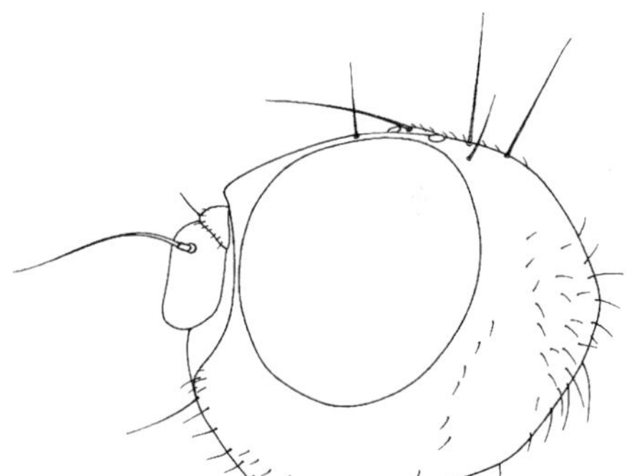

19

20

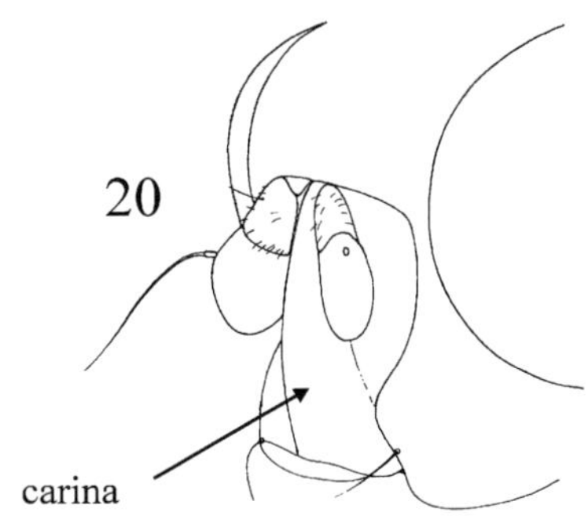

Figs 18-20. 18: *Zuskamira inexpectata* Pont, head, oblique view. 19: *Susanomira caucasica* Pont, head, lateral view. 20: *Susanomira caucasica* Pont, face, oblique view. (or – orbital seta, – poc – postocellar seta, vte – outer vertical seta, vti – inner vertical seta.)

– Orbital seta very short or absent (Figs 324-325). Basal medial cell (except in *speiseri*) and basal part of discal medial cell with normal clothing-hairs, like the rest of the wing. Katepisternum with a short seta in upper posterior corner (except in *speiseri*, which has no postpronotal seta and has katepisternum white pruinose along upper part). Fore femur in both sexes with a series of very short anteroventral spinules on apical half and, in male, with a row of widely-spaced posteroventral setae (e.g. Fig. 342). ♂: hind tibia with the osmeterium weak, scarcely perceptible, or absent .... *Nemopoda* Robineau-Desvoidy (p. 117)

9(7) Face sloping forwards, with a conspicuous facial carina, and lower margin projecting well beyond profrons (Figs 19-20). Basal scutellar seta absent. Gena broader, equal to length of antennal flagellomere (one-third of eye-depth). 1 short proepisternal seta. Wing in Fig. 294. Larger species, wing-length 4.1-5.2 mm. ♂: sternite 3 with a dense tuft of long dark setae (Fig. 295) .... *Susanomira* Pont (p. 103)

– Face receding, with a weak facial carina, and lower margin not projecting (Fig. 18). Basal scutellar seta half as long as apical one. Gena narrower, half width of antennal flagellomere. With only a proepisternal hair present. Wing unmarked, clear. Smaller species, wing-length 2.6-2.7 mm. ♂: sternite 3 without a tuft of setae, with only the usual setulae (Fig. 80)......................... .... *Zuskamira* Pont (p. 45)

Eggs (based on Meier & Pont, 2000) are known for all European genera

1 Egg without respiratory filaments (Fig. 8) .. .. *Ortalischema* Frey (p. 40)

– Egg with 1 or 2 respiratory filaments (Figs 9-10)................................... 2

2(1) Egg with 2 respiratory filaments (Fig. 9) ... . *Orygma* Meigen (p. 36)

– Egg with 1 respiratory filament (Fig. 10) 3

3(2) Egg filament only 0.6-0.8 times as long as the egg .... *Themira annulipes* (Meigen) (p. 60)

– Egg filament at least 1.5 times as long as egg 4

4(3) Filament ribbon-like and twisted around its own axis (Fig. 30)........................ .... *Susanomira* Pont (p. 103)

– Filament normally round in cross-section (Figs 21-22, 24-25), but if ribbon-like (Fig. 23) then not twisted around its own axis 5

5(4) Filament distinctly flattened in cross-section and thus looking like a ribbon (Fig. 23); egg filament 3-4 times as long as the egg .... *Nemopoda* Robineau-Desvoidy (p. 117)

– Filament perfectly round in cross section (Figs 21-22, 24-25); egg filament 1.5-4 times as long as the egg 6

6(5) Chorion along entire filament broken up into small polygons (Fig. 27).................. .... *Saltella* Robineau-Desvoidy (p. 107)

– Chorion of at least the proximal two thirds of filament more or less solid, except for at most a few rows of small pores (Figs 26, 31) 7

7(6) Respiratory filaments with irregularly arranged struts around a central canal (Fig. 21) .. most *Themira* Robineau-Desvoidy (p. 49)

– Respiratory filaments with a single layer of regularly arranged struts around the central canal (Figs 24-25)........................ 8

8(7) Struts with simple bases at the central canal and bifurcated bases at the outside chorion of the filament............................. .... *Zuskamira* Pont (p. 45)

– Struts with simple bases at the central canal and the outside chorion of the filament (Figs 24-25)................................. ...... *Themira gracilis* (Zetterstedt) (p. 72) .... *Meroplius* Rondani (p. 128) .... *Sepsis* Fallén (p. 136)

Larvae (based on Meier & Pont, 2000) are known for all European genera

1 Ventral tubercles absent (Fig. 32); stigmatophores without dorsal tubercles (Fig. 32); posterior spiracles as in Fig. 33 .... *Ortalischema* Frey (p. 40)

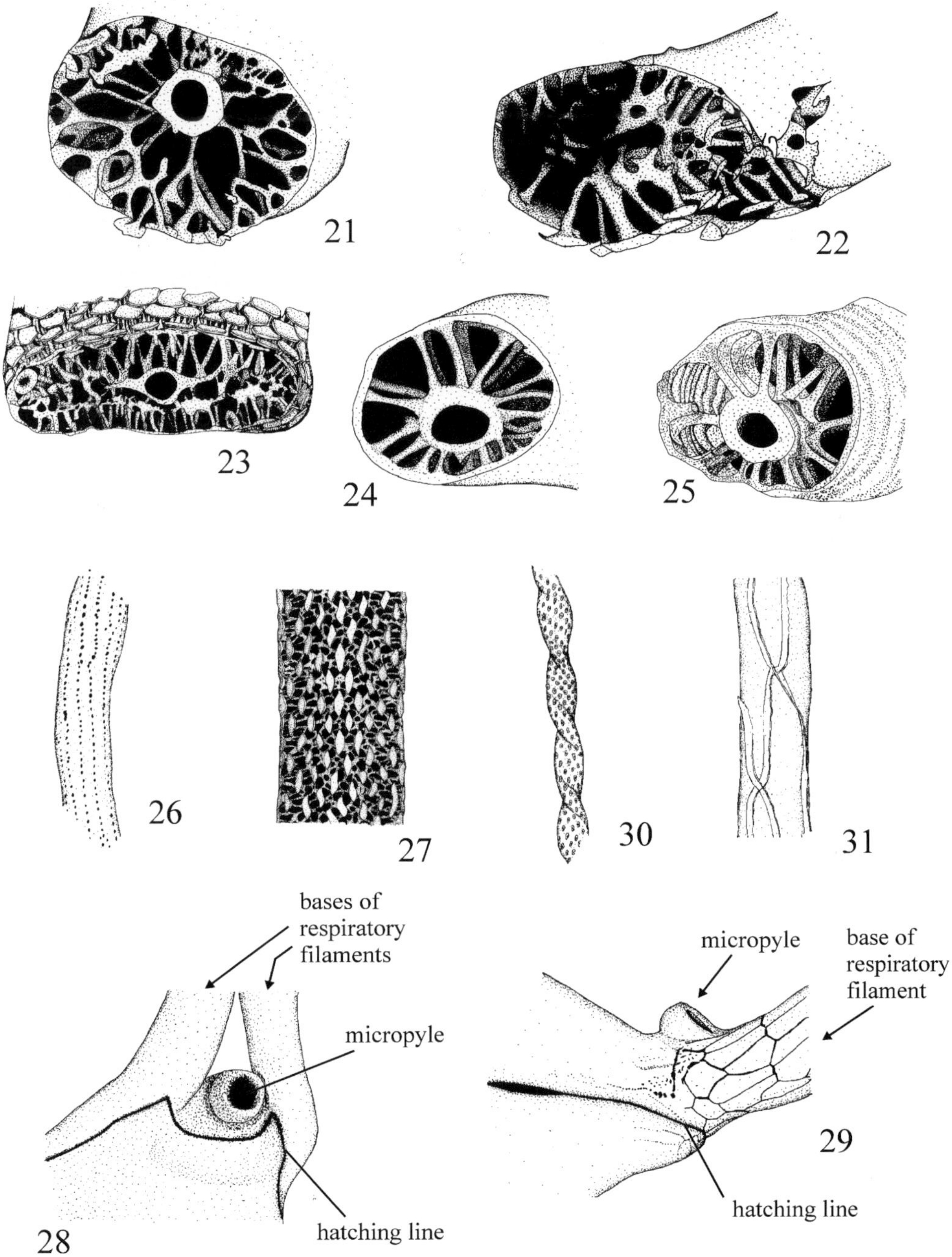

Figs 21-31. Eggs. Section through the filament. 21: *Themira annulipes* (Meigen); 22: *Orygma luctuosum* Meigen; 23: *Nemopoda nitidula* (Fallén); 24: *Sepsis orthocnemis* Frey; 25: *Sepsis cynipsea* (Linnaeus). Aeropyles of egg filament. 26: *Themira minor* (Haliday); 27: *Saltella sphondylii* (Schrank). Micropyle. 28: *Orygma luctuosum* Meigen; 29: *Nemopoda nitidula* (Fallén). Egg filament. 30: *Susanomira caucasica* Pont; 31: *Sepsis thoracica* (Robineau-Desvoidy).

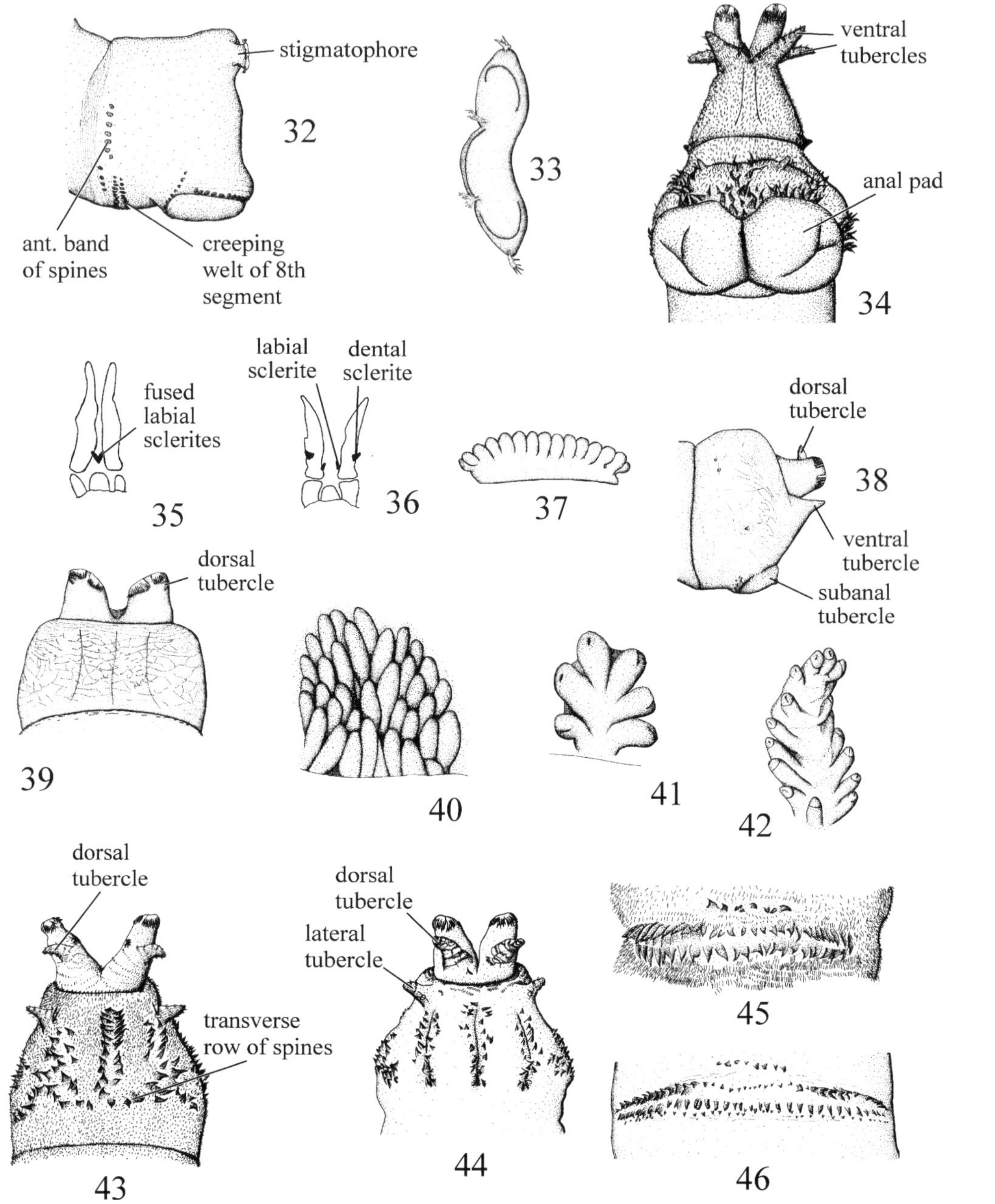

Figs 32-46. Larval morphology. 32: *Ortalischema albitarse* (Zetterstedt), anal division, lateral; 33: *Ortalischema albitarse* (Zetterstedt), posterior spiracles; 34: *Saltella sphondylii* (Schrank), anal division, ventral; 35: *Saltella sphondylii* (Schrank), mouthhooks, dorsal; 36: *Ortalischema albitarse* (Zetterstedt), mouthhooks, dorsal; 37: *Orygma luctuosum* Meigen, anterior spiracles; 38: *Orygma luctuosum* Meigen, anal division, lateral; 39: *Orygma luctuosum* Meigen, anal division, dorsal; 40: *Susanomira caucasica* Pont, anterior spiracles; 41: *Sepsis biflexuosa* Strobl, anterior spiracles; 42: *Themira lucida* (Stæger), anterior spiracles; 43: *Nemopoda pectinulata* Loew, anal division, dorsal; 44: *Themira nigricornis* (Meigen), anal division, dorsal; 45: *Nemopoda pectinulata* Loew, creeping welt; 46: *Susanomira caucasica* Pont, creeping welt.

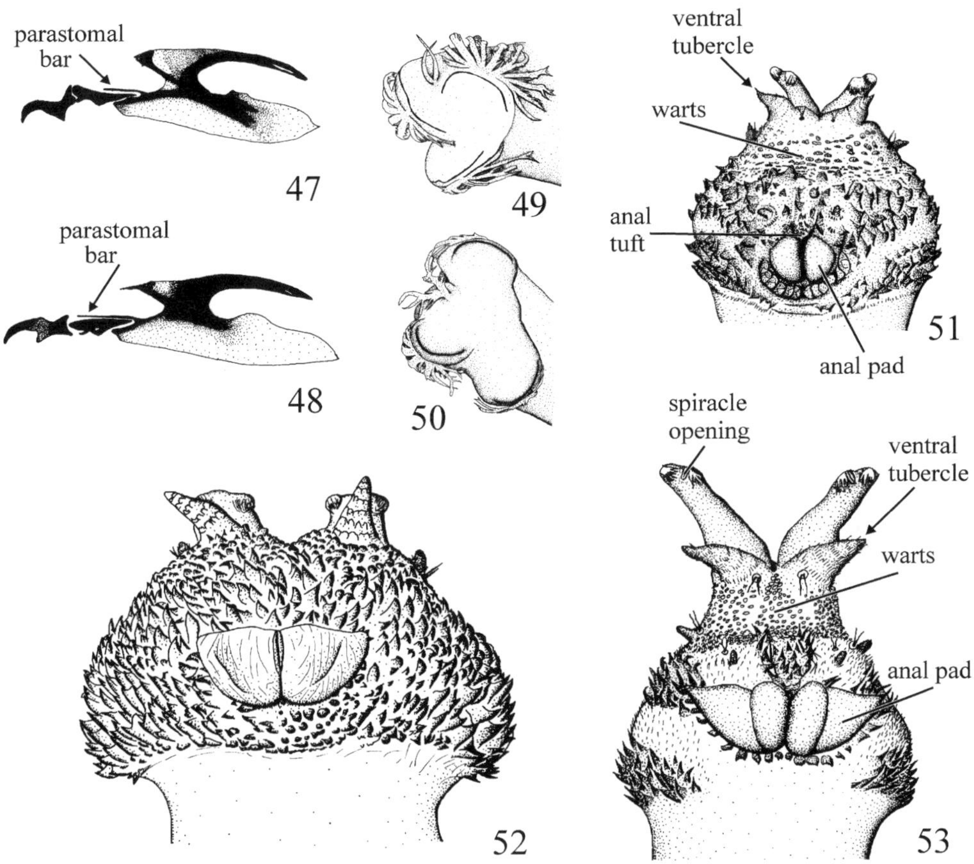

Figs 47-53. Larval morphology. 47: *Themira minor* (Haliday), cephaloskeleton, lateral; 48: *Meroplius fukuharai* (Iwasa), cephaloskeleton, lateral; 49: *Sepsis neocynipsea* Melander & Spuler, posterior spiracles; 50: *Meroplius fukuharai* (Iwasa), posterior spiracles; 51: *Meroplius fukuharai* (Iwasa), anal division, ventral; 52: *Zuskamira inexpectata* Pont, anal division, ventral; 53: *Sepsis luteipes* Melander & Spuler, anal division, ventral.

– Ventral tubercles present (e.g. Figs 34, 51-53); stigmatophores with dorsal tubercle (Figs 38, 43-44) . . . . 2[2]

2(1) Ventral tubercles clearly bifurcate (Fig. 34); labial sclerites present, fused into one V-shaped sclerite (Fig. 35) . . . . *Saltella* Robineau-Desvoidy (p. 107)

– Ventral tubercles single, not bifurcate (Figs 51-53, 132-135); labial sclerites absent or present, if present then paired (Fig. 36) . . . 3

3(2) Anterior spiracle consisting of a fan-like row of lobes (Fig. 37); without spines on lateral and dorsal sides of anal division (Figs 38-39); with labial sclerites (as in Fig. 36); spines of creeping welt dark-brown; larvae only in sea wrack . . . . *Orygma* Meigen (p. 36)

– Anterior spiracle usually tree-like (Figs 41-42), except for *Susanomira* (Fig. 40); nor-

[2] For the remaining couplets of the key, it is necessary to slide-mount the cephaloskeleton, and this will entail destroying the external features of the pseudocephalon. Before dissection, it is recommended that the pseudocephalon and thoracic segments are cut off and studied under a compound microscope, to determine (1) whether the pseudocephalon is monolobed (as in Fig. 400) or bilobed (as in Figs 59, 72, 164, 186, 199, 228, 304-305); (2) how many lobes are found on the anterior spiracles and how they are arranged (see Figs 37, 40-42); and (3) how the ridges and combs are distributed on the ventral side of the pseudocephalon (see Figs 59, 72, 199, 305, 401). This information may be needed in the keys to the species of *Themira* and/or *Sepsis*.

mally with spines on anal division (Figs 43-44, 51-53, 166); without labial sclerites; spines of creeping welt of the same colour as the cuticle 4

4(3) Anterior spiracle in a spiracle field with more than 30 lobes (Fig. 40); creeping welts with three rows of spines, the last consisting of much smaller spines than the first two (Fig. 46)......... *Susanomira* Pont (p. 103)

– Anterior spiracle more or less tree-like; with less than 30 lobes (Figs 41-42, 187); creeping welts with two rows of spines (Figs 45, 189) 5

5(4) All body segments densely hirsute (Fig. 45), integument thus with an opaque or matt appearance; anal division dorsally with three double rows and an additional transverse row of spines (Fig. 43)........................ ....*Nemopoda* Robineau-Desvoidy (p. 117)

– Body segments bare except for creeping welts and anal division (Fig. 46), integument thus translucent and shiny; anal division dorsally at most with three longitudinal double rows of spines, never with a transverse row (Figs 44, 357-358) 6

6(5) Parastomal bar short, at most reaching posterior edge of the bridge of the intermediate sclerite (Figs 47, 137) .... *Themira* Robineau-Desvoidy (p. 49)

– Parastomal bar longer, reaching beyond the bridge of the intermediate sclerite (Figs 48, 330) 7

7(6) One spiracular opening of posterior spiracle leaving the spiracular plate, extending on to the medioventral side of the stigmatophore (Figs 49, 398-399, 402); anal pad divided into two lobes on either side of anal opening ('wing-shaped': Figs 53, 395-397, 403). .... *Sepsis* Fallén (p. 136)

– Posterior spiracles with all three spiracular openings on the spiracular plate (Fig. 50); anal pad simple (Figs 51, 372) or at most with a diagonal fold (Fig. 52)............ 8

8(7) Dorsal side of anal division with bare integument or hairs between the three double rows of spines (Figs 357-358); anal pad simple (Figs 51, 372)........................... ....*Meroplius* Rondani (p. 128)

– Dorsal side of anal division with numerous large warts between the three double rows of spines (as on lateral side, see Fig. 87); anal pad with a diagonal fold (Fig. 52)......... ....*Zuskamira* Pont (p. 45)

Subfamily ORYGMATINAE Frey, 1921

Genus ***Orygma*** Meigen, 1830

Orygma Meigen, 1830: v, 6. Type-species: *Orygma luctuosum* Meigen, 1830, by monotypy. Gender: neuter.

Psalidomyia Doumerc, 1833: 92. Type-species: *Psalidomyia fucicola* Doumerc, 1833 [= *Orygma luctuosum* Meigen, 1830], by monotypy. Gender: feminine.

Adult *Orygma* can be distinguished by a number of characters, such as the large bristly appearance; small eye (Fig. 12); 2 frontal and 1 orbital setae, all strong (Fig. 12); well-developed palpus (Fig. 12a); 3 stout dorsocentrals; or densely haired anepimeron, anepisternum and katepisternum.

Adult. – Head rather triangular, facial carina not developed, face receding, eye very small and gena very broad (Fig. 12). A full complement of head setae: inner vertical, outer vertical, postocellar, 2 frontals and 1 orbital. Palpus large (Fig. 12a). Scutum and scutellum matt. Postpronotal present. 3 well-spaced dorsocentrals. 1 supraalar, 2 postalars, 1 very strong proepisternal. Proepimeral absent. Pleura matt, dusted. Anepisternal long, and 1 strong katepisternal also present; anepisternum, anepimeron and katepisternum densely setulose. Prosternum setulose. Scutellum short, 2 pairs of setae, the apical ones directed upwards; disc wholly setulose. Legs robust, with numerous setae and, in ♂, hairs as well; tibial spurs stout; tarsomeres short and stout, claws and pulvilli well-developed. ♂ fore legs not modified, and hind tibia without osmeterium. Wing without a dark spot. Tergites 1+2 to 6 with setae at sides, fewer in ♀ than in ♂.

♂ abdomen (Figs 54-56). Tergites 1+2 to 5 normal; tergite 6 present and complete between tergite 5 and tergite 7+8. Spiracles 1-3 in the intersegmental membrane; spiracles 4-6 in the tergites; spiracle 7+8 asymmetrical, between tergites 6 and 7+8 on one side, at ventral edge of tergite 7+8 on the other (Figs 55-56). Sternite 1 small, divided, setulose; sternites 2-5 large, plate-like, densely fine-setulose; sternite 6 absent; sternites 7 and 8 present, band-like, and contiguous or weakly attached at each corner to tergite 7+8. Cercal plate (Fig. 57) simple, horseshoe-shaped, without projection. Surstylus simple (Fig. 58), not articulating with epandrium. Sperm pump relatively small, narrow, tapering.

♀ abdomen. Spiracles 1-5 in the tergites; postabdominal spiracles 6 and 7 very large and placed high in tergites 6 and 7. Segments 6 and 7 with the tergites and sternites complete and very large. Segment 8 represented only by the tergite, which is weakly divided dorsally but enlarged laterally and ventrally, forming a claw-like plate on each side, thus displacing the more terminal elements of the ovipositor dorsad (Fig. 4); sternite 8 absent. Cerci small, round. Epiproct present, setose. Hypoproct weakly sclerotised, setulose. 2 spermathecae, of different sizes.

Hennig (1937: 33-34) wrote that in the ♂ sternite 7 is absent, but from the position and articulation of the post-abdominal sclerites it appears that it is sternite 6 that is absent. Hennig's 'sternite 6' is actually sternite 7, and sternite 8 (not shown in Hennig's fig. 33) is present as a ribbon-like band. For the same reason, we interpret Griffiths' (1972: fig. 95) 'sternite 6' as sternite 7. Griffiths (1972: 184) has accurately noted the asymmetry of segment 7+8 in this genus, in contrast with the Sepsinae.

Egg. – Egg of *O. luctuosum* rather slender and with two filaments which are slightly clubbed at ends (Fig. 9). Filaments dorsolateral to the micropyle and hatching line (Fig. 28). Filaments closed on the outside but medially open through gaps between irregularly shaped plates, which are internally supported by a meshwork of struts. On inner sides of the filaments, plastron therefore in direct contact with the surrounding air. In cross section, arrangement of struts irregular, usually branched (Fig. 22). Filaments without central canal. Micropyle cup-shaped and protruding (Fig. 28).

Larva. – Pseudocephalon bilobed, cephalic lobes almost entirely covered by combs; facial mask with numerous ridges in parallel arrangement (Fig. 59).

Anal division short; lateral tubercles absent (Fig. 38); stigmatophores with dorsal tubercle (Figs 38-39); ventral tubercles present, simple (Fig. 38); subanal tubercles large and on either side of anal opening; anal pad very narrow, only bordering anal opening; anal tuft and preanal protuberance absent; anal division between anterior margin of anal pad and origin of stigmatophores about as wide as long; with two rows of spines between anal pad and segment 7; remaining creeping welts of abdominal segments with three rows of dark-brown spines (Fig. 60); without spines or hairs on dorsal or lateral aspect of anal division (Figs 38-39).

Cephaloskeleton: labial sclerites present, separate; dental sclerites absent (as in Figs 47-48); intermediate sclerite anteriorly without flat plate; epistomal sclerite approximately as long as wide, with posterior bars; parastomal bars long, only slightly shorter than intermediate sclerite; basal sclerite 1.5 times as long as high, dorsal cornua shorter and approximately 1.5 times narrower than ventral cornua.

Anterior spiracles fan-like, broad, with 15 short and wide lobes; lobes 2-2.5 times as long as wide (Fig. 37).

Posterior spiracles with three almost equal, narrow, strongly arched spiracular openings, situated on one almost round plate (Fig. 61).

Discussion. – *Orygma* is the basal genus of the family Sepsidae, a placement supported by character-states in the adult and immature stages. Plesiomorphies (i.e. ground-plan characters of the family Sepsidae) are: a full complement of head setae (orbital, inner vertical, outer vertical, postocellar); palpus large; postpronotal seta present; supraalar seta present; basal scutellar seta present; prosternum setulose; ♂ fore leg without ornamentation; ♂ hind tibia without osmeterium; ♂ postabdominal tergite 6 present; ♂ postabdominal sternites 7 and 8 present as separate plates; ♂ surstylus fused seamlessly to epandrium; ♂ sperm pump narrow, relatively small; ♀ postabdominal segments 6 and 7 with large and complete tergites and sternites; ♀ postabdominal spiracles 6 and 7 in the tergites; ♀ postabdominal tergites 6 and 7 each without an anterior prolongation; ♀ postabdominal segment 8 modified into a ring; ♀ epiproct present.

Orygma is apomorphous relative to the groundplan of the Sepsinae with regard to the following characters:

1. Eye reduced in size, and consequently gena appearing very broad (also in some other Sciomyzoidea (e.g. *Coelopa*)
2. Face and epistoma strongly receding

One species, Holarctic. Adults restricted to the seashore. Larvae in wrack beds, saprophagous.

1. ***Orygma luctuosum*** Meigen, 1830

(Figs 4, 9, 12, 12a, 22, 28, 37-39, 54-61)

Orygma luctuosum Meigen, 1830: 6, plate 55 figs 13-17; Meigen plate 194 fig. 10 [Morge, 1976c]. Holotype ♂, locality not stated [probably ENGLAND], in MNHNP [see below].

Psalidomyia fucicola Doumerc, 1833: 93, plate 6 figs A1-7. Syntypes ♂♀, FRANCE, Dieppe, not in MNHNP and presumed destroyed [see below].

Diagnosis. – As for the genus.

Adult. – Colour: Frontal plate, vertex, occiput and facial ridge dark; gena, parafacial, face and lunule orange to brown. Antenna orange, flagellomere dark except at base. Palpus yellow. Thorax, scutellum and abdomen dark. Legs of ♂ mostly brown, tarsi partly and femora basally yellow, tibiae sometimes appearing translucent; in ♀ almost entirely yellow, including coxae. Wing weakly smoky. Calypter yellow, fringe-hairs yellow and brown (♂) or mostly brown (♀). Haltere yellow.

Head (Fig. 12): Frons entirely covered with short setulae. A full complement of setae present. Subvibrissal setae strong but sparse.

Thorax: Scutum matt, blackish-brown, with weak traces of 3 grey vittae along mid-line and through dorsocentrals. Scutum wholly covered with short setulae. Proepisternal surrounded by several setulae. Posterior spiracle with numerous dense setae. Scutellum short, twice as wide as long.

Legs: Fore femur with rows of posterodorsal and posteroventral setae. Fore tibia without submedian setae. Mid femur without setae except for a fine posteroventral near base. Mid tibia with 2 short anterodorsals and posterodorsals. Hind femur with irregular rows along dorsal to anterodorsal surfaces; several anteroventral setae stronger than the setulae (♂) or 4 well-spaced setae (♀). Hind tibia with 2 anterodorsal and 2 posterodorsal setae; dorsal and posterodorsal apicals very long, anteroventral apical stout.

Wings: Veins bare except for several setulae on basal part of stem-vein on ventral surface, and on apical part of vein R_{4+5} (after cross-vein r-m) on dorsal surface; subcosta often with a setula on ventral surface. Wing-surface entirely covered with microtrichiae.

Abdomen: Matt brownish-grey, not shining. Lateral setae more numerous towards tip of abdomen; syntergite 1+2 with setulae at sides. ♂ sternites 1-5 dusted, densely setulose; sternites 7 and 8 glossy, bare. ♂ genitalia very large (Figs 57-58); cercal plate small; surstylus with a small spinose process on upper anterior edge.

Wing-length: 4.8-6.3 mm (♂), 4.9-5.8 mm (♀).

Egg and larva. – See under the generic description.

Distribution. – A seashore species, sparsely but widely distributed along the coastlines of the North Sea, Norwegian Sea and Arctic Ocean. Denmark: a few old records from NEZ and NEJ; RM found one specimen in Hornbæk, North Zealand, in 1998, but attempts to establish a culture failed and the specimen was lost in the algae. Norway: Ø to F, but scarce. Sweden: Boh, Hall, Sk[3] (e.g. Zetterstedt, 1847: 2471, and 1860: 6388; Backlund, 1945: 223; Ringdahl, 1960: 20). Finland: absent. Known from the adjacent Russian coastline (Kola Peninsula and eastwards to Onega Bay, e.g. Frey, 1921: 53; Hennig, 1937: 38; Backlund, 1945: 223; Frey, 1953: 122).

Western Europe (Germany, Belgium, France, Great Britain, Ireland, Iceland) (Zuska & Pont, 1984: 157), Faroes (Lyneborg, 1968: 24), Heimaey but not Surtsey off Iceland (Lindroth et al., 1973: 36). Britanny appears to be the southern limit in Europe. Nearctic: Greenland (Vibe, 1950: 420), east coast of North America (Labrador to Massachusetts) (Steyskal, 1965: 682).

Records are from May to August, even in the northern part of the range.

Biology. – The species is restricted to the seashore and to areas where wrack beds consisting of a compressed mixture of *Fucus* and *Laminaria* are found. Adults tend to be secretive, and will not fly far when disturbed, preferring to burrow down into

[3] Backlund records the locality Vinga in the inland province Vg, but this locality is actually on the Skåne coast.

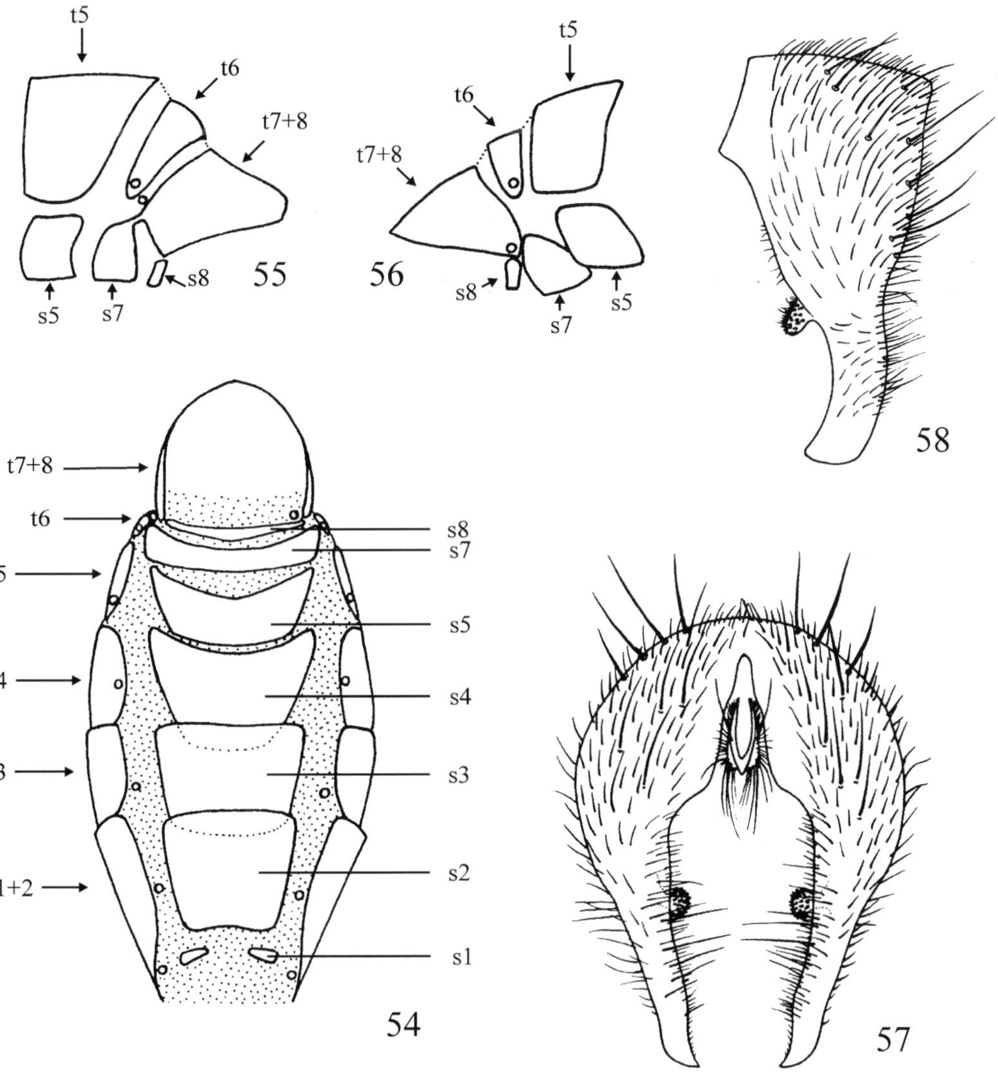

Figs 54-61. *Orygma luctuosum* Meigen, ♂. 54: abdomen, ventral view, hypopygium removed; 55: segments 5-8, left side of abdomen; 56: segments 5-8, right side of abdomen; 57: hypopygium, dorsal view; 58: hypopygium, lateral view; 59-61 larval morphology. 59: pseudocephalon, ventral; 60: creeping welt; 61: posterior spiracles. (s – sternite, t – tergite.)

the wrack. Backlund (1945: 63) considered the flattened bodies and the holding of the wings close to the abdomen to be adaptations for moving in wet and slimy surroundings. The artificially high temperature in fermenting wrack beds (Phillips & Arthur, 1994: fig. 1) enables the species to remain active throughout the year for much of its range. The flies readily copulate, often even in the collector's aspirator vial, in the course of which the fore-legs of the male are variously placed on the head or scutum of the female (RM pers. obs.).

Oviposition can occur very rapidly after eclosion and the larvae hatch after 1-4 days. Larvae live in the wrack beds, though not the wettest ones, feeding on decaying *Fucus* and *Laminaria*. They are able to tolerate up to 25% salinity. Pupariation takes place at the base of the wrack beds amongst stones, and the puparium has a white calcareous covering. The biology has been discussed in detail by Backlund (1945: passim) and Egglishaw (1960: 136-137; 1965: 195-198). The immature stages have been described by Kieffer (1898: 102-103), Scott (1920: 156, fig. 2e), Hennig (1937: 17-18, 36-37), Tsacas (1959: 87, fig. 10), Egglishaw (1960: 114-118, figs 4-14) and Meier (1995b: 427-431, figs 1, 17-18, 25; 1996: 35-37, figs 86-93). The accounts by Egglishaw and Meier are the most detailed, with descriptions and illustrations of the egg, all 3 larval instars and puparium (early instars and puparium only in Egglishaw, 1960).

Adults carry the mite *Thinoseius fucicola* Halbert (Mesostigmata, Eviphididae). Hymenopterous puparial parasitoids are *Aphaereta minuta* var. *cephalotes* Haliday (Braconidae), *Isocyrtus* sp. (Eulophidae), *Kleidotoma subaptera* Walker (Eucoilidae) and *Platymischus dilatatus* Westwood (Diapriidae). *Aleochara obscurella* Gravenhorst (syn: *algarum* Fauvel) (Coleoptera, Staphylinidae) has also been recorded as a parasitoid (Scott, 1920: 149; Backlund, 1945: 158, 161). Backlund (1945: 158) found a very high rate of hymenopterous parasitism in field populations of *luctuosum*. Adults were commonly found as prey of the asilid fly *Philonicus albiceps* (Meigen) in Ireland by Verrall (1909: 640).

Type-material. – *Orygma luctuosum*: Meigen wrote 'Hr. Dr. Leach in London schickte mir das Männchen'. ACP has seen the holotype ♂ in MNHNP, Meigen collection no. 2364; it is labelled by Meigen '*Orygma luctuosa*'. The abdomen is missing, as was also noted by Becker (1902: 250), but it clearly belongs to *Orygma luctuosum* as currently understood.

Psalidomyia fucicola: ACP has seen 1♂ and 1♀ in UMO from the Bigot collection named as *Psalidomyia fucicola* Doumerc, each with a small round label with 'TYP' printed in red (which Bigot used in his collection). These are not types as the locality is Le Havre, not Dieppe, and no type-material of *fucicola* was found in MNHNP.

Musca palpator Harris (1780b: 119, plate 35 fig. 31) has been mentioned as a possible senior synonym of *Orygma luctuosum* (e.g. Thompson & Pont, 1994: 156). However, the fly shown in Harris' illustration does not resemble *luctuosum* very closely, nor does Harris state that it was collected on the seashore. Both illustration and description best fit a species of Dryomyzidae.

Subfamily SEPSINAE Walker, 1833

Genus *Ortalischema* Frey, 1925

Ortalischema Frey, 1925: 75. Type-species: *Sepsis albitarsis* Zetterstedt, 1847, by original designation. Gender: neuter.

Protothemira Duda, 1926a: 25. Type-species: *Sepsis albitarsis* Zetterstedt, 1847, by monotypy. Gender: feminine.

Adult *Ortalischema* can be distinguished from all other European and Palaearctic genera by the absence of the postocellar seta.

Adult. – Head roundish (Fig. 16), facial carina not developed, face vertical and lower facial margin rather projecting, gena narrow. Outer vertical absent. Postocellar absent. Orbital minute. Scutum and scutellum matt. Postpronotal absent. 1 dorsocentral, 1 supraalar. 1 postalar. Proepisternal and proepimeral areas bare. Anepisternal present, short. Katepisternal absent. Anepisternum and katepisternum densely chagreened, anepimeron and meron thinly so and partly subshining. Prosternum bare. 2 pairs of scutellars (Fig. 15). Fore femur without posterodorsal setae. Tibiae without anterodorsal or posterodorsal setae. ♂ fore tibia without setae or any modifications. ♂ hind tibia without a conspicuous osmeterium, the tibia rather flattened off on inner side in apical half. Wing without a dark spot. Abdominal setae absent except for a stout seta at lower posterior corner of tergites 1+2 to 5 in ♂.

♂ abdomen (Figs 68-69). Tergites 1+2 to 5 normal; tergite 6 narrow but complete; tergite 7+8 large and long. Spiracles 1-6 in the intersegmental membrane close to the edge of the tergites; spiracles 7-8 on a weakly sclerotised patch adjacent to tergite 7+8. Sternite 1 band-like, bare; sternites 2-5 simple, plate-like, not divided or modified in any way; sternite 7+8 present as a very thin and weak ribbon ventrally between the sides of tergite 7+8. Cercal plate simple, without projections (Fig. 70). Surstylus very reduced, firmly and seamlessly attached to epandrium, without any suture or articulation; with tiny spinules at tip; with a pair of large, densely long-setose lobes, the small (anterior) lobe attached by apodemes to the epandrium and to the large lobe, the large (posterior) lobe broadly attached to the epandrium and base of surstylus and to the apodeme from the anterior lobe (Figs 70-71). Sperm pump tapering.

♀ abdomen. Sternites 2-5 broad, as in ♂. Postabdominal segments 6 and 7 with the tergites and sternites large, almost meeting or even overlapping and thus forming more or less complete rings; segment 8 with the tergite fused ring-like dorsally and continued as a broad plate laterally and ventrally, rather as in *Orygma* but not so claw-like; sternite 8 absent. Spiracles 1-5 as in ♂; spiracles 6 and 7 in the tergites. Cerci small, round. Epiproct present. Hypoproct large, cordiform. 2 spermathecae, of different sizes.

Our interpretation of the appendages on the epandrium differs from that of Hennig (1949: 33). Hennig believed that the two large lobes represented the surstylus which had been almost completely divided into two elements by a process arising from the epandrial wall. We interpret Hennig's 'process' as the true surstylus since its position, shape and lack of articulation correspond so well to the ground-plan of the family. The fact remains, however, that these two densely and long-setose lobes, whatever their interpretation, are *sui generis* elements and unique within the family Sepsidae.

Egg. – Egg without even a trace of an egg filament (Fig. 8) and almost uncurved. Anterior pole with an operculum. Hatching line unusually well developed and swollen (Fig. 8).

Larva. – Pseudocephalon bilobed, very long and pointed, cephalic lobes largely covered with combs; facial mask consisting of numerous ridges (Fig. 72).

Anal division short, lateral and ventral tubercles absent (Fig. 32); subanal tubercle moderately large; short stigmatophores without dorsal tubercle (Fig. 32); anal pad large, with deep diagonal furrow, rounded laterally; anal tuft present; preanal protuberance absent; anal division between anterior margin of anal pad and origin of stigmatophores slightly longer than wide; with a single lateral row of spines between anal pad and segment 7 (Fig. 32); two rows of creeping welt extend upwards on to lateral surface of anal division (Fig. 32); dorsal surface without spines or hairs.

Cephaloskeleton: labial sclerites present (Fig. 36), separate; dental sclerites present (Fig. 36); bridge of intermediate sclerite approximately at half length of sclerite, wide, anteriorly with a flat plate; epistomal sclerite as long as wide, without posterior bars; parastomal bars long, only slightly shorter than intermediate sclerite; basal sclerite three times as long as high, dorsal cornua shorter and four times narrower than ventral cornua.

Anterior spiracles fan-like, very short, broad, with 7-8 short, wide lobes; lobes 2-2.5 times as long as wide (Fig. 73).

Posterior spiracles with almost equal, narrow, semicircular spiracular openings on a long-oval spiracular plate (Fig. 33).

Discussion. – *Ortalischema* is also close to the base of the Sepsidae, and is the most basal genus of the Sepsinae. This is supported by character states in the adult and immature stages.

Plesiomorphies (i.e. ground-plan characters of the Sepsinae) include the following: 1 dorsocentral seta; supraalar seta present; 1 postalar seta; scutellum with basal seta present; wing with basal medial and basal radial cells distinct; ♂ fore femur with little ornamentation; ♂ hind tibia without an osmeterium; ♂ abdomen with spiracles 1-5 in the intersegmental membrane; ♂ postabdominal tergite 6 present; ♂ sperm pump relatively small, narrow; ♀ segment 8 forming a broad ring, rectum without reservoir for glandular secretion. *Ortalischema* is apomorphous relative to the ground-plan of the Sepsinae with regard to the following characters: Orbital seta minute, outer vertical seta

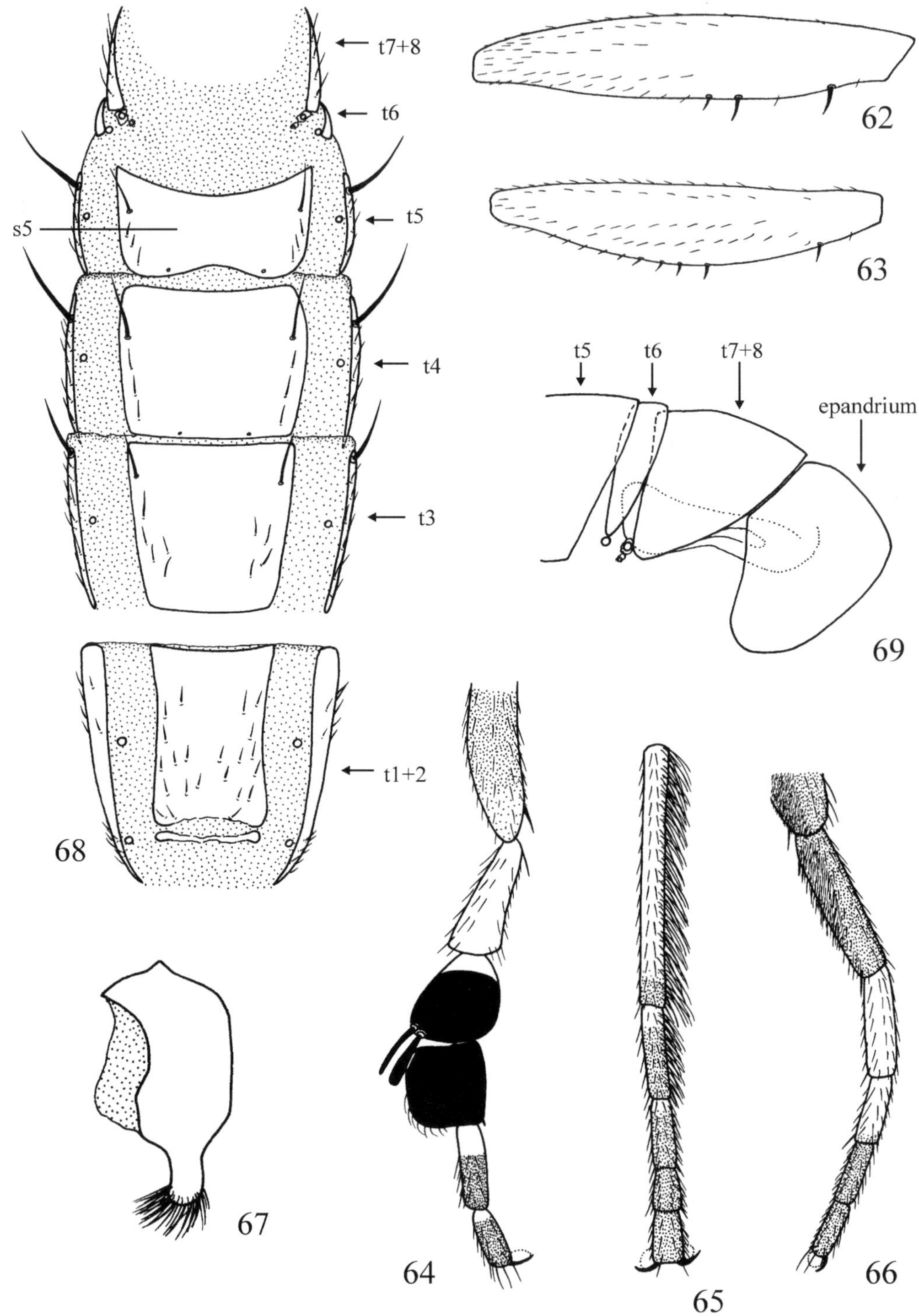

Figs 62-69. *Ortalischema albitarse* (Zetterstedt). 62: ♂ fore femur, anterior view; 63: ♂ fore femur, posterior view; 64: ♂ fore tarsus, posterior view; 65: ♂ mid tarsus, dorsal view; 66: ♀ fore tarsus, posterior view; 67: ♂ hind coxa, anterior view; 68: ♂ abdomen, ventral view, hypopygium removed; 69: tip of ♂ abdomen, lateral view. (s – sternite, t – tergite.)

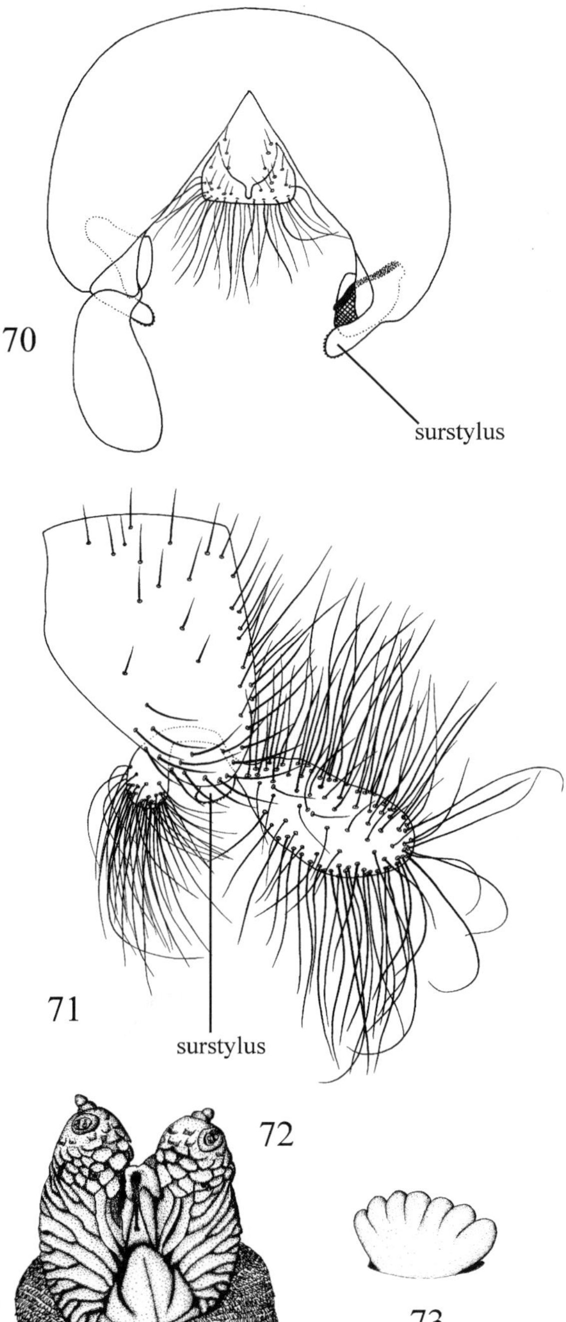

Figs 70-73. *Ortalischema albitarse* (Zetterstedt), ♂. 70: hypopygium, dorsal view, appendages on the right side and all setae on appendages and epandrium omitted; 71: hypopygium, lateral view. 72-73 larval morphology. 72: pseudocephalon, ventral; 73: anterior spiracles.

absent, postpronotal seta absent, anepisternal seta weak, prosternum bare.

Autapomorphies of *Ortalischema* are:

1. Postocellar seta absent
2. ♂ fore tarsus of unique structure and colour, adapted for presenting visual signals to the ♀
3. ♂ epandrium with a pair of large, setose, spoon-shaped appendages
4. ♀ postabdominal tergites 6 and 7 without an anterior prolongation

Two species. One in Europe and Asia (*albitarse* (Zetterstedt)), and one in East Asia (Far East province of Russia) and Japan (Hokkaido) (*maritima* Ozerov; see also Iwasa, 1995). Larva in horse dung.

2. *Ortalischema albitarse* (Zetterstedt, 1847)

(Figs 8, 15-16, 32-33, 36, 62-73)

Sepsis albitarsis Zetterstedt, 1847: 2297. Holotype ♀, 'ad diversorium Thynaes in paroecia Skogn' [NORWAY, NTI, Tynes], in ZML [see below].

Enicita rossica Frey, 1918: 23. Holotype ♀, locality not stated [from title: RUSSIA, NET, Arkhangelsk], in UZMH [see below].

Diagnosis. – As for the genus.

Adult. – Colour: Head mostly dark; anterior half of frontal plate, face, parafacial, and anterior half of gena reddish-yellow to brown. Antenna brown, tip of pedicel and basal part of flagellomere yellow. Thorax, scutellum and abdomen entirely dark. Fore coxa and trochanter yellow; fore femur dark, yellow at extreme tip; ♂ fore tibia reddish-brown, yellow at base and at tip, ♀ fore tibia dark brown; ♂ fore tarsus (Fig. 64) with tarsomere 1 yellow, tarsomeres 2 and 3 shining jet-black except for base of tarsomere 2, tarsomere 4 yellow on basal third but otherwise dark, tarsomere 5 yellow at extreme base but otherwise dark; ♀ fore tarsus (Fig. 66) dark brown except for tarsomeres 2 and 3 which are pale yellow, these rarely partly to mainly darkened; mid coxa partly yellow; hind coxa mostly dark; mid and hind trochanters yellow; mid and hind femora dark brown, reddish on basal sixth and at extreme tip; mid and hind tibiae dark brown; mid and hind tarsi mostly dark, most of tarsomere 1 and part of tarsomere 2 reddish in ♂. Wing clear. Calypter creamy, fringe-hairs white. Haltere yellow.

Head (Fig. 16): Setae very reduced: only inner vertical present. Subvibrissal setae few and weak.

Thorax: Setae very reduced, short and stout in ♂ but longer in ♀; 2 notopleurals; no trace of acrostichal setulae. Pleural setae also reduced. Katepisternum at extreme lower anterior corner (in front of mid coxa) with a stout seta, on a slight boss in ♂. Scutellum matt, grey chagreened at base; half as long as wide (Fig. 15); basal seta stout, half as long as apical seta in ♂ but rather longer and thinner in ♀.

Legs: ♂ fore femur in Figs 62-63, with a few anteroventral and posteroventral spinules in basal half; ♀ fore femur bare ventrally. ♂ fore tarsus in Fig. 64, tarsomeres 2-3 saucer-shaped, hollowed out on outer surface and glossy black here; usually 2 flattened setae on dorsal edge at tip of tarsomere 2; ♀ fore tarsus not modified (Fig. 66). Mid femur and tibia without setae. ♂ mid tarsus with some short fine curly setulae on anterodorsal to ventral surfaces of tarsomeres 1 and 2 (Fig. 65). ♂ hind coxa (Fig. 67) with a small ventral projection covered with dense setulae. Hind femur and tibia without setae; hind tibia rather curved, with the posterodorsal apical seta weak in ♂ but strong in ♀. Hind tarsus not modified.

Wings: Veins bare except for a few fine setulae on basal part of stem-vein on ventral surface. Wing-surface entirely covered with microtrichiae.

Abdomen: Subshining black with violet reflections, very weakly chagreened. Without setulae at sides of syntergite 1+2. ♂ genitalia in Figs 70-71 (also illustrated by Ozerov, 1999e: fig. 347.5).

Wing-length: 2.8-3.5 mm (♂), 2.9-3.6 mm (♀).

Body-length: 4.3-5.0 mm (♂), 4.5-5.1 mm (♀).

Egg and larva. – See under the generic description. The egg has been described in detail by Meier (1995b: 427-431, fig. 24) and the 3rd instar larva by Meier (1996: 33-35, figs 77-85).

Distribution. – Denmark: absent. Norway: only the type-locality (TRI). Sweden: SM to JÄ (Ringdahl, 1934: 3; 1948: 2). Finland: Sa (Pont, 1990: 154). Present in the adjacent area of Russia: Kr.

Widespread but highly disjunct, in Europe (Hungary, Sardinia) (Zuska & Pont, 1984: 157), Russia (Arkhangelsk area, East Siberia), Ukraine and the Caucasus Mts (e.g. Frey, 1925: 76; Ozerov, 1985: 1268; 1999e: 559; Shtakel'berg, 1958: 132), Mongolia (Soós, 1972: 353), and Japan (Hokkaido) (Iwasa, 1995: 785). Replaced in the Far East province of Russia by *O. maritima* Ozerov (1985), but the two species are sympatric (and syntopic?) on Hokkaido (Iwasa, 1995: 785).

Only collected during late June and July.

Biology. – Adults are found almost exclusively on or around horse droppings (e.g. Ringdahl, 1934: 3, 1948: 2; Pont, 1990: 154). Ringdahl (1934: 3) found it once on a flowering lime. Adult and larval habitats are discussed by Ozerov (1997d: 90).

Horse dung is the substrate in which this species breeds. Numerous eggs, lacking the egg filaments typical for all other Sepsidae, are laid in the dung with the operculum exposed to the surrounding air. The larvae are remarkably slender and pointed at the anterior end. They move with great speed through the dung. Pupariation occurs throughout the breeding substrate. Shortly before pupariation two internal, elongate, white stripes appear. The material responsible for this colouration is used for coating the puparium with very fine white pearls which harden into a hard crust, giving the puparium a white appearance. Under laboratory conditions, a culture established from females collected in Sweden (DR, Lima, June 1999) readily produced a large number of puparia in August. At room temperature no adults hatched for several months. Eclosion only occurred after storing the puparia in a refrigerator for two months. It thus appears likely that at least the Swedish populations have an obligatory diapause. A. L. Ozerov (pers. comm.) also found puparial diapause in this species in N. Ossetia.

This is an elusive species, and probably overlooked rather than genuinely scarce. The wings are strikingly short, as in *Themira leachi* and *T. annulipes*, and it is a reluctant flier. It is usually seen with the wings folded over the back, and does not indulge in wing-waving. On horse droppings, both sexes run in an agitated fashion, pausing every so often to brandish and wave the fore legs which are held forwards and outwards, forming an angle of about 90°. This behaviour is typical for a number of non-European genera of Sepsidae such as *Toxopoda* and *Paratoxopoda* (RM, pers. obs.). The modified fore tarsi of both sexes of *Ortalischema* are clearly used for visual display.

Type-material. – *Sepsis albitarsis*: The holotype ♀ was collected by Zetterstedt (1847: 2297) in

Central Norway, 4.vii.1840, 'ad diversorium Thynaes in paroecia Skogn juxta litus maris'. It is in ZML, in the Diptera Scandinaviae collection, drawer 38, where ACP has studied it. It is labelled with a small green square and Zetterstedt's label 'S. albitar- / sis Zett. ♀. / Thyns nov'. It has some strands of mould, but is otherwise in excellent condition.

Enicita rossica: ACP has seen the holotype, in drawer 104 of the Frey collection in UZMH. It is in good condition, and clearly conspecific with *albitarse*. Frey wrote that it was collected by Hellén on 14 June, close to the railway station.

Genus *Zuskamira* Pont, 1987

Zuskamira Pont, 1987c: 267. Type-species: *Zuskamira inexpectata* Pont, 1987, by original designation. Gender: feminine.

Adult *Zuskamira* can be distinguished from the other genera with a full complement of head setae (inner vertical, outer vertical, postocellar, orbital) by the bare prosternum, narrow gena, strong basal scutellar seta, and well-developed dorsocentral setae (1+3-4).

Adult. – Head roundish, facial carina small, face receding, gena very narrow. A full complement of head setae (Fig. 18): orbital, inner vertical, outer vertical, postocellar. Arista bare. Scutum and scutellum matt. Postpronotal seta present. 1+3-4 dorsocentrals, all well-developed and bristle-like; anterior postsutural pair shortest and twice as long as a ground-setula, the presutural pair and subsequent postsutural pairs becoming progressively stronger. 0 intraalar, 1 supraalar, 1 strong postalar. 1 weak proepisternal hair present, proepimeral area bare. Anepisternal present, strong. Katepisternal absent. Anepisternum shining; anepimeron and meron mostly, and katepisternum entirely, pruinose. Meral bridge absent. Prosternum bare. 2 pairs of scutellars. Fore femur without posterodorsal setae. Tibiae without anterodorsal or posterodorsal setae. ♂ fore femur and tibia modified (Figs 74-76). ♂ hind tibia without an osmeterium. Wing without a dark spot. Abdomen without dorsal setae except at posterolateral corners of tergites 4 and 5 in ♂.

♂ abdomen. Slightly constricted after syntergite 1+2, broadest at tip of tergite 4; tergite 6 absent; tergite 7+8 present but narrow (Fig. 80). Sternite 1 setulose; sternite 4 very broad, *Themira*-like, with a pair of large brushes of setae, 1 long spine-like latero-apical seta and 2 groups of median spinules, these all stout and robust (Fig. 81); sternite 5 present as an elongate Y-shaped sclerite extending below sternites 4 and 3; all vestiges of sternite 7+8 absent. Spiracles 1-5 in the intersegmental membrane; spiracles 7 and 8 on a weakly sclerotised plate adjacent to tergite 7+8. Epandrium large, as wide as widest part of tergite 4, its anterior opening (into pre-abdomen) very small. Cercal plate (Figs 82, 85) strongly sclerotised, horseshoe-shaped, with a stout spine at each tip. Surstylus (Figs 82-84) articulating with epandrium, asymmetrical, with basal processes on inner and outer surfaces. Hypandrium firmly attached at one point to tergite 7+8. Sperm pump of the 'dumb-bell' form typical of the Sepsinae (except *Ortalischema*).

♀ abdomen. Sternites 2-5 normal in size, not enlarged. Spiracles 1-5 in the intersegmental membrane; spiracles 6 and 7 at the ventral edge of the tergites. Segments 6 and 7 with the tergites and sternites well separated; the tergites narrow, and each with a median anterior prolongation. Segment 8, when not invaginated, long, extended posteriorly and ventrally, with a ventral element (sternite 8) on each side that remains separated at tip and a dorsal element (tergite 8) that forms the usual pair of ring-like bars that do not quite touch apically. Cerci round. Epiproct small, triangular. Hypoproct not detectable. 2 spermathecae, of different sizes.

Egg. – Egg with a single respiratory filament dorsal to hatching line. Micropyle fused to dorsal side of filament base (as in Fig. 29). Filament with ribbons along rows of small pores. Ribbons arranged in an irregular polygonal pattern (as in Fig. 31). Polygons very long in middle and towards end of filament, so that their parallel sides appear to form long, parallel longitudinal bands; towards the end of filament, chorion broken up into polygonal plates; end of filament slightly clubbed. In cross section, filament with a single, apparently continuous central canal and regularly arranged pillars that support the exochorion of the filament (as in Fig. 24). Pillars with bifurcated bases at chorion of filament.

Larva. – Pseudocephalon bilobed, moderately long and pointed, cephalic lobes with a few, narrow cirri with a few (2-3) long hairs each; facial

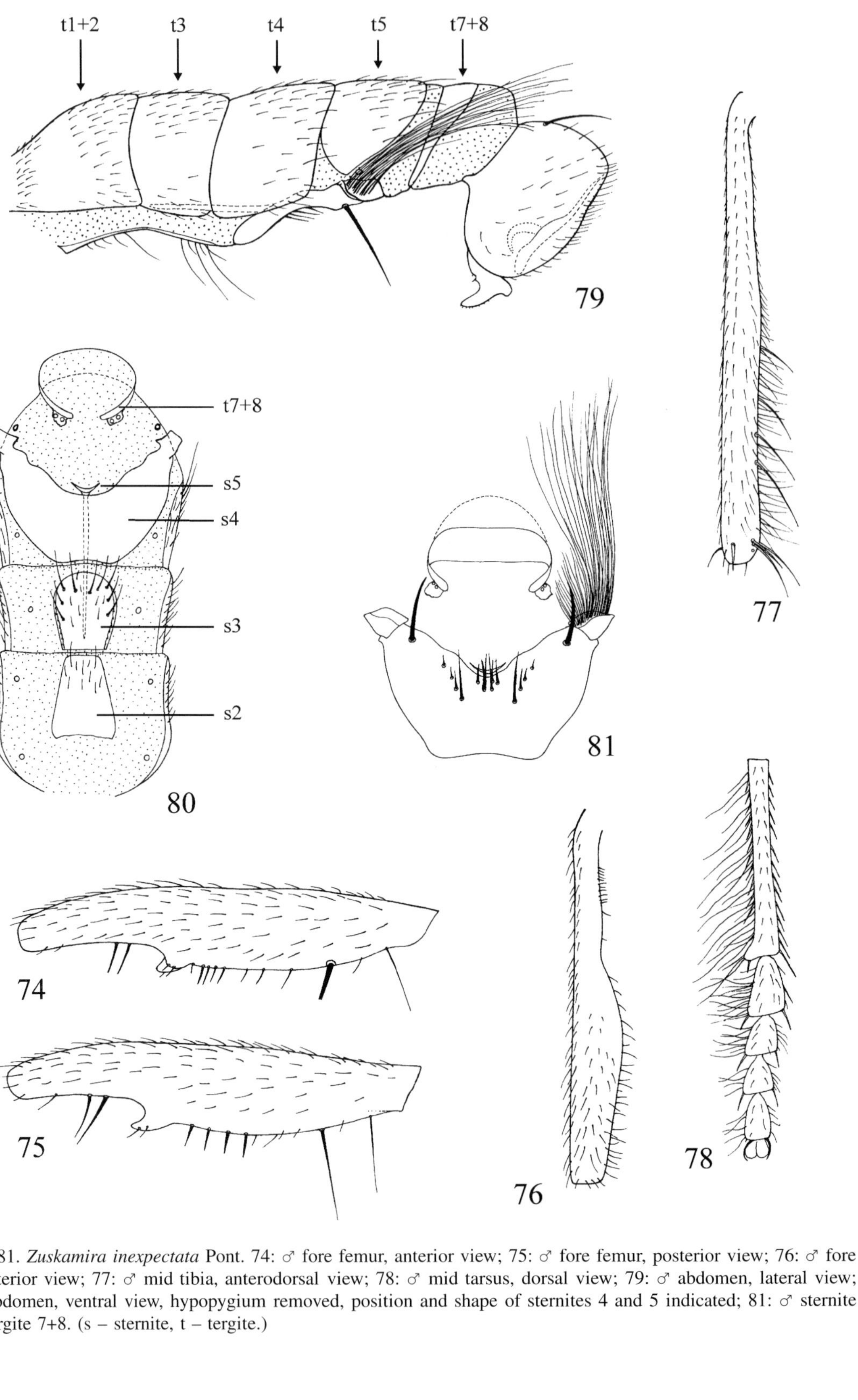

Figs 74-81. *Zuskamira inexpectata* Pont. 74: ♂ fore femur, anterior view; 75: ♂ fore femur, posterior view; 76: ♂ fore tibia, anterior view; 77: ♂ mid tibia, anterodorsal view; 78: ♂ mid tarsus, dorsal view; 79: ♂ abdomen, lateral view; 80: ♂ abdomen, ventral view, hypopygium removed, position and shape of sternites 4 and 5 indicated; 81: ♂ sternite 4 and tergite 7+8. (s – sternite, t – tergite.)

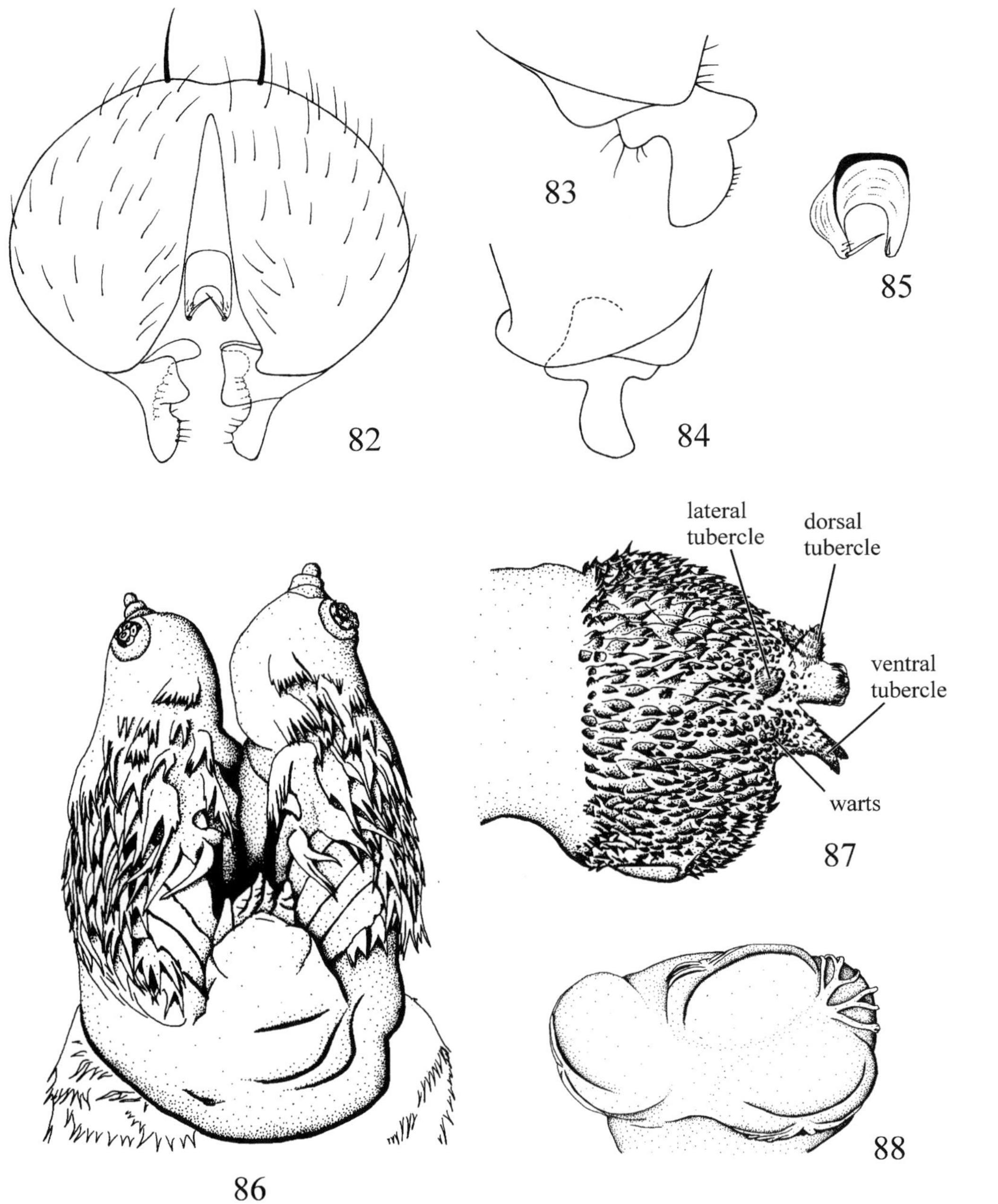

Figs 82-88. *Zuskamira inexpectata* Pont, ♂ genitalia. 82: hypopygium, dorsal view; 83: left surstylus, lateral view; 84: right surstylus, lateral view; 85: cercal plate, oblique view. 86-88. Larval morphology. 86: pseudocephalon, ventral; 87: anal division, lateral; 88: posterior spiracles.

mask with only three ridges on either side of labial lobe; labial lobe with five long projections at anterior end (Fig. 86).

Anal division short; lateral tubercles present (Fig. 87); stigmatophores with a relatively long dorsal tubercle (Fig. 87); ventral tubercles present, simple; subanal tubercles very small and bare (Figs 52, 87); anal pad small, with diagonal furrow, tapering to form a posterior angle; spines densely covering most of the anal division (Fig. 52), spines very blunt and wart-like around the ventral tubercles, anterior to the anal pad, and between the three grooves on the dorsal side of anal division.

Cephaloskeleton: labial and dental sclerites absent; bridge of intermediate sclerite at approximately half length of sclerite (as in Fig. 138), anteriorly with a flat plate; epistomal sclerite longer than wide, without posterior bars; parastomal bars only slightly longer than bridge of intermediate sclerite; basal sclerite 2-2.5 times as long as high, dorsal cornua shorter and 4-5 times narrower than ventral cornua.

Anterior spiracles tree-like, somewhat wider at base than at tip, with 11 lobes; lobes 1.5-2.5 times as long as wide (spiracle as in Fig. 42).

Posterior spiracles with almost equal, narrow, somewhat arched spiracular openings on a single plate; two spiracular openings, mirror-images of each other, located on a slightly separated extension of the spiracular plate, third one perpendicular to the other two on a more strongly separated extension (Fig. 88).

Discussion. – *Zuskamira* is also one of the more basal clades of the Sepsinae. Plesiomorphies in terms of the Sepsinae include: a full complement of head setae (orbital, inner vertical, outer vertical, postocellar); postpronotal seta present; supraalar seta present; postalar seta present; anepisternal seta present; scutellum reduced in size; ♂ hind tibia without osmeterium. *Zuskamira* is apomorphous relative to the ground-plan of the Sepsinae with regard to the following characters: dorsocentrals 1+3-4, well-developed, ♂ surstyli articulating with epandrium, not fused to epandrium, ♂ sperm pump shaped like a dumb-bell.

Autapomorphies of the genus are:

1. ♂ sternite 4 heavily spinose
2. ♂ without any vestiges of sternite 7+8
3. ♂ epandrium very enlarged
4. ♂ hypandrium firmly attached to tergite 7+8 at one point

One species, from Sweden. Larva in horse dung.

3. ***Zuskamira inexpectata*** Pont, 1987
(Figs 18, 52, 74-88)

Zuskamira inexpectata Pont, 1987c: 269, figs 12-25. Holotype ♂, SWEDEN: Västmanland, Klacka Lerberg, in BMNH.

Diagnosis. – As for the genus.

Adult. – Colour: Frons, occiput and lower margin of head black; face, facial carina, parafacial and gena orange. Antenna and arista dark brown, only the articulation between antennal scape and pedicel orange. Thorax, scutellum and abdomen black. Legs black, but most of fore coxa, most or all of trochanters, and basal quarter (♂) or extreme base (♀) of fore tibia yellow. Wing clear, brown only in basal costal cell. Calypter white, with creamy margin and fringe. Haltere yellow.

Head (Fig. 18): A row of tiny setulae present below orbital seta. Fronto-orbital plate shining from some angles, the shining area greatly expanded above, frontal vitta matt, ocellar triangle and occiput weakly grey pruinose. Gena moderate, subequal to half width of antennal flagellomere; with a row of short setae.

Thorax: Scutum, postpronotal lobe, notopleuron and scutellum matt. 2 strong subequal notopleurals; acrostichal setulae very short, in 1-2 irregular rows in front. Proepisternal depression and anepisternum shining black; katepisternum whitish-grey pruinose; anepimeron and meron mostly grey pruinose, with only a little shine on upper anterior margin of anepimeron and at upper posterior corner of meron. Scutellum short, almost twice as wide as long; with a pair of basal lateral setae, over half as long as apical pair.

Legs: Fore coxa and trochanters shining, mid and hind coxae grey dusted; femora and tibiae subshining. ♂ fore legs modified, in Figs 74-76, femur with a long strong anteroventral seta at basal third. ♀ fore femur bare ventrally. ♂ mid coxa with a strong seta on outer surface. Mid femur without setae. ♂ mid tibia (Fig. 77) in apical half with short dense hairs between anteroventral and posteroventral surfaces, and amongst them with 2 strong ventral and 1-2 anteroventral setae; apical setae: short dorsal and ventral, long anterior and anteroventral, and some short fine posteroventral and posterior. ♀ mid tibia simple, with 2 anteroventral setae. ♂ mid tarsus (Fig. 78) with tarsomere 1 attenuated and with a slight lobe at anterior tip, and tarsomeres 2-5 short and round; tarsomere

1 1.12 times as long as combined length of tarsomeres 2-5; all tarsomeres with fine curled hairs on anterior surface and with strong anteroventral and posteroventral apical setae. ♀ mid tarsus normal. ♂ hind coxa and trochanter without modifications. Hind femur with 3 anterodorsals on third quarter. Hind tibia without setae except at apex; in ♂ with several fine erect dorsal hairs on apical fifth, and some curled anterodorsals on apical quarter; ♂ without an osmeterium, but with a pigmented streak on third quarter of anterodorsal surface.

Wing: Veins bare except for a few setulae in basal part of stem-vein on ventral surface. Entirely covered with microtrichiae.

Abdomen: Shining black, ♂ hypopygium large and shining black. Syntergite 1+2 with only short setulae on sides. ♂ genitalia: Figs 82-85.

Wing-length: 2.6-2.7 mm (♂), 2.6-2.7 mm (♀).

Body-length: 3.2 mm (♂), 2.9 mm (♀).

Egg and larva. – See under the generic description.

Distribution. – Known only from the Swedish provinces of VS, DR and UP. Since the original description, this species has only been found by RM, at Lima in eastern DR, close to the border with VR.

Only collected in late June and early July.

Biology. – Another elusive species. Adults are closely associated with horse droppings and corrals. Eggs are laid in horse dung where the larvae also develop. In the laboratory, pupariation occurs after 10-18 days, within the substrate. Under laboratory conditions, a culture established from females collected in Sweden (DR, Lima, June 1999) produced puparia in August. At room temperature no adults hatched for several months. Eclosion only occurred after storing the puparia in a refrigerator for approximately two months. It thus appears likely that *Zuskamira inexpectata* has a winter diapause (see also *Ortalischema*).

Genus ***Themira*** Robineau-Desvoidy, 1830

Themira Robineau-Desvoidy, 1830: 745. Type-species: *Themira pilosa* Robineau-Desvoidy, 1830 [= *Musca putris* Linnaeus, 1758], by designation of Rondani, 1874: 170, 178. Gender: feminine.

Enicopus Walker, 1833: 253. [Preocc. by *Enicopus* Stephens, 1830.] Type-species: *Sepsis annulipes* Meigen, 1826, by monotypy. Gender: masculine.

Cheligaster Macquart, 1835: 479. Type-species: *Musca putris* Linnaeus, 1758, by designation of Duponchel in d'Orbigny, 1843: 453 [see Evenhuis & Thompson, 1990: 234]. Gender: feminine.

Enicita Westwood, 1840: 148. Replacement name for *Enicopus* Walker. Type-species: *Sepsis annulipes* Meigen, 1826, automatic. Gender: feminine.

Halidaya Rondani, 1856: 117. Type-species: '*Themira setosa* Desv.' [error for *Themira pilosa* Robineau-Desvoidy, 1830, = *Musca putris* Linnaeus, 1758], by original designation. Gender: feminine.

Enicomira Duda, 1926a: 27 (as subgenus of *Themira* Robineau-Desvoidy). Type-species: *Sepsis minor* Haliday, 1833, by monotypy. Gender: feminine.

Cheligastrula Strand, 1928: 73. Unjustified replacement name for *Cheligaster* Macquart. Type-species: *Musca putris* Linnaeus, 1758, automatic. Gender: feminine.

Annamira Ozerov, 1999d: 195 (as subgenus of *Themira* Robineau-Desvoidy). Type-species: *Sepsis leachi* Meigen, 1826, by original designation. Gender: feminine. **Syn. n.**

Nadezhdamira Ozerov, 1999d: 199 (as subgenus of *Themira* Robineau-Desvoidy). Type-species: *Sepsis superba* Haliday, 1833, by original designation. Gender: feminine. **Syn. n.**

Adult *Themira* can be recognised in both sexes by the absence of the supraalar seta, and by the short or absent anepisternal seta. Males have ornate fore legs, modified sternite 4 with long bushy setae issuing from the apical angles, and an osmeterium on hind tibia.

Adult. – Head roundish, facial carina small, face vertical or slightly receding, gena narrow to quite broad. Outer vertical absent. Postocellar present. Orbital present, sometimes short. Scutum and scutellum wholly glossy to matt or pruinose. Postpronotal seta present or absent. 1 dorsocentral, but a few additional short setae present in *biloba* and rarely also in *putris*. Intraalar absent, supraalar absent, 1 strong postalar. Proepisternal hairs present, only strong in *nigricornis*; proepimeral area usually with some tiny hairs. Anepisternal short but distinct, or not distinguishable from the hind-marginal setulae. Pleura mostly glossy to mostly pruinose; pleura with varying amounts of silvery-grey pruinosity. Prosternum setulose. 1 scutellar (apical), but absent in *simplicipes*. Fore femur with or without posterodorsal setae. Tibiae without anterodorsal (except on hind tibia in ♂ *leachi*) or posterodorsal setae. ♂ fore tibia (and femur) modified. ♂ hind tibia with the osmeterium on anterodorsal surface, field-like, varying in position and extent. Wing without a dark spot. Abdominal setae absent except on tergites 4 and 5.

♂ abdomen. Tergites 1+2 to 4 normal; tergite 5 normal, often with a trend towards reduction in species with tergite 4 greatly expanded (*superba*, *malformans*, *annulipes*; not *leachi*). Tergite 6 absent; tergite 7+8 with 2 spiracles in lower corner. Spiracles 1-5 in the intersegmental membrane (except spiracle 5 in *annulipes* which may be in tergite 5). Sternite 1 complete, band-like and setulose, reduced to two lateral plates in *minor*, *paludosa* and *leachi*, bare in *leachi*; sternites 2 and 3 complete; sternites 4 and 5 highly modified, sternite 4 greatly expanded and with tufts of bushy setae in posterior corners, sternite 5 large or small, a weakly sclerotised plate confluent with sternite 4 or partly or mainly concealed beneath it; sternite 7+8 usually present as a very narrow band connecting tergite 7+8 across the ventral part of the abdomen. Cercal plate simple at tip, weakly bilobed, or produced into one or two long teat-like projections, usually with a pair of setae at tip. Surstylus usually simple, elongate or short, often asymmetrical. Hypandrium attached only to the epandrium, by two rods, one of which connects with the anterior epandrial wall and one of which connects with the surstylus. Sperm-pump large, 'dumb-bell' in shape.

♀ abdomen (based on *putris*, *minor*, *superba*, *annulipes*). Sternites 2-5 normal in size, not enlarged. Spiracles 1-5 in the intersegmental membrane at the edge of the tergites; spiracles 6 and 7 in the tergites. Postabdominal segments 6 and 7 with the tergites and sternites hardly to narrowly separated, sometimes partly overlapping, the tergites each with an anterior extension, the sternites broad and strongly sclerotised. Segment 8, when not invaginated, long, extended posteriorly and ventrally, with a ventral element (sternite 8) on each side that remains separated at tip and a dorsal element (tergite 8) that forms the usual pair of ring-like bars that do not quite touch apically. Cerci small, round. Epiproct usually absent (present in *annulipes*). Hypoproct present. 2 spermathecae, of different sizes.

Egg. – Description based on *annulipes*, *biloba*, *gracilis*, *lucida*, *minor*, *putris*; a single filament also present in *nigricornis*.

Themira eggs with a single filament with a micropyle as in Fig. 10. In cross section, filament with a central canal and struts that are more or less irregularly arranged (Fig. 21), except for *gracilis* with single layer of regularly-arranged struts between central canal and chorion (as in Fig. 24). Surface of filament with regularly arranged rows of aeropyles (Fig. 26). Arrangement of aeropyles at base of filament in polygons (similar to Fig. 29); in the middle and at the end of the filament, aeropyles arranged in parallel, longitudinal rows (Fig. 26). Filament of *annulipes* unusually short, and this is the only sepsid with a single filament that is shorter than the main body of the egg. Proximal end of filament of *gracilis* lacking aeropyles and aeropyles sparse on the remaining filament.

Larva. – Pseudocephalon bilobed, cephalic lobes usually with numerous combs (Figs 164, 186, 199, 228); facial mask from largely covered with ridges (Fig. 228) to entirely covered with combs (Fig. 199); in *annulipes* and *simplicipes* (former subgenus *Enicita*) with facial mask reduced in size.

Anal division short; lateral tubercles present; stigmatophores with a dorsal tubercle (Figs 44, 166); ventral tubercles present, simple; subanal tubercles small to large; anal pad small to moderately large, with diagonal furrow, rounded or angled laterally; anal tuft present; preanal protuberance present or absent (Figs 132-135, 165, 188); in most *Themira* spp. distance between anterior margin of anal pad and posterior margin of segment 7 from half to a quarter width of segment 7, but in *annulipes* and *simplicipes* (former subgenus *Enicita*) anterior margin of anal pad and posterior margin of segment 7 almost in contact with each other; *leachi* with a double row of spines on abdominal segment 7 (Fig. 11); arrangement of spines on anal division variable, from laterally with numerous spines in two bands and three double rows on dorsal surface (Figs 44, 166) to few spines laterally and spines dorsally in no apparent pattern (Fig. 209).

Cephaloskeleton: labial sclerites absent; in most *Themira* spp. dental sclerites completely fused to mouthhooks (Fig. 47); bridge of intermediate sclerite at approximately half length of sclerite (somewhat more anterior in *gracilis*), anteriorly with a flat plate (Fig. 138); but in *annulipes* and *simplicipes* dental sclerites present, partially fused to mouthhooks (Fig. 137), and bridge of intermediate sclerite situated in anterior third of sclerite, thin; bridge anteriorly without a flat plate (Fig. 136); epistomal sclerite longer than wide, with posterior bars; parastomal bars short, at most reaching posterior edge of bridge of intermediate sclerite (Figs 47, 137), basal sclerite 2-2.5 times as long as high, dorsal cornua shorter and 1.5-2 times (in *annulipes* and *simplicipes*) or 2.5-4 times (in all other *Themira* spp.) narrower than ventral cornua.

Anterior spiracles tree-like, thin and long (*superba*) or more or less wide (e.g. *leachi*), basally with additional lobes, with 10-18 lobes; lobes 2-4 times as long as wide (Figs 42, 187).

Posterior spiracles with almost equal, narrow, weakly arched spiracular openings on an oval plate; two spiracular openings parallel or almost parallel, the third one perpendicular to the other two (Figs 167, 190).

Discussion. – Plesiomorphies (i.e. ground-plan characters of the Sepsinae) include the following: orbital setae present; postvertical setae present; scutellum reduced in size; basal radial and medial wing-cells distinct; hind tibia without anterodorsal or posterodorsal seta; ♀ abdominal spiracles 1-5 situated in the intersegmental membrane; ♂ sternite 7+8 present as a narrow band; ♀ with postabdominal tergites 6 and 7 each with an anterior prolongation. *Themira* is apomorphous relative to the ground-plan of the Sepsinae with regard to the following characters: outer vertical seta absent (shared with *Ortalischema*), anepisternal seta very small or absent (very small only in *Ortalischema*), ♂ hind tibia with an osmeterium, ♂ surstylus articulating with epandrium (except in *germanica*, *gracilis*, *leachi*), ♂ sperm-pump shaped like a 'dumbbell'.

Autapomorphies of the genus are:

1. Absence of the supraalar seta
2. Lobe on anterior side of ♂ fore tibia (weak in *nigricornis*)
3. Parastomal bar of cephaloskeleton half the length of the intermediate sclerite

On adult characters, *Themira* is a well-founded monophyletic group. Because of the varying development of a number of 'significant' characters such as postpronotal seta, katepisternal and other pleural pruinosity, ♂ surstylus articulation and shape, ♂ fore leg ornamentation, there have been consistent attempts to divide the genus up into named subgenera (Ozerov, 1999d) or even to split it into several genera (Mangan, 1976; Steyskal, 1965, 1987). With 32 described species, *Themira* is not a large genus and these attempts seem to us to be both unnecessary and flawed: there has been no cladistic analysis made of the genus (ideally combining egg, larval and adult characters) and to pick out single species or small species-groups on the basis of characters whose polarities are not stated or even known merely creates artificial groups and paraphyletic rest-groups.

Holarctic, impinging on the Neotropical region in Mexico and the Oriental region in N India, Nepal, China (Jiangsu). 32 species worldwide, of which 16 (plus one new species, described here) occur in Europe. A key to Russian species (22 species) was given by Ozerov (1999e), and the same author has given an excellent and detailed review of the World species (Ozerov, 1999d).

Key to European species of *Themira* Robineau-Desvoidy

Adults

1 Postpronotal seta absent. Syntergite 1+2 posteriorly at sides without longer setulae (as in *Sepsis fulgens*, Fig. 386). Fore femur without posterodorsal setae (1-2 very short ones present in *minor* and *paludosa*). ♀: fore femur without stout anteroventral spinules (sometimes 2-3 short ones present in *leachi*) 2

– Postpronotal seta present and strong, weak only in *gracilis*. (Care is needed with this character as the seta is sometimes knocked off and the scar is difficult to detect[4]). Syntergite 1+2 posteriorly at sides with several longer stronger setulae (as in *Sepsis violacea*, Fig. 385). Fore femur with several

[4] The species *minor* and *lucida,* and even *paludosa* and *germanica,* are easily confused.

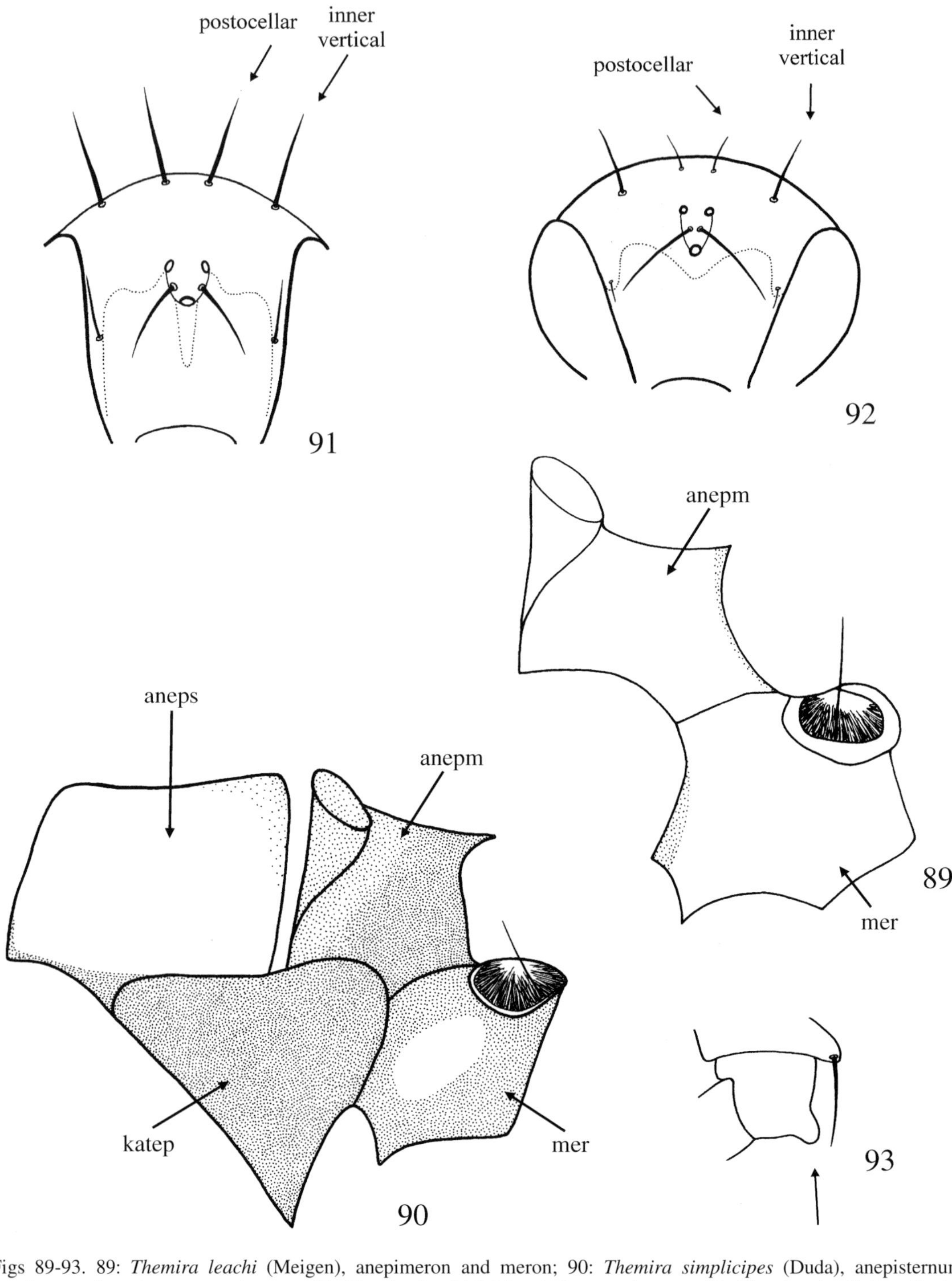

Figs 89-93. 89: *Themira leachi* (Meigen), anepimeron and meron; 90: *Themira simplicipes* (Duda), anepisternum, anepimeron, katepisternum and meron; 91: *Themira leachi* (Meigen), frons and vertex; 92: *Themira simplicipes* (Duda), frons and vertex; 93: *Themira leachi* (Meigen), ♂ hind trochanter, lateral view. (anepm – anepimeron, aneps – anepisternum, katep – katepisternum, mer – meron.)

strong posterodorsal setae in apical half (very short in ♀ *superba*). ♀: fore femur with 4-8 stout anteroventral spinules..... 6

2(1) Katepisternum and meron glossy black, with virtually no white pruinosity (Fig. 89). Postocellar seta subequal to inner vertical (Fig. 91). Anterior notopleural about half as long as posterior one. ♂: sternite 4 with a pair of long dense broad tufts of setae issuing from from the posterior corners, that are always visible in dried specimens (Fig. 196); sternites 2-4 each with a conspicuous median tubercle, best seen in lateral view (Fig. 194); hind trochanter with a lobe-like process on inner surface (Fig. 93); fore legs in Figs 191-192; hind tibia expanded at middle, where the osmeterium ends (Fig. 193), and twisted along its longitudinal axis; genitalia in Figs 197-198. ♀: wings distinctly smoky; mid and hind tibiae with an anteroventral seta.... .... 9. *leachi* (Meigen)

– Katepisternum entirely white pruinose, meron pruinose except on an oval median patch (Fig. 90). Postocellar seta only half as long as inner vertical (Fig. 92). Anterior notopleural less than a third as long as posterior one. ♂: sternite 4 with the tufts of setae in posterior corners shorter and narrower, not always visible in dried specimens; sternites 2-4 without tubercles; hind trochanter simple. ♀: wings clear; hind tibia, and usually mid tibia, without anteroventral setae 3

3(2) Apical scutellar seta absent. Anepimeron entirely grey pruinose (Fig. 90). Fore femur largely, and mid and hind femora broadly, yellow. Mid tibia with a short anteroventral seta. Calypters dark brown. Scutum wholly matt, not shining. ♂: osmeterium on hind tibia short, only one fifth of tibial length (Fig. 275); fore leg in Figs 272-273, tarsomeres 2 and 3 subequal; genitalia in Figs 277-278......... .... 18. *simplicipes* (Duda)

– Apical scutellar seta present and strong. Anepimeron glossy except for a very narrow posterior band. Fore femur mostly dark, at most yellow at base and tip. ♂: osmeterium on hind tibia longer, one third of tibial length; fore leg with tarsomeres 2 and 3 not subequal (Figs 98-99)......... 4

4(3) Legs almost wholly black, except for the coxae and trochanters. Scutum mainly glossy black, except posteriorly, and postpronotal lobes glossy black; but if more extensively matt then at least anterior area of scutum and most of postpronotal lobes glossy black. Orbital subequal to postocellar. (Head more round in profile, the occiput at middle just under half of eye-width (Fig. 94); calypters white) ♂: mid tarsus black, tarsomeres 4 and 5 rather enlarged and dorso-ventrally flattened, at tip of tarsomere 4 with a short but expanded posterior seta (Fig. 100); fore leg with tarsomere 3 shorter than tarsomere 2 (Fig. 98); fore legs in Figs 219-220; hind tibia in Fig. 221; sternite 4 and genitalia Figs 224-227 .... 12. *minor* (Haliday)

– Legs partly yellow, in addition to coxae and trochanters the bases of the mid and hind femora yellow and (*annulipes*) also parts of the tibiae and tarsi. Scutum entirely and evenly matt, without any trace of shine; postpronotal lobes matt (*annulipes*) or shining (*paludosa*). Orbital half as long as postocellar. ♂: mid tarsus different.... 5

5(4) Postpronotal lobes matt. Meron glossy on extreme posterior part as well as medially (Fig. 97). Calypters brown. Mid tibia often with an anteroventral seta in both sexes. ♂: mid tarsus (Fig. 101) with tarsomere 1 elongated, tarsomeres 2-4 short and laterally compressed, tarsomere 1 mainly yellow with a black tip, tarsomere 2 black with a pale base, tarsomeres 3-5 black; fore tarsus with tarsomere 3 longer than tarsomere 2 (Fig. 99); fore legs in Figs 140-141; hind tibia in Fig. 142; sternite 4 and genitalia in Figs 145-147 .... 4. *annulipes* (Meigen)

– Postpronotal lobes partly glossy. Meron only glossy medially, pruinose posteriorly (as in Fig. 96). Calypters white. Mid tibia without anteroventral seta. ♂: mid tarsus simple, tarsomeres not as described above; fore tarsus with tarsomere 3 shorter than tarsomere 2 (as in Fig. 98); fore legs in Figs 239-240; hind tibia in Fig. 241; sternite 4 and genitalia in Figs 244-246..... .... 14. *paludosa* Elberg

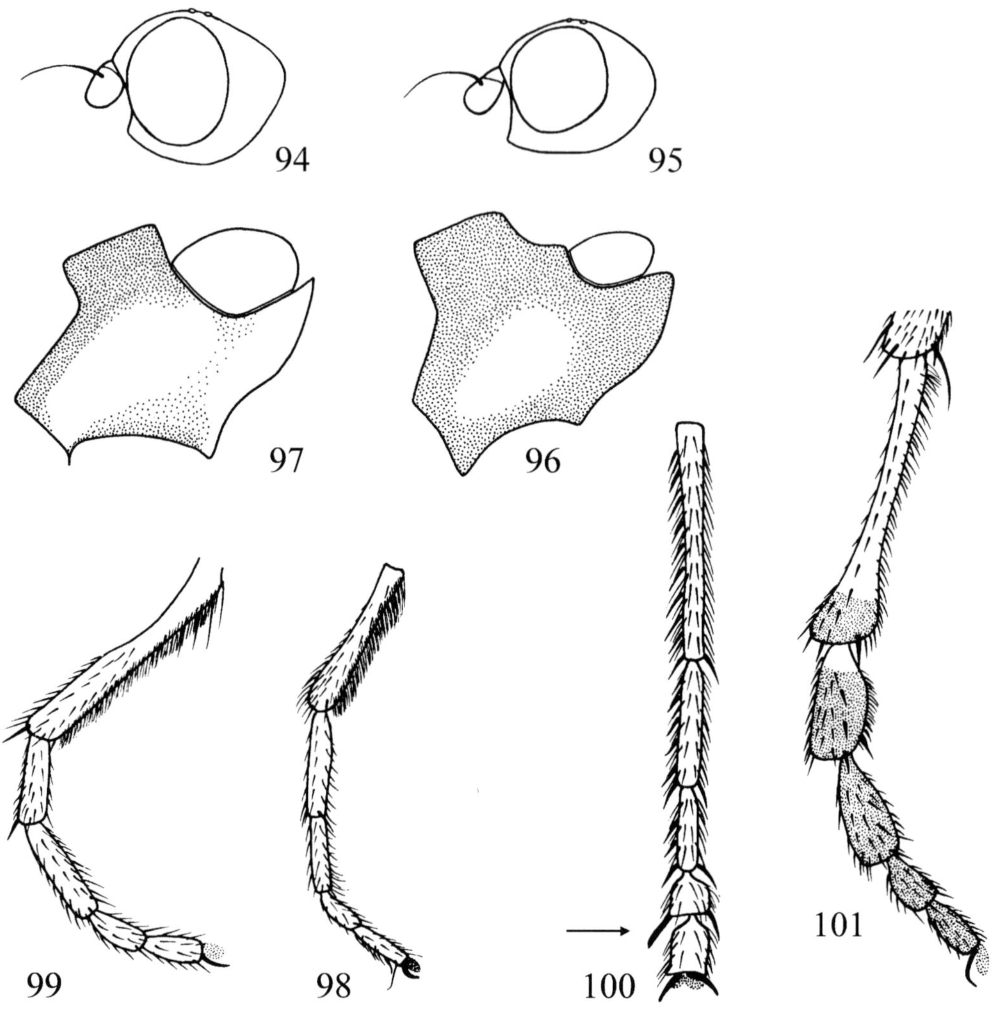

Figs 94-101. Head in lateral view. 94: *Themira minor* (Haliday); 95: *Themira annulipes* (Meigen). Meron. 96: *Themira minor* (Haliday); 97: *Themira annulipes* (Meigen). ♂ fore tarsus, lateral view. 98: *Themira minor* (Haliday); 99: *Themira annulipes* (Meigen). ♂ mid tarsus. 100: *Themira minor* (Haliday), dorsal view; 101: *Themira annulipes* (Meigen), lateral view.

6(1) Katepisternum mainly or entirely glossy black, at most with a white pruinose band along upper part and entire lower part glossy black (Figs 102-103, 105) 7

– Katepisternum entirely white pruinose (Fig. 121) 11

7(6) Meron entirely pruinose, without a glossy median patch (Fig. 102). Katepisternum with a continuous pruinose band along upper third (Fig. 102). Anepisternum weakly pruinose in upper posterior corner (Fig. 102). Legs partly yellow: coxae, trochanters, mid and hind femora at base, fore tibia, and mid and hind tibiae partly yellow. Hind femur without trace of anterodorsal setae around middle, or at most with 1 seta. Articulation of antennal pedicel and flagellomere orange. Groundcolour of gena and facial carina orange. ♂: fore legs in Figs 177-178; hind tibia and osmeterium in Fig. 179; sternite 4 and genitalia in Figs 182-185. ♀: fore femur with 4

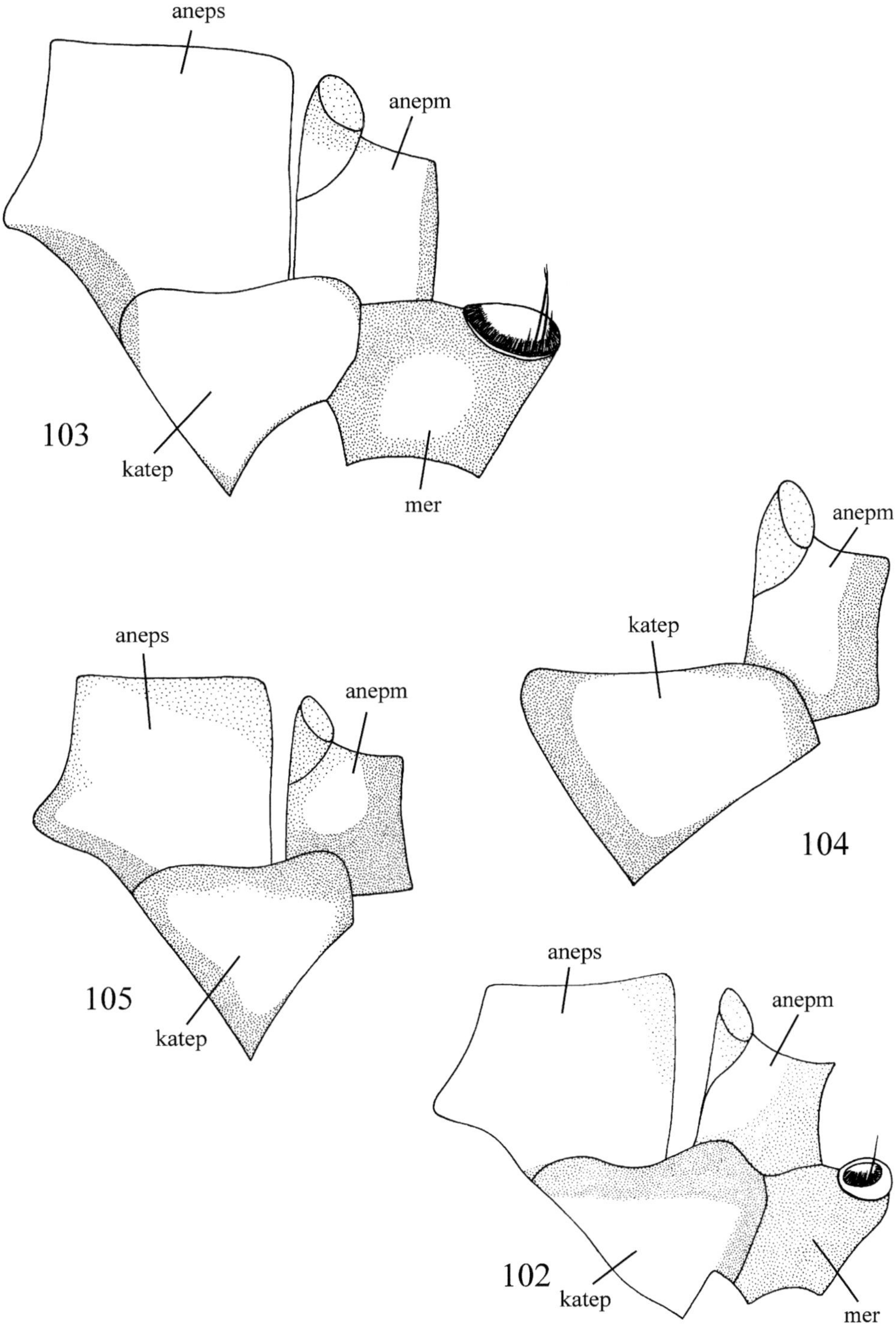

Figs 102-105. Pleural plates. 102: *Themira gracilis* (Zetterstedt), anepisternum, anepimeron, katepisternum, meron. 103: *Themira superba* (Haliday), anepisternum, anepimeron, katepisternum, meron. 104: *Themira malformans* Melander & Spuler, anepimeron, katepisternum. 105: *Themira arctica* (Becker), anepisternum, anepimeron, katepisternum. (anepm – anepimeron, aneps – anepisternum, katep – katepisternum, mer – meron.)

weak anteroventral spinules; mid and hind femora bare ventrally .... 8. *gracilis* (Zetterstedt)

– Meron with a glossy oval median patch, pruinose around the edges (Fig. 103). Katepisternum much more narrowly pruinose along upper margin, if at all (Figs 103-105). Anepisternum entirely glossy black (except *arctica*, which is weakly pruinose in upper posterior corner, Fig. 105). Legs entirely black. Hind femur with at least weak anterodorsal setae around middle. Antennae black. Gena and facial carina brown or black 8

8(7) Katepisternal pruinosity very reduced along upper margin, restricted at the most to a small patch in posterior and/or anterior corner (Fig. 103). Scutum mostly glossy black (*superba*) or only thinly pruinose and subshining (*malformans*). Mid and hind coxae glossy to subshining outside. Gena and facial carina black. ♂: fore femur very curved, almost S-shaped (Figs 210, 279); mid femur rather curved, in apical half with either posteroventral (*superba*, Fig. 110) or anterior (*malformans*, Fig. 111) setae; hind femur with fine posterior setae on at least basal half. ♀: hind femur with some fine posterior setae on basal quarter to third 9

– Katepisternum with a complete though narrow band of pruinosity along upper margin (Fig. 105). Scutum more densely pruinose. Mid and hind coxae pruinose. Gena and facial carina usually brown. ♂: fore femur more or less straight; mid femur straight, without long setae; hind femur without posterior setae; hind coxa and trochanter without modifications. ♀: hind femur without posterior setae 10

9(8) Katepisternum without any pruinosity in upper posterior corner, lower anterior edge of katepisternum almost devoid of pruinosity except above and below (Fig. 103). Anepimeron with a narrow pruinose band on posterior margin (Fig. 103). Fore coxa entirely glossy black. Scutum and postpronotal lobes largely glossy black. Anterior notopleural usually absent. ♂: fore tarsus with tarsomere 1 not modified and tarsomere 2 shorter than tarsomere 3 (Fig. 116); hind tibia not modified dorsally, without anteroventral setae; hind trochanter with a downwardly-directed thorn-like process on inner surface (Figs 108-109); hind coxa not modified (Fig. 106); mid femur with fine curled posteroventral setae on apical half (Fig. 110); hind femur with rather longer fine posterior setae on all but apical third; mid tibia with only short ventral apical setae; fore legs in Figs 279-280; hind tibia and osmeterium in Fig. 281; sternite 4 and genitalia in Figs 284-286. ♀: mid femur without anterior setae in apical half .... 19. *superba* (Haliday)

– Katepisternum with a small band of pruinosity in upper posterior corner, lower anterior edge of katepisternum with a broad uninterrupted pruinose band (Fig. 104). Anepimeron usually narrowly pruinose on lower as well as posterior margins (Fig. 104). Fore coxa narrowly pruinose above. Scutum and postpronotal lobes thinly pruinose, subshining. Anterior notopleural short and fine. ♂: fore tarsus with tarsomere 1 produced apically into a short blunt dorsal projection and tarsomere 2 longer than tarsomere 3 (Fig. 117); hind tibia at tip of anterodorsal surface with a short comb of erect setulae and 1 stout curled anterior seta, and with several fine setae on apical 3/5 of anteroventral surface (Fig. 112); hind trochanter not modified; hind coxa with a short stout spine in front towards middle (Fig. 107); mid femur with straight sparse anterior setae on apical half (Fig. 111); hind femur with rather shorter fine posterior setae on basal half; mid tibia with the anteroventral and ventral apical setae long and fine (Fig. 113); fore legs in Figs 210-212; hind tibia and osmeterium in Fig. 213; sternite 4 and genitalia in Figs 216-218. ♀: mid femur with some short anterior setae in apical half .... 11. *malformans* Melander & Spuler

10(8) Anepisternum with a small pruinose area in upper posterior corner; katepisternum only narrowly pruinose on posterior margin (Fig. 105). Anterior notopleural seta subequal to or only slightly shorter than posterior seta. ♂: fore tarsus with tarsomere 1 shorter, 1.7 times as long as

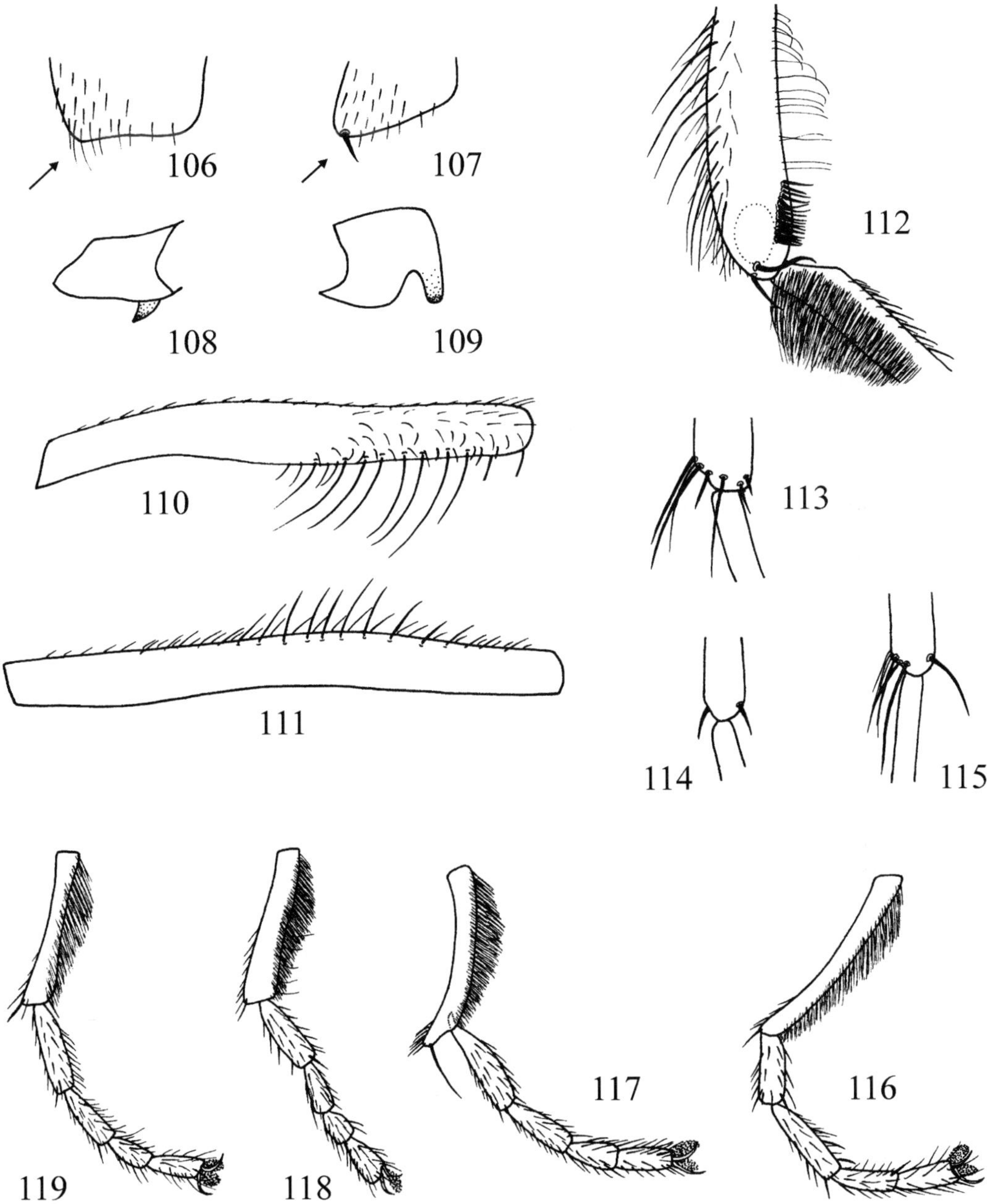

Figs 106-119. Tip of ♂ hind coxa, lateral view. 106: *Themira superba* (Haliday); 107: *Themira malformans* Melander & Spuler. 108, 109: *Themira superba* (Haliday), ♂ hind trochanter, anterior and posterior views. 110: *Themira superba* (Haliday), ♂ mid femur, posterior view. 111: *Themira malformans* Melander & Spuler, ♂ mid femur, ventral view. 112: *Themira malformans* Melander & Spuler, apex of ♂ hind tibia, anterodorsal view. Tip of ♂ mid tibia, anterior view. 113: *Themira malformans* Melander & Spuler; 114: *Themira arctica* (Becker); 115: *Themira ringdahli* Pont, sp. n. ♂ fore tarsus, anterior view. 116: *Themira superba* (Haliday); 117: *Themira malformans* Melander & Spuler; 118: *Themira arctica* (Becker); 119: *Themira ringdahli* Pont, sp. n.

tarsomere 2 (Fig. 118); hind tibia with the osmeterium situated on basal 2/5 of anterodorsal surface (Fig. 150); mid tibia with the anteroventral and ventral apical setae minute (Fig. 114); fore legs in Figs 148-149; sternite 4 and genitalia in Figs 153-155. ♀: fore femur with 4-5 anteroventral spinules .... 5. *arctica* (Becker)

– Anepisternum entirely glossy; katepisternum more broadly pruinose on posterior margin. Anterior notopleural short and fine, no more than half as long as posterior seta. ♂: fore tarsus with tarsomere 1 longer, twice as long as tarsomere 2 (Fig. 119); hind tibia with the osmeterium situated on median third of anterodorsal surface (Fig. 266); mid tibia with the anteroventral and ventral apical setae long and fine (Fig. 115); fore legs in Figs 263-265; sternite 4 and genitalia in Figs 269-271. ♀: not known...................... .... 17. *ringdahli* Pont, sp. n.

11(6) Mid tibia with 1 anteroventral seta. Prosternum broader, dilated anteriorly (Fig. 122). Fore coxa pruinose on entire anterior surface and broadly at base on outer surface. Gena broader, equal to at

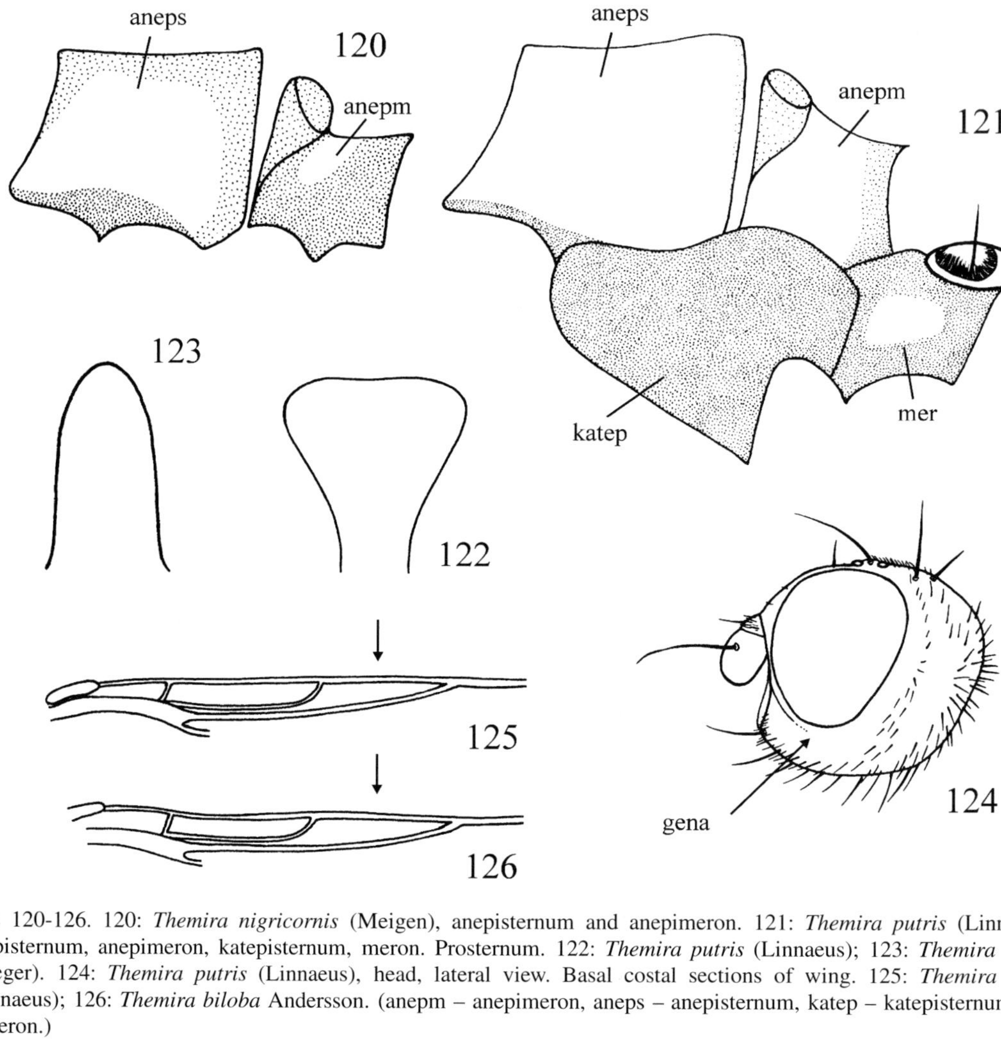

Figs 120-126. 120: *Themira nigricornis* (Meigen), anepisternum and anepimeron. 121: *Themira putris* (Linnaeus), anepisternum, anepimeron, katepisternum, meron. Prosternum. 122: *Themira putris* (Linnaeus); 123: *Themira lucida* (Stæger). 124: *Themira putris* (Linnaeus), head, lateral view. Basal costal sections of wing. 125: *Themira putris* (Linnaeus); 126: *Themira biloba* Andersson. (anepm – anepimeron, aneps – anepisternum, katep – katepisternum, mer – meron.)

least 2/3 length of antennal flagellomere (Fig. 124). Anepisternal setula distinct. Several tiny proepimeral setulae present. 12

– Mid tibia without anteroventral seta. Prosternum narrower, not dilated anteriorly (Fig. 123). Fore coxa entirely glossy, except very narrowly at base on outer surface. Gena narrower, at middle equal to no more than half width of antennal flagellomere (as in Fig. 94). Anepisternal setula absent. Proepimeral setulae absent 14

12(11) Anepisternum mainly glossy but pruinose around edges; anepimeron pruinose except on a median patch (Fig. 120); meron entirely pruinose. Scape, pedicel, and sometimes base of flagellomere orange to yellow. The entire genal area (below eye) orange to yellow. Legs partly yellow: coxae, trochanters, mid and hind femora basally, knees, fore tibia mainly, and mid and hind tibiae in part. Scutellum without basal setula. Proepisternal setula short but conspicuous. ♂: fore tarsus with tarsomere 1 well over twice as long as tarsomere 2 (10 : 3.5-4.5); fore tibia without a submedian anterior lobe, and fore legs in Figs 229-230; hind tibia in Fig. 231; sternite 4 and genitalia in Figs 234-237. ♀: anterior notopleural subequal to posterior one; fore femur with 4-6 anteroventral spinules . . . . 13. *nigricornis* (Meigen)

– Anepisternum virtually entirely glossy; anepimeron glossy except on posterior margin; meron pruinose except on an oval median patch (Fig. 121). Antennae entirely black. The entire genal area (below eye) black, at most partly brown. Legs virtually entirely black. Scutellum with a pair of small basal setulae. Proepisternal setula minute and inconspicuous. ♂: fore tarsus with tarsomere 1 shorter, less than twice as long as tarsomere 2; fore tibia with a median anterior lobe, and fore legs in Figs 156-158, 255-256. ♀: anterior notopleural half to two-thirds length of posterior one; fore femur usually with 5-7 anteroventral spinules 13

13(12) Section of costa between humerus and subcosta much longer than that between subcosta and vein R_1(10 : 7) (Fig. 125). All dorsocentral setulae tiny, except for the usual pair of prescutellar setae. ♂: osmeterium shorter, occupying almost a third of hind tibial length, the tibia not expanded medially (Fig. 257); fore legs in Figs 255-256, fore femur with the more apical of the ventral pegs bent at tip, tarsomere 1 subequal to tarsomere 2; sternite 4 in Fig. 260, the lobes with a brush of very long setae; genitalia in Figs 261-262, cercal plate with a pair of long setae and surstylus notched on inner surface . . . . 16. *putris* (Linnaeus)

– Section of costa between humerus and subcosta subequal to that between subcosta and vein R_1(10 : 10) (Fig. 126). Several strong and conspicuous dorsocentral setulae or setae present before the prescutellar pair. ♂: osmeterium longer, occupying almost half of hind tibial length, the tibia itself distinctly expanded medially (Fig. 159); fore legs in Figs 156-158, fore femur with the more apical of the ventral pegs entirely straight, tarsomere 1 longer than tarsomere 2 (10 : 7); sternite 4 in Fig. 161, the lobes with numerous shorter setae in addition to the brush of long setae; genitalia in Figs 162-163, cercal plate with a pair of short setae and surstylus distinctly bilobed on inner surface . . . . 6. *biloba* Andersson

14(11) Postpronotal lobes and scutum glossy black, the scutum usually only pruinose before scutellum, sometimes more extensively so. Legs wholly black, at most knees and trochanters reddish. ♂: fore tarsus with tarsomere 2 longer than tarsomere 3, tarsomere 1 of normal structure (Fig. 131); hind tibia with the osmeterium long (Fig. 202), occupying median half of tibia; fore femur (Fig. 200) straight, without a strong anteroventral seta at base; fore tibia (Fig. 201) with the tooth short, and with (2-) 3 setae slightly exceeding tibial depth; mid femur, hind femur and hind tibia without setae; sternite 4 and genitalia in Figs 205-208 . . . . 10. *lucida* (Stæger)

– Postpronotal lobes and scutum matt. ♂: fore tarsus with tarsomere 2 shorter than tarsomere 3, tarsomere 1 curved and constricted (Fig. 130); hind tibia with the osmeterium short (Figs 171, 249), occupying a quarter to a third of tibial length; fore femur (Figs 169, 247) with a strong basal anteroventral seta 15

15(14) Legs wholly black, including coxae and trochanters, at most the joints obscurely reddish. ♂: fore femur (Fig. 247) strongly curved, with a submedian anteroventral projection bearing 2 heavy spines in addition to the usual anteroventral peg; fore tibia (Fig. 248) with the submedian anterior tooth short, bearing 4-5 setae the longest of which reach almost to tip of tibia, the tooth preceded by 2 stout anterior setae; hind tibia (Figs 129, 249) with a ventral excision near middle, and the tibia bent at this point, without erect dorsal to anterodorsal hairs on apical third; hind femur (Fig. 127) with 3-4 erect posteroventral setae at middle and a row of short curled anteroventrals on basal half; mid tibia with several apical posterior setae and a row of short posteroventral to posterior setulae on apical third (Fig. 128); sternite 4 and genitalia in Figs 252-254. ♀: fore femur with 2-3, rarely as many as 5, anteroventral spinules .... 15. *pusilla* (Zetterstedt)

– Fore coxa and all trochanters yellow, and in female bases of femora and often also knees yellowish, otherwise legs black. ♂: fore femur (Fig. 169) straight, without a submedian anteroventral projection, with 2 moderate anteroventral spines in addition to the usual anteroventral peg; fore tibia (Fig. 170) with the submedian anterior tooth long, not bearing long setae or spines, and not preceded by strong anterior setae; hind tibia straight and without a median ventral excision, with fine erect dorsal to anterodorsal hairs on apical third (Fig. 171); hind femur bare ventrally, and mid tibia bare posteriorly; sternite 4 and genitalia in Figs 174-176. ♀: fore femur with 5-7 anteroventral spinules .... 7. *germanica* Duda

Larvae

1 Larva with 7 spiny creeping welts on abdominal segments (Fig. 11); anal pad strongly extended and drawn out into a long lateral and a shorter posteromedian corner (Figs 11, 132) .... 9. *leachi* (Meigen)

– Larva without creeping welt on 7th abdominal segment, i.e. with only 6 welts (Fig. 168); anal pad either round or only weakly drawn out into corners (Figs 133-135, 139, 188, 238) 2

2(1) Bridge of intermediate sclerite in anterior third of sclerite (Fig. 136); dental sclerites not completely fused to mouthhooks (Fig. 137); anal pad laterally rounded and almost as wide as anal division (Fig. 133) .... 4. *annulipes* (Meigen) .... 18. *simplicipes* (Duda)

– Bridge of intermediate sclerite at approximately half the length of sclerite (Fig. 138), in *gracilis* in a slightly more anterior position, but not in anterior third; dental sclerite fused to mouthhook (Fig. 47); anal pad laterally either somewhat angular (Fig. 135) or if rounded then at most 2/3 width of anal division (Figs 134, 139) .... remaining *Themira* spp.

4. *Themira annulipes* (Meigen, 1826)

(Figs 21, 95, 97, 99, 101, 133, 137, 140-147)

Sepsis annulipes Meigen, 1826: 292; Meigen plate 268 fig. 3 [Morge, 1976c]. Lectotype ♂, locality not stated [GERMANY: probably Hamburg], in NMW [see designation below].

Nemopoda brunicosa Robineau-Desvoidy, 1830: 745. Holotype/syntypes ♀, FRANCE, Saint-Sauveur, not in MNHNP and presumed destroyed. **Syn. n.**

Nemopoda varipes Meigen, 1838: 351; Meigen plate 157 fig. 11 [Morge, 1976a]. Holotype ♀, 'Baiern' [GERMANY, Bavaria], in MNHNP [see below].

Sepsis marginata Hoffmeister, 1844: 13. Holotype ♂, locality not stated [GERMANY, Hessen: Kassel district], not located and presumed destroyed [see below]. **Syn. n.**

Enicita annulipes var. *crassiseta* Duda, 1926a: 33, 72. Lectotype ♀, FRANCE, Chamonix, in ZMHU [by designation of Ozerov, 1997a: 479].

Enicita nigra Duda. 1926a: 72, 88. Unavailable name; citation of a Meigen MS name, in synonymy with *Enicita annulipes* and *Themira superba*.

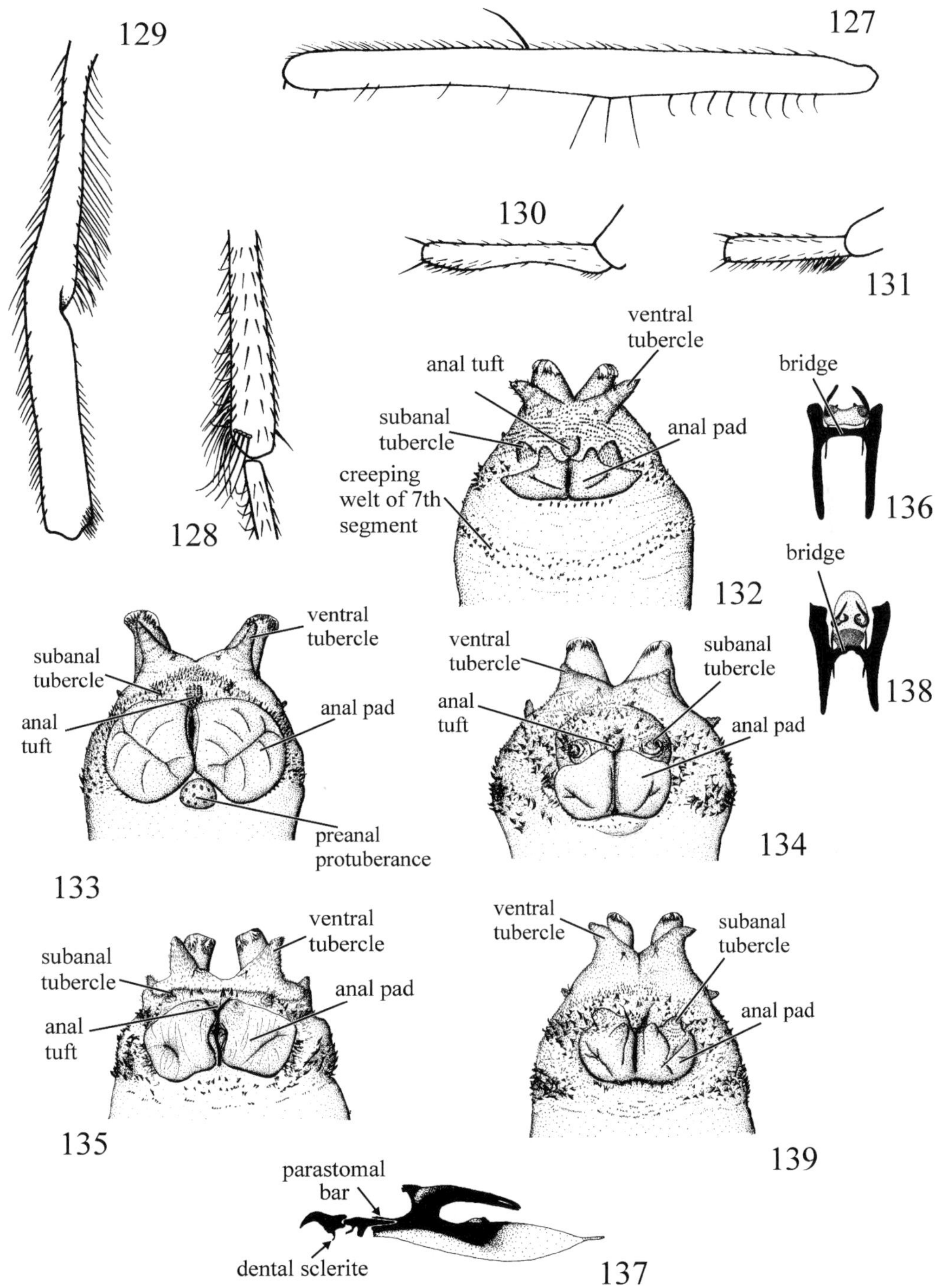

Figs 127-139. *Themira pusilla* (Zetterstedt). 127: ♂ hind femur, anterior view; 128: tip of ♂ mid tibia, posterior view; 129: ♂ hind tibia, anterior view. ♂ fore tarsomere 1, anterior view: 130: *Themira germanica* Duda; 131: *Themira lucida* (Stæger). 132-139. Larval morphology. Anal division, ventral. 132: *Themira leachi* (Meigen); 133: *Themira annulipes* (Meigen); 134: *Themira minor* (Haliday); 135: *Themira lucida* (Stæger); 136: *Themira simplicipes* (Duda), intermediate sclerite, dorsal; 137: *Themira annulipes* (Meigen), cephaloskeleton, lateral; 138: *Themira minor* (Haliday), intermediate sclerite, dorsal; 139: *Themira putris* (Linnaeus), anal division, ventral.

Diagnosis. – Adult ♂ recognised by the remarkable form and colour of the mid tarsomeres (Fig. 101). ♀ recognised by the absence of postpronotal setae, entirely pruinose katepisternum, and partly yellow legs; it is most similar to *paludosa*, but can be reliably distinguished by the characters given in the key (p. 53).

Larval pseudocephalon with a very small facial mask consisting of combs only; cephaloskeleton with dental sclerites only partially fused to mouthhooks (Fig. 137) and bridge of intermediate sclerite in its anterior third (as in Fig. 136); anal division with a very small subanal tubercle and a very large laterally rounded anal pad with diagonal fold (Fig. 133); anterior spiracles with 12-14 lobes; with six abdominal creeping welts; anal pad about as wide as the anal division (Fig. 133); with few spines on lateral side of anal division, on dorsum with three double rows of spines and no spines between the rows (as in Fig. 44).

Adult. – Colour: Head mostly black; parafacial, gena and face reddish. Antenna dark brown, inner surface of flagellomere and articulation of the segments mostly orange-yellow. Thorax, scutellum and abdomen entirely black. Fore coxa except dorsal tip and trochanters yellow, mid and hind coxae mostly dark brown. Fore legs black, extreme base of femur, knees and tip of tibia yellow. Mid and hind legs dark, but femora broadly yellow at base and on knees, and tip of mid tibia and often tip of hind tibia and hind tarsomere 1 yellow, less conspicuously so in ♀ than in ♂; ♂ mid tarsus black but a basal ring on tarsomere 2 and most of tarsomere 1 pale (Fig. 101), ♀ mid tarsus dark but tarsomere 1 largely yellow. Wing rather smoky (♂) or almost clear (♀); calypter brown, with dark brown margin and fringe-hairs; haltere creamy-yellow.

Head (Fig. 95): Frons mostly shining, occiput weakly brownish-grey pruinose. Postocellar about half length of inner vertical; orbital very short, only half as long as postocellar. Gena at middle much less than half length of antennal flagellomere; subvibrissal setae numerous but very short.

Thorax: Scutum, postpronotal lobes and scutellum brownish-grey matt and nowhere shining. Setae: postpronotal absent; anterior notopleural very short, hardly 1/3 length of posterior one; acrostichal setulae minute. Pleural pruinosity: anepisternum glossy; katepisternum entirely white pruinose; anepimeron thinly pruinose on upper anterior corner and on posterior half or less, otherwise glossy; meron thinly pruinose above, in front and below (Fig. 97), otherwise glossy but this glossy patch partly or completely divided by a vertical pruinose bar descending from lower margin of posterior spiracle, posterior margin therefore glossy. Setae reduced; no katepisternal setula. Scutellum short, twice as wide as long, without basal setulae.

Legs: Fore coxa apparently mostly shining, except basally, mid and hind coxae mostly pruinose; trochanters shining, femora and tibiae subshining. Fore femur without posterodorsal setae; ♂ fore legs in Figs 140-141; ♂ fore tarsal ratio 10-3-4-2.3-2.3, i.e. tarsomere 3 longer than tarsomere 2; ♀ fore legs simple, femur without any anteroventral spinules. ♂ mid legs straight; tibia with the apical setae normal; 1 anteroventral seta in both sexes; tarsomere 1 elongated, tip of tarsomere 1 and tarsomeres 2-4 enlarged and compressed (Fig. 101). ♂ hind coxa and trochanter without modifications. Hind femur without anterodorsal setae. Hind tibia without setae; ♂ osmeterium in Fig. 142, the tibia without erect fine anterodorsal setulae.

Wings: Rather short, narrower in ♂ than in ♀. Veins bare except for a few setulae in basal part of stem-vein on ventral surface. Entirely covered with microtrichiae.

Abdomen: Almost wholly matt, with some shining areas on syntergite 1+2 and on mid-line of tergite 3. ♂ tergites 1+2, 3 and 4 only weakly expanded behind (Figs 143-144). Without setae, except at sides of hind-margin of tergite 5 in ♂; syntergite 1+2 without any erect setulae at sides. ♂ sternites with weak setulae (Fig. 144); sternite 4 and genitalia in Figs 145-147.

Wing-length: 1.9-2.8 mm (♂), 2.3-3.2 mm (♀).

Remark. – *T. annulipes* was fully described and illustrated by Ozerov (1999d: 187, fig. 17); the ♂ genitalia were also illustrated by Ozerov (1999e: fig. 348.1).

Egg. – The eggs have the shortest known egg filaments in Sepsidae, but are otherwise morphologically typical for the genus (Meier, 1995b: 427-432).

Larva. – The larva has been included in a key by Mangan (1977: 339), who also illustrated the cephalopharyngeal skeleton (op. cit.: fig. 4). It has been fully described by Meier (1996: 105-107, figs 364-372).

Distribution. – Widespread and common. Denmark: almost all provinces (e.g. Stæger, 1844: 35;

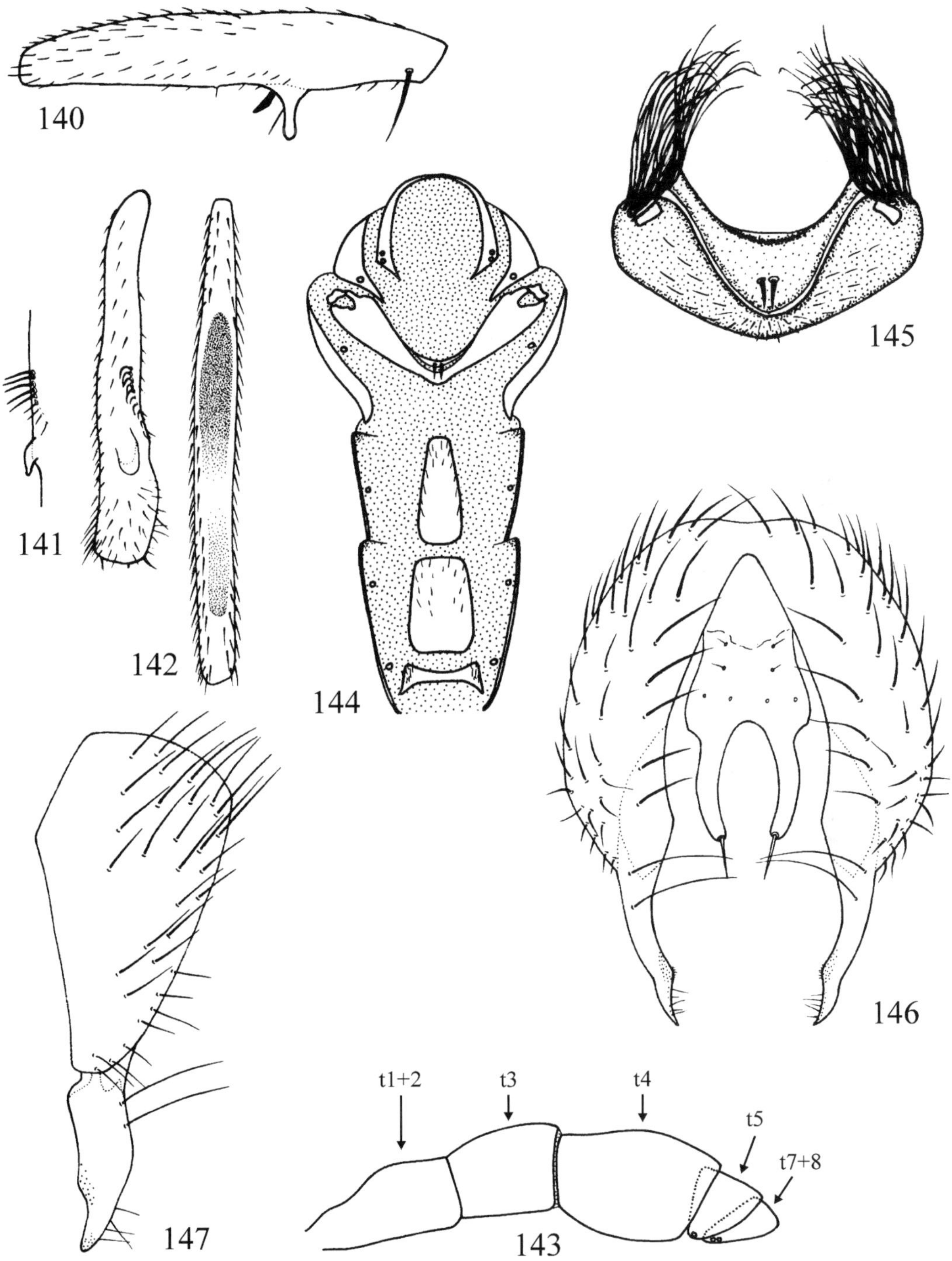

Figs 140-147. *Themira annulipes* (Meigen), ♂. 140: fore femur, anterior view; 141: fore tibia, anterior view (detail: ventral view of the projection); 142: hind tibia, dorsal view; 143: abdomen, lateral view; 144: abdomen, ventral view, hypopygium removed, position and shape of sternites 4 and 5 indicated; 145: sternites 4 and 5; 146: hypopygium, dorsal view; 147: hypopygium, lateral view. (t – tergite.)

Lyneborg, 1965: 231). Norway: northwards to TR. Sweden: SK to TO (e.g. Zetterstedt, 1847: 2304; Ringdahl, 1951: 143). Finland: northwards to LkE (Frey, 1908: 585).

Mainland Europe and the British Isles, but absent from the Mediterranean area, eastwards to China and the Far East province of Russia (Zuska & Pont, 1984: 164; Ozerov, 1999e: 561). Nearctic: from Alaska, British Columbia and Utah, eastwards to Quebec and Virginia (Mangan, 1977: 338; Ozerov, 1999d: 206-207).

Records are from mid April to early September.

Biology. – Adults are eurytopic (Minder, 1963: 388), but most common in meadows and damp muddy habitats (Bährmann, 1983a: 104-105) close to ponds and mires, especially where the mud is enriched by cattle or waterfowl droppings. They are often attracted to freshly trampled grass, and can be trapped in small numbers with faeces and fruits (Gregor, 1966: 82). They may be the dominant species of Sepsidae in damp meadows and copses (Naswaty, 1960: 89-90). They are strikingly short-winged, and reluctant to fly more than a short distance when disturbed. Territorial behaviour and the courtship display of the male, in which the mid tarsomeres are displayed prior to any any attempts at genital contact, have been studied by Mangan (1976: 1026).

Mangan (l.c.) thought that cattle feeding pens or sewage beds were the most likely larval breeding-sites, although Meier (1996: 107) found that the mortality rates of larvae feeding on cattle dung were unusually high. Larvae have been found in goose and duck droppings (Ozerov, 1999d: 187), and in high-water flood debris (ZMUC; Boness, 1975: 390). The species overwinters as a puparium (Lobanov, 1962a: 37).

Type-material. – *Sepsis annulipes* was described without the locality, number or sex given, 'in dem Museum des Hrn. v. Winthem in Hamburg'. ACP has seen 4♂ syntypes in NMW, which were also seen by Duda (1926a: 72). Each is labelled 'annulipes [handwritten] / Coll. Winthem [printed]', each is in excellent condition, and each is *annulipes* as currently understood. One has an additional old label 'annulipes'. ACP has labelled this ♂ and designates it herewith as lectotype, and the remaining 3♂ as paralectotypes. 1♂ in MNHNP, Meigen collection no. 2246 (see also Becker, 1902: 230), also seen by ACP, cannot be a syntype since it is labelled by Meigen 'Sepsis annulipes Baiern'. Although badly damaged, it can be recognised as *Themira annulipes*.

Nemopoda varipes: Meigen described this species from 'Baiern', without giving the number or sex. The holotype ♀ is in MNHNP, Meigen collection no. 2245 (see also Becker, 1902: 230), where ACP has studied it. It is labelled 'Sepsis varipes Baiern' by Meigen, and is clearly *Themira annulipes*.

Sepsis marginata: Nothing is known about the fate of Hoffmeister's collection, and enquiries by ACP to Professor R. Bährmann and through him to the curator of the Kassel Museum have been unsuccessful; the collection must now be assumed to have been destroyed. Hoffmeister compared his new species with *annulipes*, but his notes (♂ mid tarsus with tarsomeres 1 and 2 each with a silver-white ring before the black part, the tip of tarsomere 1 and all of tarsomere 2 expanded) clearly indicate that he was in fact dealing with *annulipes*.

5. ***Themira arctica*** (Becker, 1915)
(Figs 105, 114, 118, 148-155)

Cheligaster arctica Becker *in* Becker et al., 1915: 67, fig. 7 [February 1915]. Holotype ♂, 'aus dem polaren Ural' [RUSSIA, Ural Mts, northern Tyumen' district], in ASSP [see below].
Themira dampfi Becker, 1915: 128, fig. A [8 October 1915]. Lectotype ♂, FAROE IS: Sand, in ZMHU [by designation of Hennig, 1949: figs 13, 23, 97, 153; by inference, see discussion below].

There has been some confusion over the name and identity of this species. *T. arctica* was unknown to Duda (1926a) and Hennig (1949), but Hennig (1949: 44) suggested that *arctica* and *dampfi* might be synonyms. In fact, as Ringdahl (1934) was able to show, there are two good species here. One is indeed *arctica*, of which *dampfi* is a synonym. This is the species previously known as *dampfi* by European authors, and (correctly) as *arctica* by North American authors and also by Frey (1925), Chernov (1965) and Ozerov (1999d, e). The name *arctica* is slightly older than *dampfi* (see above). The other species, called *arctica* by Ringdahl, proves to be an undescribed species and is formally described and named below (p. 97).

Diagnosis. – Adult ♂ recognised by the partly shining katepisternum (Fig. 105), presence of a postpronotal seta, and simple apical setae on mid tibia (Fig. 114). It is most similar to *ringdahli* Pont, sp. n. and best recognised by the features given in the key (p. 58).

Adult. – Colour: Head mostly black, gena and facial carina often brown. Antenna black. Thorax, scutellum and abdomen entirely black. Legs wholly black. Wing clear; calypter creamy, margins and fringe-hairs yellow; haltere yellow.

Head: Frons mostly shining, the rest brownish-grey pruinose. Postocellar subequal to inner vertical; orbital only slightly shorter than this. Gena just over half length of antennal flagellomere; subvibrissal setae numerous but very short.

Thorax: Scutum, postpronotal lobes and scutellum thinly brownish-grey pruinose. Setae: postpronotal present; anterior notopleural strong, hardly shorter than posterior one or 3/4 of its length; acrostichal setulae minute. Pleural pruinosity in Fig. 105: anepisternum with an inconspicuous matt patch in upper posterior corner, pruinose in proepimeral area; katepisternum with little pruinosity on posterior margin; anepimeron and meron pruinose, each with a central undusted oval patch. Setae reduced; no katepisternal setula. Scutellum short, twice as wide as long, with or without a tiny basal setula.

Legs: Fore coxa shining except above, mid and hind coxae pruinose; trochanters and femora shining. Fore femur with some short posterodorsal setae before tip; ♂ fore legs in Figs 148-149, tibia with 1 or 2 setae on anterior surface close to the projection; ♀ fore legs simple, femur with 3-5 anteroventral spinules around middle. ♂ mid legs straight. Mid tibia with or without a short anteroventral seta, in ♂ the apical setae normal (Fig. 114). ♂ hind coxa and trochanter without modifications. Hind femur with the anterodorsal setae very fine. Hind tibia without setae; ♂ with the osmeterium in Fig. 150, the tibia with the dorsal ground-setulae fine and erect in apical third.

Wings: Veins bare except for several setulae in basal part of stem-vein on ventral surface. Entirely covered with microtrichiae.

Abdomen: Glossy black. ♂ tergites 1+2, 3 and 4 moderately expanded behind (Figs 151-152). Without setae except at posterior lateral corners of tergites 4 and 5 in ♂; syntergite 1+2 with erect setulae at sides. ♂ sternites with weak setulae (Fig. 152); sternite 4 and genitalia in Figs 153-155.

Wing-length: 2.3-3.0 mm (♂), 2.7-3.4 mm (♀).

Remarks. – *T. arctica* was fully described and illustrated by Ozerov (1999d: 173-174, fig. 1); the ♂ genitalia were also illustrated by Ozerov (1999e: fig. 348.2). Note that Ozerov's illustration of ♂ sternites 4 and 5 (1999d: fig. 1.7) resembles that shown here for the new species (Fig. 269) rather than *arctica* (Fig. 153).

Egg and larva. – Not known.

Distribution. – An arctic-subarctic species, widespread but rare in the north. Denmark: absent; Ozerov's (1986a: 640) record refers to its occurrence in the Faroes. Norway: TR. Sweden: LY, LU, TO. Finland: LkW, Le, Li.

Circumboreal, but records sparse. Iceland and Faroe Is (Zuska & Pont, 1984: 164), including the new island of Surtsey (Lindroth et al., 1973: 35). Russia: Lr (UZMH; Frey, 1925: 76); WS, Tyumen' (type-locality); ES, Taymyr Peninsula (Chernov, 1965: 77; Ozerov, 1986a: 640; 1999d: 174); possibly Kirghiziya, Tien Shan Mts, over 2555 m (Sychevskaya & Vtorov, 1969: 820, but identification queried). Nearctic: Greenland (Vibe, 1950: 420), North-West Territories & Labrador (Steyskal, 1965: 684; Ozerov, 1999d: 174).

Only collected in July and August.

Biology. – Almost nothing known. Adults occur in open tundra areas, around pools and in mires. Chernov (1965: 77) found them around greatly decomposed offal and carcasses. The larvae may breed in the droppings of waterfowl.

Type-material. – *Cheligaster arctica*: the holotype ♂ is in ASSP, and has been seen by ACP. It is labelled in Russian 'Polar Urals / north of Tobolsk Government / F. Zaytsev 6.vii.09' [ACP translation]. It is damp, only just on the pin, and with head and right fore tibia+tarsus missing. It belongs to *arctica* as defined here. According to A. L. Ozerov (pers. comm.), the 'Tobolsk government' is now Tyumen' district, and the type-locality is most probably in the Urals to the west of Salekhard near the mouth of the River Ob.

Themira dampfi: Becker described this species from 3♂, and these are in ZMHU, where ACP has studied them. Duda (1926a: 92) saw two of them, Gregor (1966: 78) saw all three, and Ozerov (1986a: 640) saw one; the ♀ mentioned by Ozerov is actually a syntype of *roseni* Becker and not of *dampfi* (see discussion under *lucida*, p. 81). Hennig (1949) also saw some of the syntypes. His figures (op. cit.: figs 13, 23, 97, 153) are of the

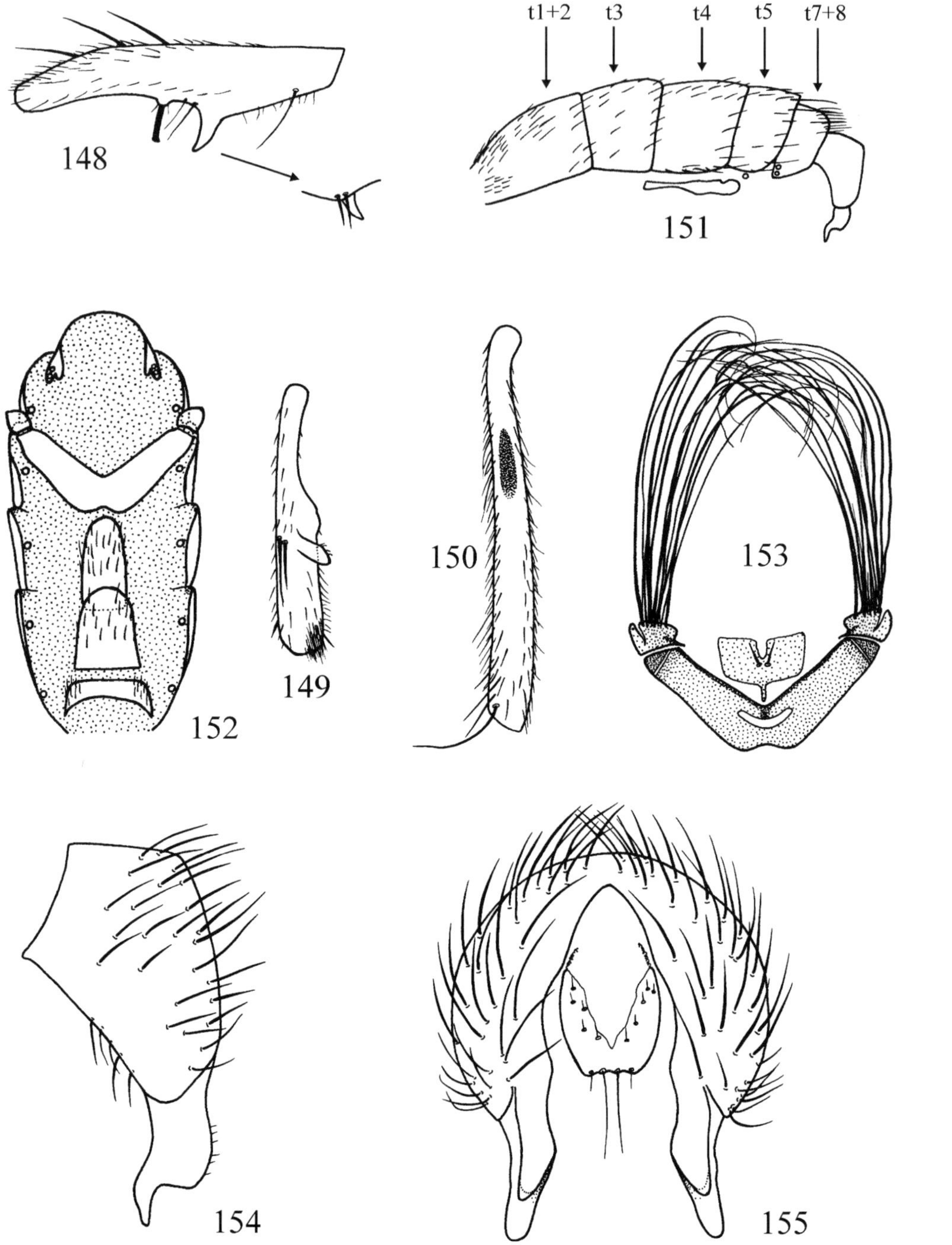

Figs 148-155. *Themira arctica* (Becker), ♂. 148: fore femur, anterior view (detail: posterior view of the peg); 149: fore tibia, anterior view; 150: hind tibia, dorsal view; 151: abdomen, lateral view; 152: abdomen, ventral view, hypopygium removed, position and shape of sternite 4 indicated; 153: sternites 4 and 5; 154: hypopygium, dorsal view; 155: hypopygium, lateral view. (t – tergite.)

'Typus', and this action effectively fixes the specimen illustrated as the lectotype. ACP has formally labelled this ♂ as lectotype. It is labelled 'Sand. / a. Wiesen. [in meadows] / 31.8.12', and is now in very poor condition. It can be recognised without doubt as the present species. Unfortunately, the slides that Hennig prepared of the fore legs and the genitalia cannot be found, in either DEI or in ZMHU, and must now be presumed lost (H. Wendt and J. Ziegler, pers. comm.). Becker's second ♂, also from Sand, 25.8.12, has been labelled as paralectotype; it also belongs to this species. The third ♂, from Kvalvig, has not been found.

6. ***Themira biloba*** Andersson, 1975

(Figs 126, 156-168)

Themira biloba Andersson, 1975: 57, figs 2, 4, 6, 8, 10, 12. Holotype ♂, SWEDEN: Skåne, Krankesjön, Revinge, in ZML.

Diagnosis. – Adult *biloba* is a relatively large species with broad gena, very similar to *putris* and best distinguished by the characters given in the key (p. 59).

Larval morphology very similar to that of *minor*, except for an anal division that is more hairy; anterior spiracles with 16-23 lobes.

Adult. – Colour: Head mostly black, part of frons, parafacial, gena and face brown. Antenna black. Thorax, scutellum and abdomen entirely black. Wing clear; calypter white, with creamy margin and white fringe-hairs; haltere creamy.

Head: Frons mostly shining, occiput weakly brownish-grey pruinose. Postocellar equal to inner vertical; orbital half as long as inner vertical. Gena broad, at middle rather less than length of antennal flagellomere (about 2/3); subvibrissal setae numerous but short.

Thorax: Scutum and postpronotal lobes thinly brownish-grey matt, and subshining from most points-of-view; scutellum more densely matt. Setae: postpronotal present; anterior notopleural 1/3 to 1/2 (♂) or 2/3 (♀) length of posterior one; acrostichal setulae minute; middle of scutum with several additional short but distinct dorsocentral setae. Pleural pruinosity: anepisternum almost wholly glossy, thinly pruinose along hind margin; katepisternum entirely whitish-grey pruinose; anepimeron thinly pruinose on upper anterior corner and on posterior 1/4 to 1/3, otherwise glossy; meron pruinose, with a central glossy oval patch. Setae reduced; anepisternal setula distinct; no katepisternal setula. Scutellum short, twice as wide as long, with a pair of short basal setulae.

Legs: Fore coxa shining except basally and in front, mid and hind coxae pruinose; trochanters shining, femora subshining, tibiae pruinose. Fore femur with several posterodorsal setae in apical half; ♂ fore legs in Figs 156-158; fore tarsal ratio 10-6.5-4-2.5-3.3, i.e. tarsomere 1 rather longer than tarsomere 2; ♀ fore legs simple, femur with 5-6 anteroventral spinules. ♂ mid femur weakly curved; tibia with the apical setae normal; tarsomeres 4 and 5 rather flattened and expanded; tibia with one anteroventral seta in both sexes. ♂ hind coxa and trochanter without modifications. Hind femur with a few short anterodorsal setae. Hind tibia without setae; rather expanded around middle in ♂; ♂ osmeterium in Fig. 159, occupying half of tibial length; the tibia with fine anterodorsal setulae, especially before apex.

Wings: Veins bare except for a few setulae in basal part of stem-vein on ventral surface. Entirely covered with microtrichiae. Section of costa between humeral cross-vein and subcosta subequal to that between subcosta and vein R_1 (10 : 10) (Fig. 126).

Abdomen: Glossy black. ♂ tergites 1+2, 3 and 4 moderately expanded behind. Without setae; syntergite 1+2 with erect setulae at sides. ♂ sternites with weak setulae (Fig. 160); sternite 4 and genitalia in Figs 161-163.

Wing-length: 3.1-3.4 mm (♂), 3.4-3.8 mm (♀).

Remark. – *T. biloba* was fully described and illustrated by Ozerov (1999d: 175, fig. 3).

Larva. – Pseudocephalon bilobed, cephalic lobes with numerous combs (Fig. 164); facial mask consisting of ridges surrounded by combs.

Anal division short; lateral tubercles present; stigmatophores with a dorsal tubercle (Fig. 166); ventral tubercles present, simple; subanal tubercles hairy and moderately large; anal pad large, with a diagonal furrow, somewhat angled laterally; anal tuft present; preanal protuberance present (Fig. 165); distance between anterior margin of anal pad and base of ventral tubercles similar to width of segment 7, with an anterior and posterior band of spines laterally and three double rows of spines dorsally; anal division densely hirsute between spinous areas.

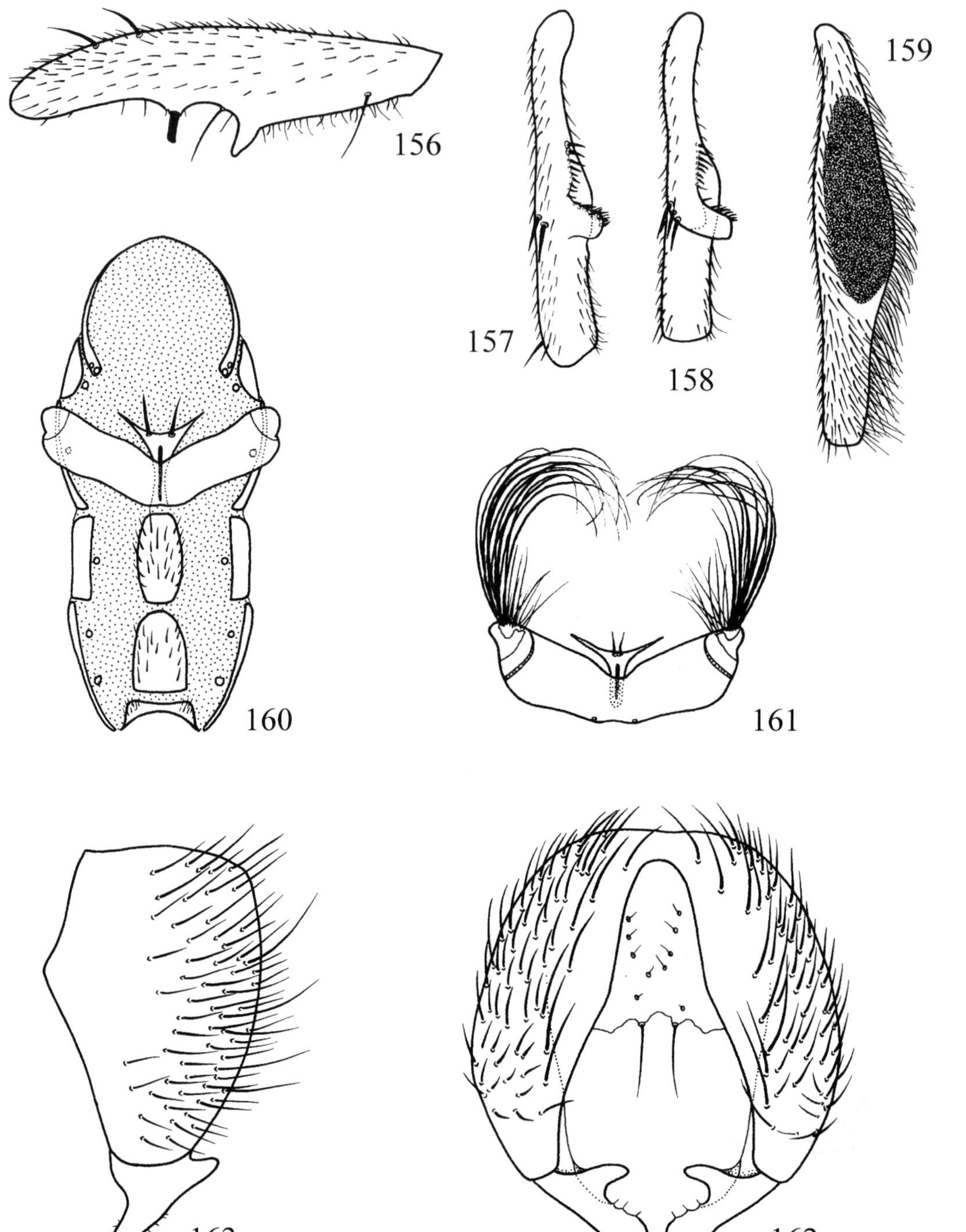

Figs 156-163. *Themira biloba* Andersson, ♂. 156: fore femur, anterior view; 157, 158: fore tibia, anterior view; 159: hind tibia, dorsal view; 160: abdomen, ventral view, hypopygium removed, position and shape of sternites 4 and 5 indicated; 161: sternites 4 and 5; 162: hypopygium, dorsal view; 163: hypopygium, lateral view.

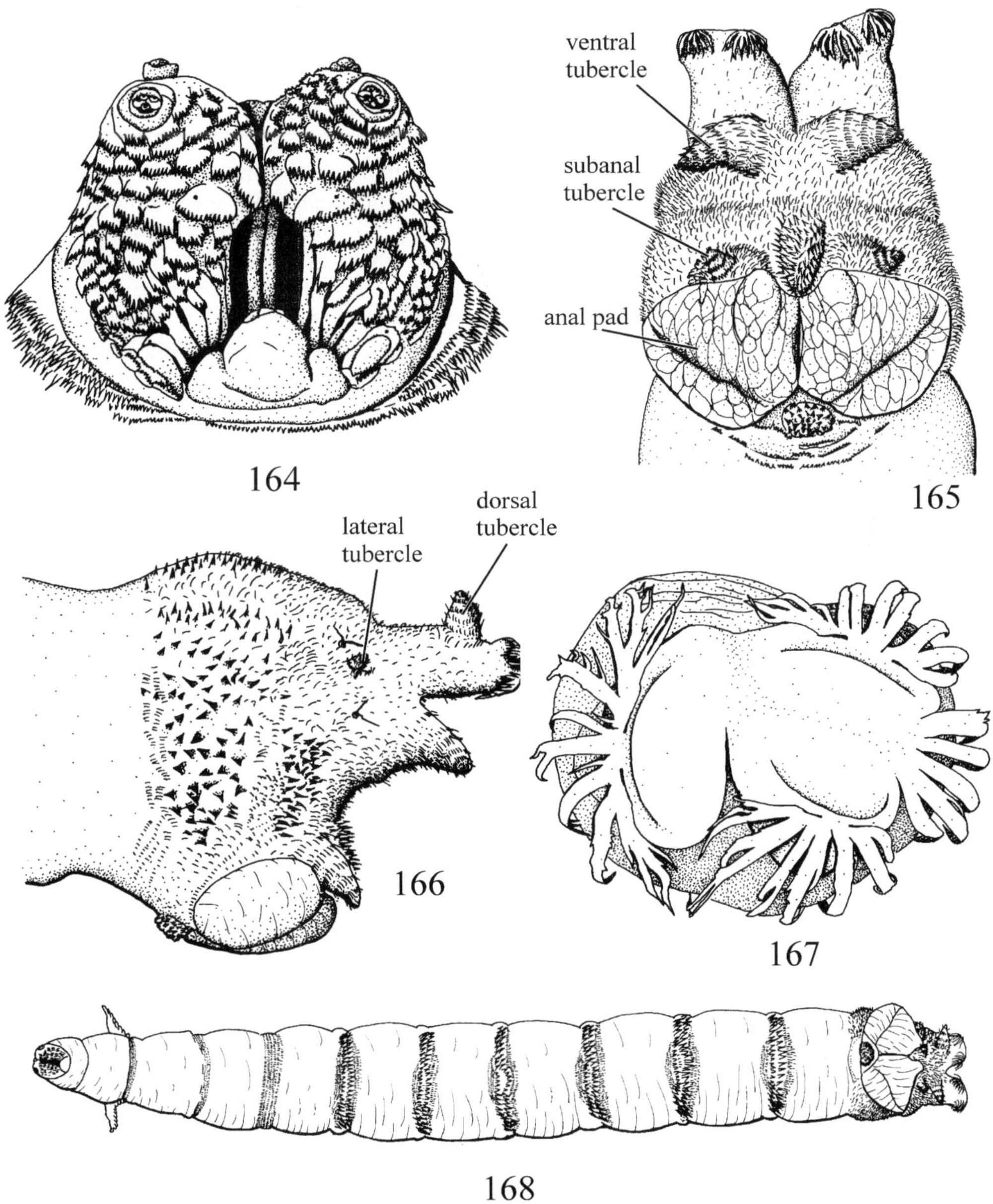

Figs 164-168. *Themira biloba* Andersson. Larval morphology. 164: pseudocephalon, ventral; 165: anal division, ventral; 166: anal division, lateral; 167: posterior spiracles; 168: larval habitus, ventral.

Cephaloskeleton: labial sclerites absent; dental sclerites fused to mouthhooks (as in Fig. 47); bridge of intermediate sclerite at approximately half length of sclerite, anteriorly with a flat plate (as in Fig. 138); epistomal sclerite longer than wide, with posterior bars; parastomal bars short, barely reaching posterior edge of bridge of intermediate sclerite (as in Fig. 47), basal sclerite 2-2.5 times as long as high, dorsal cornua shorter and 3-4 times narrower than ventral cornua.

Anterior spiracles tree-like, with wide base; basally with additional lobes, with 16-23 lobes; lobes 2-4 times as long as wide (as in Fig. 187).

Posterior spiracles with almost equal, narrow, weakly arched spiracular openings on an oval plate; two spiracular openings parallel, the third one perpendicular to the other two (Fig. 167).

Distribution. – Scarce, but possibly overlooked. Denmark: NEZ. Norway: absent. Sweden: SK. Finland: N, Ka (Pont, 1990: 154).

Otherwise only known from eastern England (Pont, 1986a: 91), the Netherlands (Van der Goot, 1989: 36), Germany (Meier, 1999: 190), and the Czech Republic (Zuska & Barták, 1997: 74; Barták et al., 1997: 227).

Only collected in July and August.

Biology. – Adults have been found only near water, and may be associated with the droppings of waterfowl and river birds. In England they have been collected off guano on the nest of a coot (*Fulica atra*), in Germany the species was also collected at a muddy spot littered with waterfowl droppings along a lakeshore in Götz (Brandenburg), and in Bohemia the species was found on a 'shore of a small drainage pond' (Barták et al., 1997: 227). Under laboratory conditions the flies readily oviposited in cattle dung, and the larvae developed very well and in large numbers in this substrate.

7. ***Themira germanica*** Duda, 1926

(Figs 130, 169-176)

Themira germanica Duda, 1926a: 83. Holotype ♂, 'Zehlau' [RUSSIA, CET, Kaliningrad district], in ZMHU [Ozerov, 1986a: 641; 1997a: 480].

Diagnosis. – Adult *germanica* is recognised among the species with a postpronotal seta by the characteristic and large genitalia, and by the external characters given in the key (p. 60). It is very similar to *paludosa*, but differs by the presence of a postpronotal seta and in details of the fore leg ornamentation.

Adult. – Colour: Head black, sometimes brown on parafacial, face and gena. Antenna black. Thorax, scutellum and abdomen entirely black. Legs black, but fore coxa and all trochanters mostly yellowish (sometimes brownish), and in ♀ also bases of femora and often also knees. Wing clear; calypter and fringe-hairs creamy, margin yellow; haltere yellow.

Head: Frons mostly shining, occiput thinly brownish-grey pruinose. Postocellar subequal to inner vertical; orbital half length of inner vertical. Gena at middle hardly half width of antennal flagellomere; several short subvibrissal setae present.

Thorax: Scutum and postpronotal lobes entirely though thinly brownish-grey pruinose, subshining from some points-of-view but nowhere glossy; scutellum more densely brown pruinose. Setae: postpronotal present; anterior notopleural weak, hardly 1/4 length of posterior one; acrostichal setulae minute. Pleural pruinosity: anepisternum entirely glossy; katepisternum entirely whitish-grey pruinose; anepimeron thinly pruinose on upper anterior part and along hind margin; meron pruinose, with a central glossy oval patch. Setae reduced; no katepisternal setula. Scutellum short, twice as wide as long, without basal setulae.

Legs: Fore coxa shining except at extreme base, mid and hind coxae pruinose; trochanters and femora shining; tibiae shining to subshining. Fore femur with 1 or more posterodorsal setae in apical half; ♂ fore legs in Figs 169-170; ♂ fore tarsal ratio 10-3.3-4-2.8-3, i.e. tarsomere 3 slightly longer than tarsomere 2; ♀ fore legs simple, femur with 5-7 anteroventral spinules. ♂ mid femur slightly curved; tibia with the apical setae normal. ♂ hind coxa and trochanter without modifications. Hind femur with several short anterodorsal setae. Hind tibia without setae; ♂ osmeterium in Fig. 171, short and narrow; the tibia with fine erect anterodorsal setulae on apical third or more.

Wings: Veins bare except for a few setulae in basal part of stem-vein on ventral surface. Entirely covered with microtrichiae.

Abdomen: Glossy black. ♂ tergites 1+2, 3 and 4 hardly expanded behind (Figs 172-173). Without setae, except for some weak ones at sides of hindmargin of tergite 5 in ♂; syntergite 1+2 with erect setulae at sides. ♂ sternites with weak setulae (Fig. 173); sternite 4 and genitalia in Figs 174-176.

Wing-length: 2.4-3.2 mm (♂), 2.9-3.7 mm (♀).

Remarks. – *T. germanica* was fully described and illustrated by Ozerov (1999d: 176-177, fig. 5); the ♂ genitalia were also illustrated by Ozerov (1999e: fig. 348.3).

Egg and larva. – Not known.

Distribution. – Uncommon. Denmark: absent. Norway: TR. Sweden: SK (e.g. Ringdahl, 1941: 16) to TO. Finland: only in the southern provinces (Pont, 1990: 154).

Uncommon and known only from a few localities in Belgium (Grootaert, 1991: 135), the Netherlands (Van der Goot, 1985b: 178; 1986a: 2),

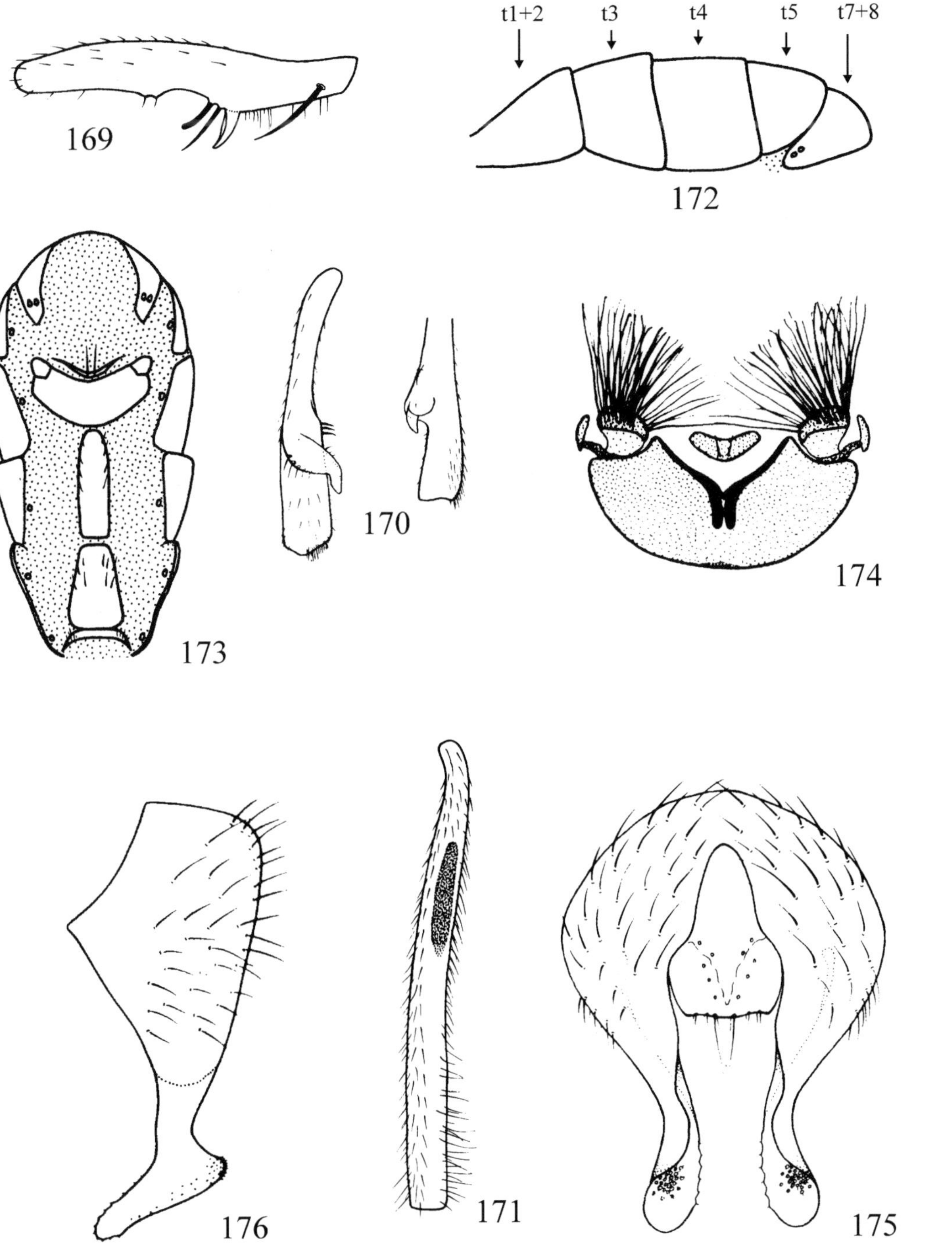

Figs 169-176. *Themira germanica* Duda, ♂. 169: fore femur, anterior view; 170: fore tibia, anterior view (detail: ventral view of apical half); 171: hind tibia, dorsal view; 172: abdomen, lateral view; 173: abdomen, ventral view, hypopygium removed, position and shape of sternites 4 and 5 indicated; 174: sternites 4 and 5; 175: hypopygium, dorsal view; 176: hypopygium, lateral view. (t – tergite.)

Switzerland (Haenni, 1997: 73), Great Britain and Germany (Zuska & Pont, 1984: 164), North Italy (Munari, 1993: 12). Russia: Krasnoyarsk Territory (Ozerov, 1986a: 641) and Transbaikalia (Ozerov, 1999e: 563); the type-locality is now in Russia (CET, Zehlau Bruch, south-east of Kaliningrad).

Collected in July, August and early September.

Biology. – Adults have been collected around open tarns and pools; Ringdahl (1941: 20) listed it as a hydrophilous species. Nothing is known of the life-cycle or immature stages.

8. ***Themira gracilis*** (Zetterstedt, 1847)

(Figs 102, 177-190)

Sepsis gracilis Zetterstedt, 1847: 2300. Lectotype ♂, 'in graminosis ad diversorium Skalstugan, ad radicem jugi alpini situm' [SWEDEN: Jämtland, Skalstugan], in ZML [by designation of Gregor, 1966: 78; see below].

Meroplius plurisetosa Duda, 1926a: 98. Holotype ♀, UKRAINE: Körösmezö [now = Jasiňa], in ZMHU [Ozerov, 1997a: 483].

Diagnosis. – Adult recognised by the partly glossy katepisternum together with partly yellow legs.

Larval morphology similar to that of *minor*, but pseudocephalon smaller and facial mask mostly consisting of cirri, anal division with more hairs, and double rows of spines on dorsum more distinct than in *minor*; anterior spiracles with 7-10 lobes.

Adult. – Colour: Occiput, vertex and most of frons black; lunule, face, parafacial and gena orange. Antenna black, inner edge of flagellomere and articulation between flagellomere and pedicel orange. Thorax, scutellum and abdomen entirely black. Fore coxa mostly yellow, mid and hind coxae mostly dark; trochanters yellow; fore femur dark, narrowly yellow at base and apex, tibia yellow, tarsus dark; mid and hind legs mostly dark, the femora broadly yellow at base, knees yellow, tips of tibiae and basal tarsomeres yellow. Wing clear; calypter creamy, margin and fringe-hairs creamy to yellow; haltere yellow.

Head: Frons subshining, the rest brownish-grey to grey pruinose. Postocellar subequal to inner vertical; orbital half as long as this. Gena narrow, not even half length of antennal flagellomere; subvibrissal setae very short.

Thorax: Scutum, postpronotal lobes and scutellum entirely matt, brownish-grey pruinose. Setae: postpronotal present but weak; anterior notopleural just over half as long as posterior one; acrostichal setulae minute. Pleural pruinosity as in Fig. 102: anepisternum thinly pruinose on upper posterior margin; katepisternum pruinose along upper and posterior margins; anepimeron pruinose except for an oval glossy patch; meron entirely pruinose. Setae reduced; no katepisternal setula. Scutellum short, twice as wide as long, without a basal setula.

Legs: Fore coxa shining except above, mid and hind coxae pruinose; trochanters and femora shining. Fore femur with a short posterodorsal seta before tip; ♂ fore legs in Figs 177-178; ♀ fore legs simple, femur with 4 strong anteroventral spinules around middle. ♂ mid legs straight. Mid tibia without an anteroventral seta, the apical setae normal in ♂. Hind coxa and hind trochanter without modifications. Hind femur without distinct anterodorsal setae. Hind tibia without setae; ♂ with the osmeterium in Fig. 179, the tibia without any erect dorsal ground-setulae.

Wings: Veins bare; basal part of stem-vein on ventral surface sometimes with a few setulae, sometimes bare. Entirely covered with microtrichiae.

Abdomen: Glossy black. ♂ tergites 1+2, 3 and 4 moderately expanded behind (Figs 180-181). Without setae except in posterior lateral corner of tergite 5 in ♂; syntergite 1+2 with erect setulae at sides. ♂ sternites with virtually no setulae (Fig. 181); sternite 4 and genitalia in Figs 182-185.

Wing-length: 2.1-3.0 mm (♂), 2.7-2.9 mm (♀).

Remarks. – *T. gracilis* was fully described and illustrated by Ozerov (1999d: 177-178, fig. 6); the ♂ genitalia were also illustrated by Ozerov (1999e: figs 349.2, 6). The species has also been discussed in detail by Soós (1958: 207), Gregor (1966: 78), who gave a wealth of excellent illustrations, and Ozerov (1983: 636).

Larva. – Pseudocephalon bilobed, cephalic lobes with a moderate number of cirri; facial mask small, mostly with cirri, with only very few ridges on either side of labial lobe (Fig. 186).

Anal division short; lateral tubercles present; stigmatophores with a dorsal tubercle; ventral tubercles present, simple; subanal tubercles hairy and rather small; anal pad small, with a diagonal furrow, somewhat angled laterally; anal tuft present; preanal protuberance present (Fig. 188); distance between anterior margin of anal pad and

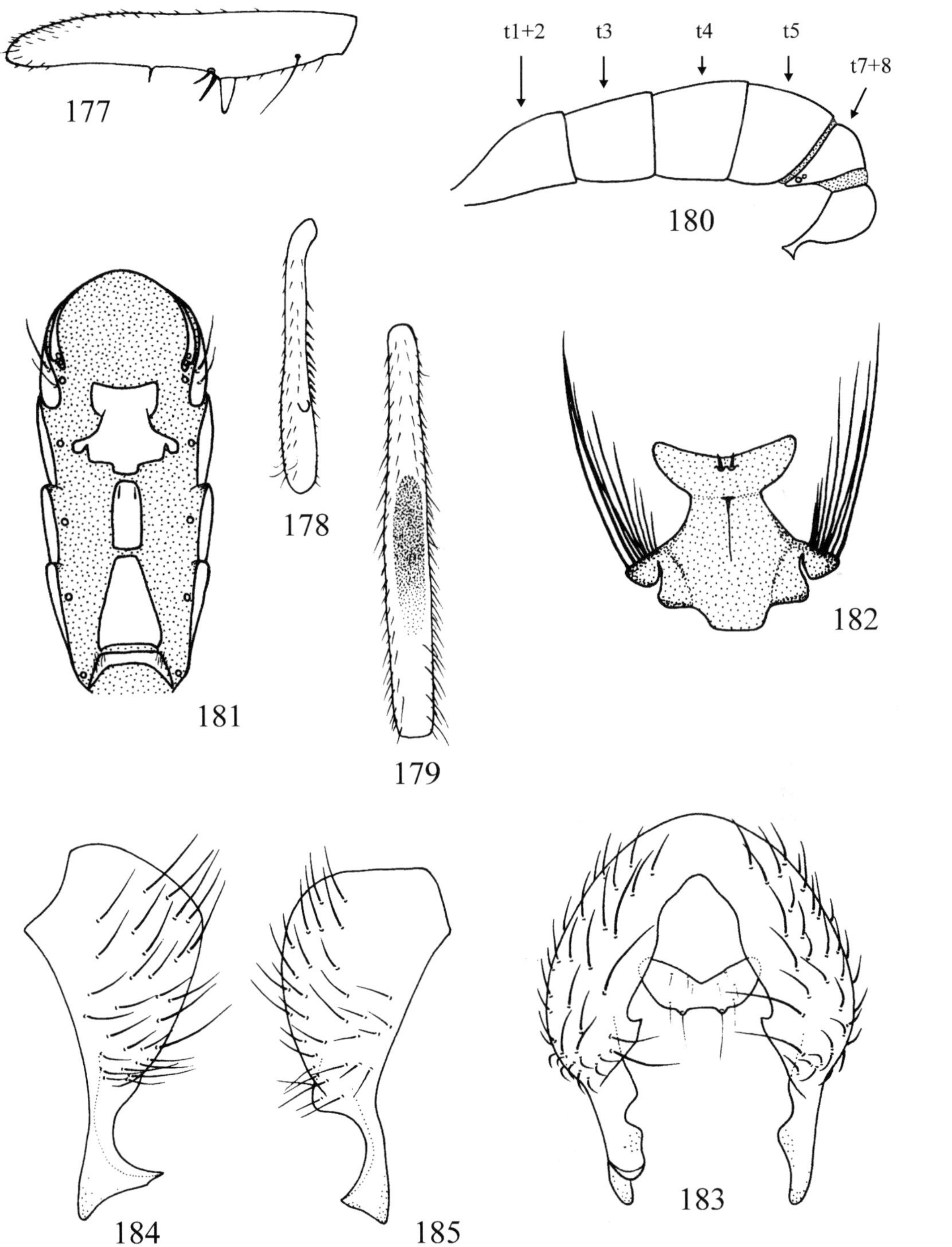

Figs 177-185. *Themira gracilis* (Zetterstedt), ♂. 177: fore femur, anterior view; 178: fore tibia, anterior view; 179: hind tibia, dorsal view; 180: abdomen, lateral view; 181: abdomen, ventral view, hypopygium removed, position and shape of sternites 4 and 5 indicated; 182: sternites 4 and 5; 183: hypopygium, dorsal view; 184: hypopygium, lateral view, left side; 185: hypopygium, lateral view, right side. (t – tergite.)

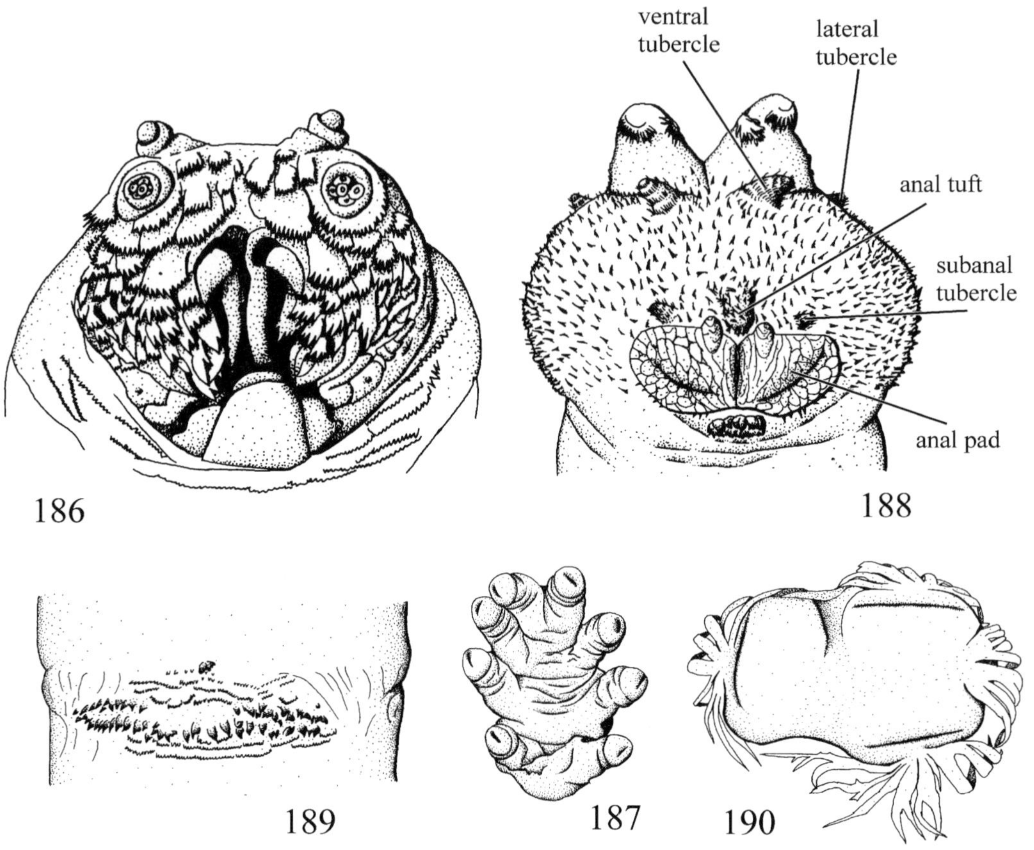

Figs 186-190. *Themira gracilis* (Zetterstedt). Larval morphology. 186: pseudocephalon, ventral; 187: anterior spiracles; 188: anal division, ventral; 189: creeping welt; 190: posterior spiracles.

base of ventral tubercles two-thirds width of segment 7, with an anterior and posterior band of spines laterally, and three double rows of spines dorsally, with few hairs between double rows; anal division densely hirsute between spinous areas.

Cephaloskeleton: labial sclerites absent; dental sclerites fused to mouthhooks (as in Fig. 47); bridge of intermediate sclerite barely in anterior half of sclerite, anteriorly with a flat plate (as in Fig. 138); epistomal sclerite longer than wide, with posterior bars; parastomal bars short, less than half length of intermediate sclerite (as in Fig. 47), basal sclerite 2-2.5 times as long as high, dorsal cornua shorter and 2.5-3 times narrower than ventral cornua.

Anterior spiracles tree-like, with 7-10 lobes; lobes 2-4 times as long as wide (Fig. 187).

Posterior spiracles with almost equal, narrow, weakly arched spiracular openings on an oval plate; two spiracular openings parallel, the third one perpendicular to the other two (Fig. 190).

Distribution. – A little-known but locally common species. Denmark: 1 specimen, without locality but maybe NEZ (Stæger). Norway: R and TR. Sweden: SK to JÄ (Ringdahl, 1934: 3; 1951: 143). Finland: only the southern provinces. Frey's (1908: 587) ♂ from Valamo actually belongs to *paludosa*; Siebke's (1877: 153) material belongs to *annulipes*.

The middle belt of Europe eastwards (Great Britain, Germany, Austria, Slovakia, Hungary, Romania) (Zuska & Pont, 1984: 164), Latvia (Karps, 1986: 6), Belgium (Grootaert, 1991: 135), Switzerland (Haenni, 1997: 73), North Italy (Munari, 1984: 48); Russia: Kr (UZMH), Leningrad district (Shtakel'berg, 1958: 132), Moscow district and North Ossetia (Ozerov, 1999d: 178), Amur district and Far East province (Ozerov, 1983: 636; 1989: 842; 1999d: 178); Ukraine (Zuska & Pont, 1984: 164), Mongolia (Soós, 1972: 357).

Collected in June, July and August.

Biology. – Adults have been collected on a sphagnum moor (Soós, 1958: 209), on a dead sheep in a mountain pasture (Gregor, 1966: 81), and on a dead cow and in grasses near a cesspool (Ozerov, 1999d: 178). It is often associated with cow dung (e.g. Zetterstedt, 1855: 4782). It has been collected in horse corrals, farmyards where horses and cattle are kept, and, abundantly, in a copse by a stream where the grass and mud were heavily trampled and be-dunged by cattle (ACP, pers. obs.).

Females collected on horse dung at Lima, DR (June 1999) laid eggs in this substrate. The larvae developed slowly and pupariated after approximately two weeks. As in *Zuskamira* and *Ortalischema* from the same locality, no adults emerged unless the puparia were kept in the refrigerator for approximately two months. At least the Swedish populations of this species appear to have a diapause.

Type-material. – *Sepsis gracilis*: Gregor (1966: 78) studied Zetterstedt's ♂ syntype from his Diptera Scandinaviae collection, drawer 38 (as did Ringdahl, 1934: 5), and designated it as lectotype. It has also been studied by ACP. It is labelled by Zetterstedt 'S. gracilis ♂. / Zett. Skalstug.'; the head and right mid leg are missing. The ♀ syntype, with Zetterstedt's label 'S. gracilis ♀. / Zett. Skalst.', has been labelled by ACP as paralectotype; it is conspecific with the ♂.

9. ***Themira leachi*** (Meigen, 1826)

(Figs 11, 89, 91, 93, 132, 191-199)

Sepsis leachi Meigen, 1826: 291; Meigen plate 157 fig. 9 [Morge, 1976a]. Holotype ♂, 'aus England' [ENGLAND], in MNHNP [see below].

Themira phantasma Robineau-Desvoidy, 1830: 746. Holotype/syntypes ♂, FRANCE, Saint-Sauveur, not in MNHNP and presumed destroyed.

Nemopoda tarsalis Walker, 1833: 252. Syntypes ♀ [not ♂ as stated], ENGLAND: near London, and near Ambleside, Westmoreland, not in BMNH and presumed destroyed [see below].

Nemopoda fumipennis Walker, 1833: 253. Holotype/syntypes ♀, ENGLAND: near London, not in BMNH and presumed destroyed [see below].

Sepsis coxarum Zetterstedt, 1837: 51. Nomen nudum.

Sepsis coxarum Zetterstedt, 1838: 748. Lectotype ♀, 'Hab. in Lapponia Tornensi rarissime' [SWEDEN, TO], in ZML [see designation below].

Themira curvipes Van der Wulp, 1864: 137, plate 8 figs 9-14. Syntypes ♂♀, NETHERLANDS: The Hague, not in ZMAN [see below].

Nemopoda umbripennis Van der Wulp, 1871: 190. Syntype 1♀, NETHERLANDS: 'Oosterdoksdijk te Amsterdam', in ZMAN [see below].

Diagnosis. – Adult *leachi* is a striking species, recognised by the glossy katepisternum and absence of postpronotal seta, and in the ♂ by the striking modifications to the fore leg and hind tibia (Figs 191-193).

Larval pseudocephalon with a facial mask consisting of numerous combs arranged in rows (Fig. 199); cephaloskeleton without a separate dental sclerite (as in Fig. 47), and bridge of intermediate sclerite at half length of the sclerite (as in Fig. 138); anal division with a medium-size subanal tubercle and a distinctively shaped anal pad with a diagonal fold (Figs 11, 132); anterior spiracles with 8-11 lobes; with seven abdominal creeping welts; with few spines on lateral side of anal division, on dorsum with three double rows of spines and no spines between the rows (as in Fig. 44).

Adult. – Colour: Head mostly black; parafacial, gena and face reddish. Antenna dark brown, inner surface of flagellomere and articulation of flagellomere and pedicel mostly orange-yellow. Thorax, scutellum and abdomen entirely black. Coxae and trochanters yellow, at most coxae infuscated basally. Fore legs black, but ♂ femur yellow at extreme base and on apical 1/4, and tibia yellow on rather more than basal half; in ♀, base of femur, knees, and tip of tibia yellow. Mid and hind legs black, but basal 1/4 of femora, knees, tip of mid tibia, and in ♂ most of mid tarsus yellow. Wing distinctly smoky, less so towards base; calypter dirty white (♂) to brown (♀), margin and fringe-hairs brownish-yellow (♂) to brown (♀); haltere whitish to creamy.

Head: Frons mostly shining, occiput weakly brownish-grey pruinose. Postocellar subequal to inner vertical; orbital half as long as this (Fig.

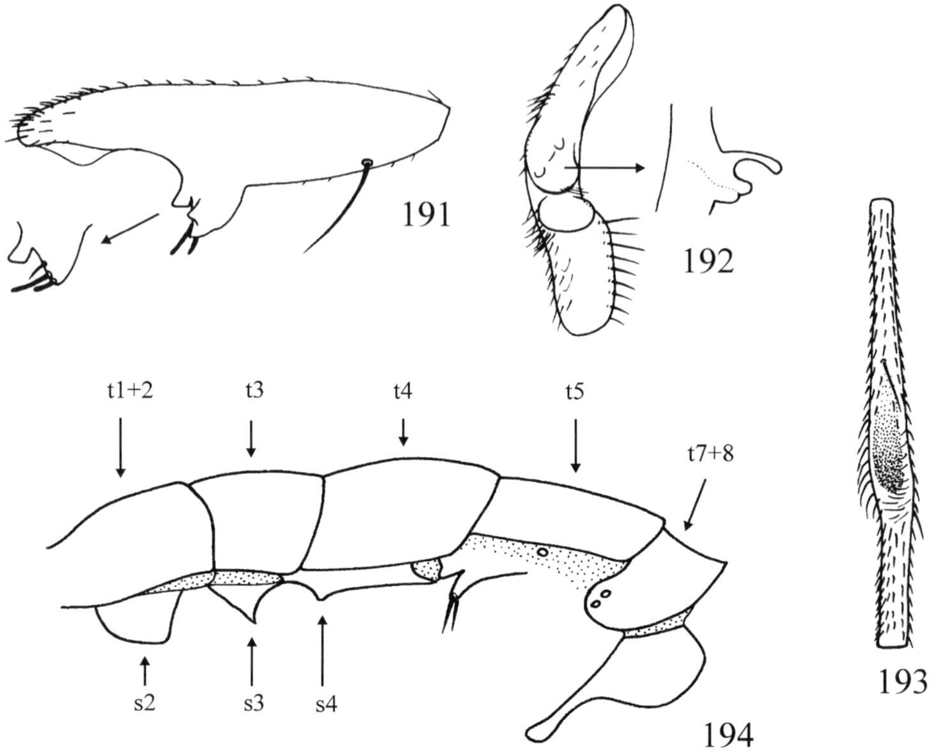

Figs 191-194. *Themira leachi* (Meigen), ♂. 191: fore femur, anterior view (detail: posterior view of the protuberance); 192: fore tibia, anterior view (detail: dorsal view of the protuberance); 193: hind tibia, dorsal view; 194: abdomen, lateral view. (s – sternite, t – tergite.)

91). Gena at middle hardly half length of antennal flagellomere; subvibrissal setae numerous and moderately long.

Thorax: Scutum, with postpronotal lobes and scutellum, brownish-grey matt and nowhere shining. Setae: postpronotal absent; anterior notopleural short, half length of posterior one; acrostichal setulae minute. Pleural pruinosity: anepisternum glossy; katepisternum glossy, except very narrowly in front and behind mid coxa; anepimeron glossy except very narrowly on hind-margin; meron also glossy except very narrowly alongside mid coxa (Fig. 89). Setae reduced; no katepisternal setula. Scutellum short, twice as wide as long, without basal setulae.

Legs: Fore coxa apparently mostly shining except basally, mid and hind coxae mostly pruinose; trochanters shining, femora and fore tibia mostly subshining, other tibiae shining. Fore femur without posterodorsal setae; ♂ fore legs in Figs 191-192; ♂ fore tibia short, only about half length of fore tarsus; ♂ fore tarsal ratio 10-6-3.75-2.5-2.5; ♀ fore legs simple, femur without any anteroventral spinules. ♂ mid legs straight; tibia with the apical setae normal; tarsi not modified; 1 anteroventral seta in both sexes. ♂ hind coxa without modifications. ♂ hind trochanter with a downwardly-directed lobe-like process on inner surface (Fig. 93). Hind femur without anterodorsal setae. Hind tibia with 1 anteroventral and, in ♂, with 1 strong anterodorsal towards base; ♂ hind tibia rather expanded at middle, where the osmeterium ends, and twisted along its longitudinal axis (Fig. 193), the tibia without erect anterodorsal setae or setulae apart from the seta near base.

Wings: Rather short and narrow. Veins bare except for a few setulae in basal part of stem-vein on ventral surface, sometimes absent in ♂. Entirely covered with microtrichiae.

Abdomen: Shining black but not glossy, the posterior segments in particular appearing rather shagreened. ♂ tergites 1+2, 3 and 4 expanded behind (Figs 194-195). Without setae, except along sides of tergite 5 in ♂; tergites 1+2 without any

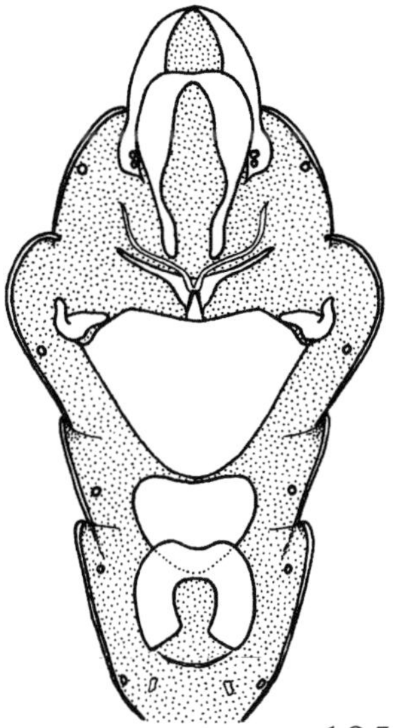

195

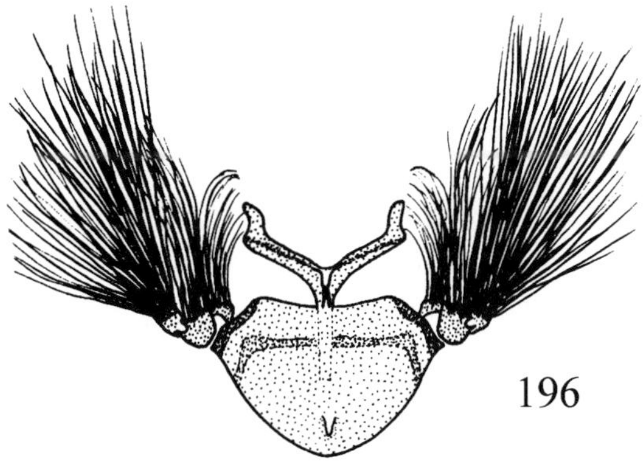

196

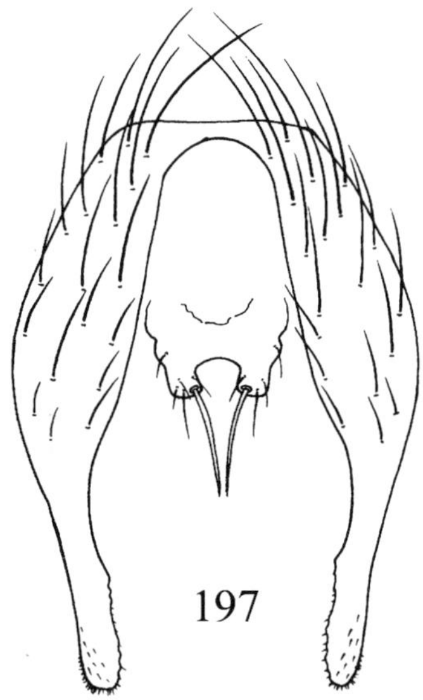

197

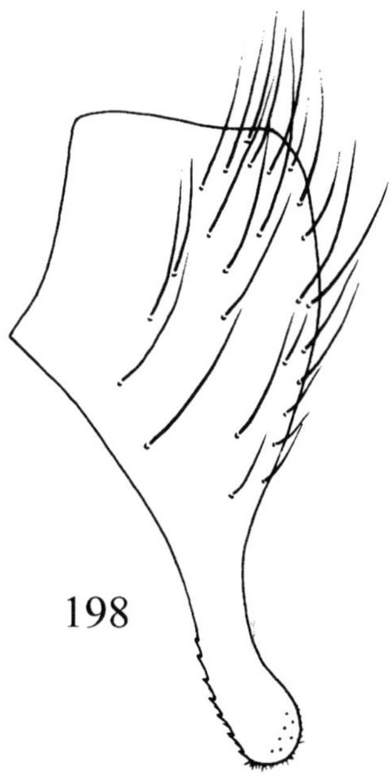

198

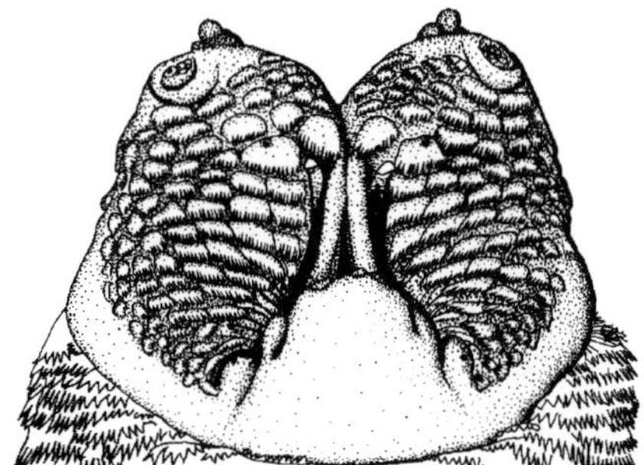

199

Figs 195-199. *Themira leachi* (Meigen), ♂. 195: abdomen, ventral view hypopygium removed, position and shape of sternites 4 and 5 indicated; 196: sternites 4 and 5; 197: hypopygium, dorsal view; 198: hypopygium, lateral view. 199: larval pseudocephalon, ventral.

erect setulae at sides. ♂ sternites 2-4 each with a conspicuous median tubercle, best seen in lateral view (Fig. 194), without any setulae; sternite 4 and genitalia in Figs 196-198, surstyli fused to hypandrium.

Wing-length: 2.4-2.8 mm (♂), 2.4-3.0 mm (♀).

Remarks. –*T. leachi* was fully described and illustrated by Ozerov (1999d: 195-197, fig. 26); the ♂ genitalia were also illustrated by Ozerov (1999e: figs 348.6).

Larva. – The first larval description was by Bouché (1834: 96, plate 6, fig. 12). It has been fully described by Meier (1996: 109, figs 382-390).

Distribution. – Widespread and rather common. Denmark: most provinces. Norway: widespread (e.g. Siebke, 1873: 93; 1877: 153), and common in the north (Zetterstedt, 1847: 2293). Sweden: surprisingly sparse, ÖG northwards. Ringdahl (1951: 143) gives SK to Lapland, but it is not mentioned in any of his faunistic papers on south Swedish Diptera, nor is there any material in his collection; Zetterstedt (1855: 4781) recorded it from SK. Finland: most provinces (Frey, 1908: 586; 1950: 11).

Common in Central and North Europe, from Great Britain and Ireland eastwards through Russia to the Far East province (Ozerov, 1999d: 197; 1999e: 559), and Mongolia (Zuska & Pont, 1984: 165).

Collected from late May to mid September.

Biology. – Adults occur amongst ground-vegetation, usually in damp or shaded habitats (Bährmann, 1993a: 104-105), often near dung or faeces. They are often attracted to freshly trampled grass, and occasionally visit traps baited with faeces or fruits (Gregor, 1966: 82). They have short wings and are unwilling to fly, as already noted by Zetterstedt (1847: 2293).

Larvae have been found in human excrement (Bouché, 1834: 96, plate 6 fig. 12; Gregor, 1966: 81), in the decaying stem of an umbelliferous plant (Ozerov, 1999d: 197), and in high-water flood refuse (ZMUC). Lobanov (1962a: 37) reported that the puparia hibernate in rotting grass.

Type-material. – *Sepsis leachi*: Meigen described this from an unspecified number of males: 'Das Männchen aus England, von Dr. Leach'. ACP has studied the holotype ♂ in MNHNP, Meigen collection no. 2243 (see also Becker, 1902: 230). It is labelled 'Sepsis leachi ♂' by Meigen. It lacks head and abdomen, but clearly belongs to *leachi* as currently understood. A ♀ with it is not a syntype and is *Meroplius minutus*.

No material of Walker's *tarsalis* and *fumipennis* has survived in any of the sources for Walker material known to ACP. However, the descriptions indicate that he had females of this very characteristic species before him. He attributed dark legs and smoky wings to his *fumipennis*, and this can only be *Themira leachi*. His description of *tarsalis* as a ♂ must be an error: the combination of dark and simple legs does not occur in males of *Themira* or *Nemopoda*. His specimens must have been ♀ and, again, can only have been *Themira leachi*. In Curtis' collection (NMV), there is no material of *tarsalis*, but there are 2♀ of *fumipennis* without data which are *leachi* (seen by ACP).

Sepsis coxarum: Zetterstedt described *coxarum* from both sexes, with no indication of any precise locality. In ZML, in the Insecta Lapponica collection, Diptera box 17, there are 2♀ syntypes which have been studied by ACP. One is labelled by Zetterstedt 'S. coxarum Zett. / Ins. L. ♂. S. Le= / achi ♀ ... [illegible] / Lapp. Torn.'; the other is labelled by Zetterstedt 'S. Leachi. ♀ / coxarum Zett. / Ins. Lapp. / Torneå'. ACP has labelled and designates herewith the first of these as lectotype; it is in excellent condition. The Torneå ♀ has been labelled as paralectotype. Both are *leachi*. There is also a ♂ with these 2♀, but the only locality data are a small green tag, which refers to material collected by Zetterstedt in 1840 in Jämtland, and so it cannot be a syntype; it is also *leachi*.

Themira curvipes and *umbripennis*: As pointed out by Duda (1926a: 76, 79), Van der Wulp had difficulty in sexing his *Themira* material correctly, and he described the ♂ and ♀ of *leachi* as separate species. According to H. de Jong (pers. comm.), Van der Wulp's two syntypes of *curvipes* are not in ZMAN and are presumed lost: Den Haag, 11.viii, Van der Wulp, ♂; Den Haag, June, Snellen van Vollenhoven, ♀. On the other hand, one syntype of *umbripennis* is in ZMAN, but (as also noted by Duda, 1926a: 76) it is a ♀ not a ♂: Amsterdam, 31.viii.1867, Kinker.

10. *Themira lucida* (Stæger, 1844)

(Figs 42, 123, 131, 135, 200-209)

Sepsis lucida Stæger *in* Schiødte, 1844: 32. Lectotype ♂, DENMARK: Amager I., in ZML [see designation below].

Themira roseni Becker, 1915: 129, fig. B. Lectotype ♀, FAROE IS: Sand, in ZMHU [see designation below]. **Syn. n.**

Diagnosis. – Amongst the species with postpronotal seta and pruinose katepisternum, adult *lucida* is best recognised by the wholly dark legs, glossy scutum and, in the ♂, by characters of the fore legs and genitalia (Figs 200-201, 205-208).

Larval pseudocephalon as in *leachi* (see Fig. 199); anal division with a small subanal tubercle and a relatively large anal pad with a diagonal fold (Fig. 135); cephaloskeleton as in *leachi*; anterior spiracles with 13-17 lobes; with six abdominal creeping welts; with relatively numerous spines on lateral side of anal division, on dorsum with three double rows of small spines, with hairs between rows (Fig. 209).

Adult. – Colour: Head mostly black, dull brownish on parafacial, face and gena. Antenna black. Thorax, scutellum and abdomen entirely black. Legs black. Wing clear; calypter and fringe-hairs creamy, margin yellow; haltere yellow.

Head: Frons mostly shining, occiput thinly brownish-grey pruinose. Postocellar subequal to inner vertical; orbital only 1/3 length of inner vertical. Gena at middle hardly half width of antennal flagellomere; several short subvibrissal setae present.

Thorax: Scutum and postpronotal lobes shining black, with some weak brownish pruinosity at neck and between dorsocentral setae and scutellum; scutellum more densely brownish pruinose. Setae: postpronotal present; anterior notopleural weak, hardly 1/4 length of posterior one; acrostichal setulae minute. Pleural pruinosity: anepisternum entirely glossy; katepisternum entirely whitish-grey pruinose; anepimeron thinly pruinose on upper anterior part and very narrowly along hind-margin; meron pruinose, with a central glossy oval patch. Setae reduced; no katepisternal setula. Scutellum short, twice as wide as long, with a pair of short basal setulae.

Legs: Fore coxa shining except at extreme base, mid and hind coxae pruinose; trochanters and femora shining; tibiae shining to subshining. Fore femur with 1 or more posterodorsal setae in apical half; ♂ fore legs in Figs 200-201; fore tarsal ratio 10-6.5-4.5-3-3.3, i.e. tarsomere 2 longer than tarsomere 3; ♀ fore legs simple, femur with 5-6 anteroventral spinules. ♂ mid legs straight, tibia with the apical setae normal. ♂ hind coxa and trochanter without modifications. Hind femur with 2-3 short anterodorsal setae. Hind tibia without setae; ♂ osmeterium in Fig. 202, broad, the tibia with fine erect anterodorsal setulae on apical third.

Wings: Veins bare except for a few setulae in basal part of stem-vein on ventral surface. Entirely covered with microtrichiae.

Abdomen: Glossy black. ♂ tergites 1+2, 3 and 4 hardly expanded behind (Figs 203-204). Without setae, except at sides of hind-margin of tergite 5 in ♂; syntergite 1+2 with erect setulae at sides. ♂ sternites with weak setulae (Fig. 204); sternite 4 and genitalia in Figs 205-208.

Wing-length: 2.8-3.4 mm (♂), 2.7-3.5 mm (♀).

Remarks. – *T. lucida* was fully described and illustrated by Ozerov (1999d: 178-179, fig. 7); the ♂ genitalia were also illustrated by Ozerov (1999e: fig. 348.7).

Males of *lucida* are most often confused with *minor* in collections. If the postpronotal seta is broken off, then the species are very similar in external features, fore legs and genitalia. *T. lucida*, however, lacks the short but expanded posterior seta on mid tarsomere 4 that is present in *minor* (Fig. 100).

Females of *lucida* and *pusilla* are very difficult to separate, and little reliance should be placed on identifications of females without associated males. The dusting on the scutum is very variable in *lucida*: sometimes only the posterior part of the postpronotal lobes and adjacent parts of the scutum are glossy, but usually the glossy areas are much more extensive. The scutum and postpronotal lobes are usually wholly pruinose in *pusilla*, but sometimes there may appear to be some shine around the posterior part of the postpronotal lobes. Such females are impossible to separate, but additional assistance may be found with the following: there are (3-) 4-5 or even 6 anteroventral spines on fore femur in *lucida*, but only 2-3, rarely 5, in *pusilla*; generally, *lucida* is more southern in its distribution.

Egg. – The egg was described by Meier (1995b: 427-432).

Larva. – A full description of the 3rd instar larva was given by Meier (1996: 111, figs 391-399).

Distribution. – Rather common and widespread in the south. Denmark: most provinces (e.g. Lyneborg, 1965: 231). Norway: surprisingly few records, R, N and TR (e.g. Ringdahl, 1954a: 52). Sweden: SK to TO (e.g. Ringdahl, 1934: 4; 1951: 143). Finland: Al to Li (Pont, 1990: 154).

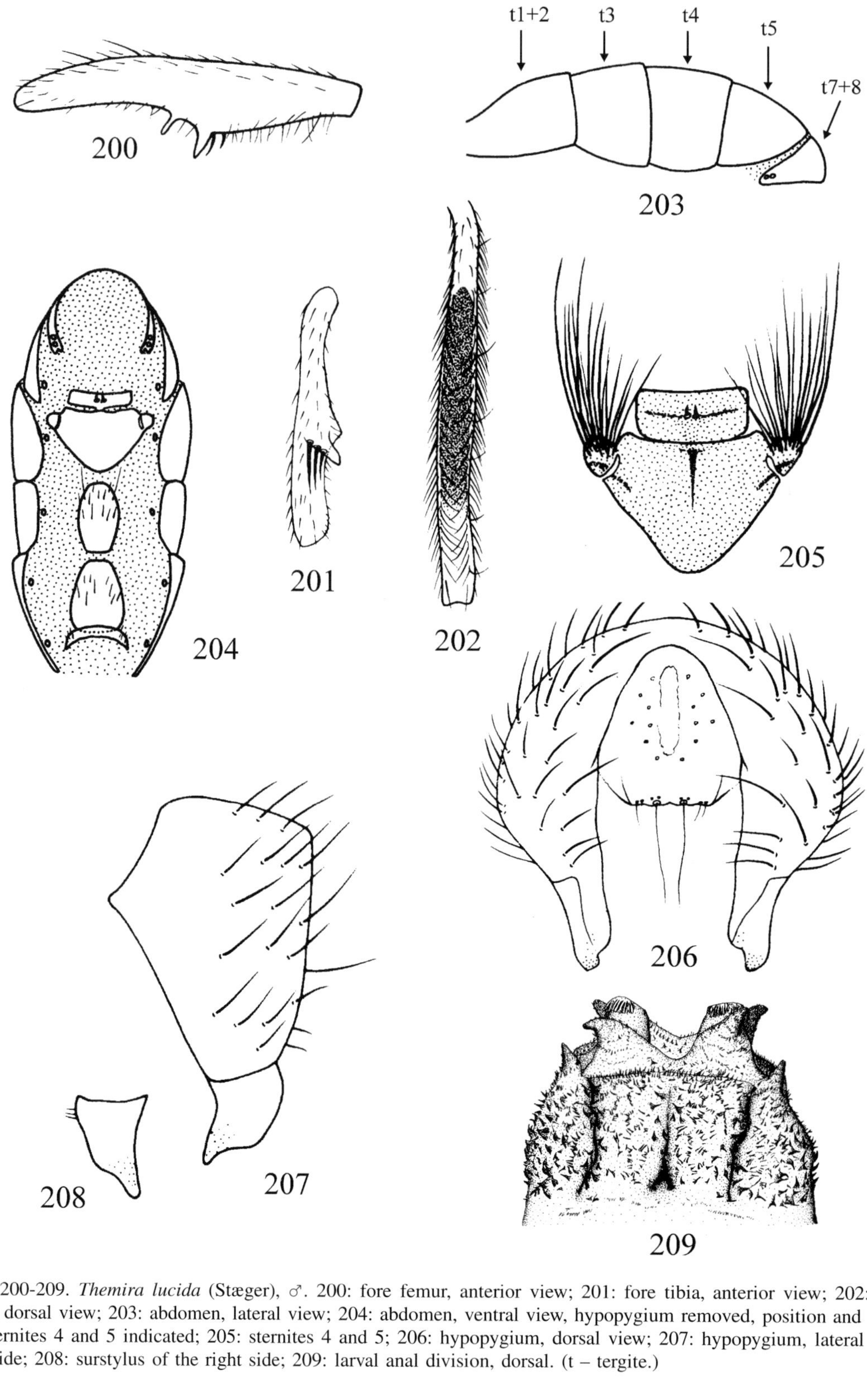

Figs 200-209. *Themira lucida* (Stæger), ♂. 200: fore femur, anterior view; 201: fore tibia, anterior view; 202: hind tibia, dorsal view; 203: abdomen, lateral view; 204: abdomen, ventral view, hypopygium removed, position and shape of sternites 4 and 5 indicated; 205: sternites 4 and 5; 206: hypopygium, dorsal view; 207: hypopygium, lateral view, left side; 208: surstylus of the right side; 209: larval anal division, dorsal. (t – tergite.)

Most of Europe, from Ireland and the Faroes east to Russia (Moscow) and Ukraine (Zuska & Pont, 1984: 165), and throughout Asiatic Russia to the Far East province (Ozerov, 1999d: 179). Nearctic: Alaska (Ozerov, 1999d: 179). Some of the more southern records (e.g. Algeria in Zuska & Pont, l.c.) may be based on misidentifications of *minor*, which has a pronounced Mediterranean distribution.

Collected from late May to mid September.

Biology. – Adults are found in damp meadows, around tarns and pools, and on river banks (Duda, 1926a: 86), especially on waterfowl droppings (RM, pers. obs.; F. Püchel, pers. comm.), but also on rabbit droppings (Minder, 1963: 388) and they have also been reported from *Vespula vulgaris* nests (O'Toole, 1978: 163). They are also attracted to traps baited with fish meat, faeces, or fruits (Gregor, 1966: 82). Adults have been found carrying two unidentified species of mite (Parasitidae).

It has been reared from high-water flood refuse (ZMUC) as well as goose and duck droppings (RM, pers. obs.). The flies were reared twice in the laboratory: once from an unknown substrate (resembling goose dung) on which they were also caught and once from duck dung. In both substrates the flies develop very rapidly (10 days from egg to adult at 24°C).

Type-material. – *Sepsis lucida*: ACP found 1♂ syntype in ZML, in the Diptera Scandinaviae collection, drawer 38, given to Zetterstedt by Stæger, labelled by Stæger 'Amager / 17/8.42'. This is in excellent condition, and belongs to *lucida* as currently understood; it has been labelled and is herewith designated as lectotype. In ZMUC are 2♂ 1♀ syntypes of *lucida*, which Duda (1926a: 79, 86) also studied. ACP has labelled them as paralectotypes. As pointed out by Duda, the 2♂ are actually *minor*, whereas only the ♀ is *lucida*, for which reason the ZML ♂ is preferred for lectotype designation.

Themira roseni: Becker described *roseni* from 1♂ and 5♀. The ♂ has not been seen since, but Duda (1926a: 80) followed by Hennig (1949: 38) and Zuska & Pont (1984: 165) surmised that it might be *minor*. Duda also saw 3♀ syntypes, of which he identified two as *lucida* and one as *dampfi* (i.e. *arctica* of the present work). However, Lyneborg (1968: 24) was convinced that *minor* did not occur in the Faroes and that *roseni* ♂ would probably prove to be *lucida*. We agree with Lyneborg, especially as Becker's figure of the ♂ fore leg of *roseni* does actually fit *lucida* better than *minor*.

ACP has seen 4♀ syntypes of *roseni* from ZMHU, but renewed search for the ♂ has been unsuccessful. 1♀ is labelled 'Sand. / 30.viii.12', and it has been labelled and is herewith designated as lectotype; it is in good condition and belongs to *lucida* as defined in this paper. The presence of some stout anteroventral spinules on fore femur confirms that it is *lucida* and not *minor*. 2♀ on one pin are labelled 'Sand. / Küste. [= coast] / 30.8.12', and have been labelled as paralectotypes; both are *lucida*. 1♀, labelled 'Vaag / Kirchhof / 21.9.12', has been labelled as paralectotype, and is *arctica*.

11. ***Themira malformans*** Melander & Spuler, 1917
(Figs 104, 107, 111-113, 117, 210-218)

Themira malformans Melander & Spuler, 1917: 46. Lectotype ♂, CANADA: Hudson Bay Territory, in USNM [by designation of Ozerov, 1998a: 86].

Diagnosis. – Adult recognised by the glossy katepisternum and presence of postpronotal seta, together with the highly ornamented ♂ fore and hind legs (Figs 210-213).

Adult. – Colour: Head and antenna black. Thorax, scutellum, legs and abdomen entirely black, sometimes mid tarsomere 1 orange. Wing clear; calypter creamy, margin and fringe-hairs yellow; haltere creamy to yellow.

Head: Frons mostly shining, fronto-orbital plate and rest of head dark brown pruinose. Postocellar subequal to inner vertical; orbital half as long as this. Gena over half length of antennal flagellomere; subvibrissal setae numerous and very short.

Thorax: Scutum thinly pruinose (♂) or subshining (♀), narrowly but more densely brown pruinose along notopleural, prealar, postalar and prescutellar areas. Postpronotal lobes thinly pruinose to subshining. Scutellum pruinose. Setae: postpronotal present; anterior notopleural fine but distinct, half as long as posterior one; acrostichal setulae minute. Pleural pruinosity dull- to brownish-grey (Fig. 104): anepisternum with only tiny pruinose areas; katepisternum pruinose along anterior edge and in upper posterior corner, i.e. more extensively than in *superba* but not as much as in *arctica* or *ringdahli*; anepimeron moderately pruinose behind, meron pruinose except for an oval central area. Setae reduced; no katepisternal setula. Scutellum short, twice as wide as long, without basal setulae.

Legs: Fore coxa shining except above, mid coxa partly shining outside, hind coxa subshining; trochanters and femora shining; mid tibia pigmented in basal half dorsally and hind tibia pigmented all along anterodorsal surface beyond osmeterium. Fore femur with some posterodorsal setae before tip, these numerous and long in ♂; ♂ fore legs in Figs 210-212; ♀ fore legs simple, femur with 4-5 strong anteroventral spinules around middle. ♂ mid femur curved, with a series of sparse straight anterior setae in apical half (Fig. 111); normal and bare in ♀ except for some short anterior setae in apical half. ♂ mid tibia rather curved, flattened in basal half, without ventral seta, apical setae fine and elongate (Fig. 113); ♀ with 1-2 short anteroventral setae. ♂ hind coxa with a short spine in front towards middle (Fig. 107). ♂ hind trochanter without modifications. Hind femur with some anterodorsal setae around middle; in ♀ with short fine posterior setae in basal quarter. ♂ hind tibia with the osmeterium in Fig. 213, with some fine erect dorsal setulae on apical 2/5 and some fine but longer anteroventrals on apical 3/5; at tip rather hollowed out on anterodorsal surface, and with a short comb of erect setulae and a stout curled anterior seta (Fig. 112).

Wings: Veins bare except for several setulae on basal part of stem-vein on ventral surface. Entirely covered with microtrichiae.

Abdomen: Glossy black. ♂ tergites 1+2, 3 and 4 greatly expanded behind (Figs 214-215), tergite 4 very broad and largely concealing the following tergites. Without setae except at the sides of tergite 4 in ♂; syntergite 1+2 with erect setulae at sides; ♂ tergite 5 slightly reduced, tergite 7+8 with setulae dense and tufted (Fig. 214). ♂ sternites with weak setulae (Fig. 215); sternite 4 and genitalia in Figs 216-218.

Wing-length: 3.0 mm-3.7 mm (♂), 3.1-3.9 mm (♀).

Remarks. – *T. malformans* was fully described and illustrated by Ozerov (1999d: 202, fig. 31); the ♂ genitalia were also illustrated by Ozerov (1999e: fig. 348.9).

Egg and larva. – Not known.

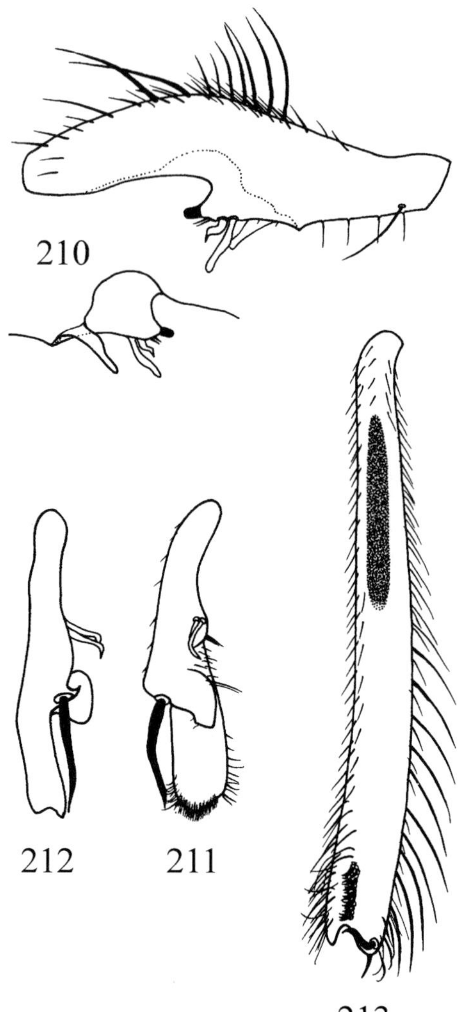

Figs 210-213. *Themira malformans* Melander & Spuler, ♂. 210: fore femur, anterior view (detail: posterior view of the protuberance); 211: fore tibia, anterior view; 212: fore tibia, dorsal view; 213: hind tibia, dorsal view.

Distribution. – A rare northern species. Denmark: absent. Norway: O (1000 m, ZMUC), Nn (Ringdahl, 1954a: 52) and TR (ZML). Sweden: LY (ZML) and TO (Ringdahl, 1931: 25; 1934: 5, who recorded it as new to Sweden and to Europe). Finland: LkW and Li (Pont, 1990: 154). Also from the adjacent part of Russia: Lr, Kola Peninsula (UZMH; Ozerov, 1999e: 563).

Otherwise known only from Russia: W Siberia, Tyumen' province (Ozerov, 1986a: 642), and

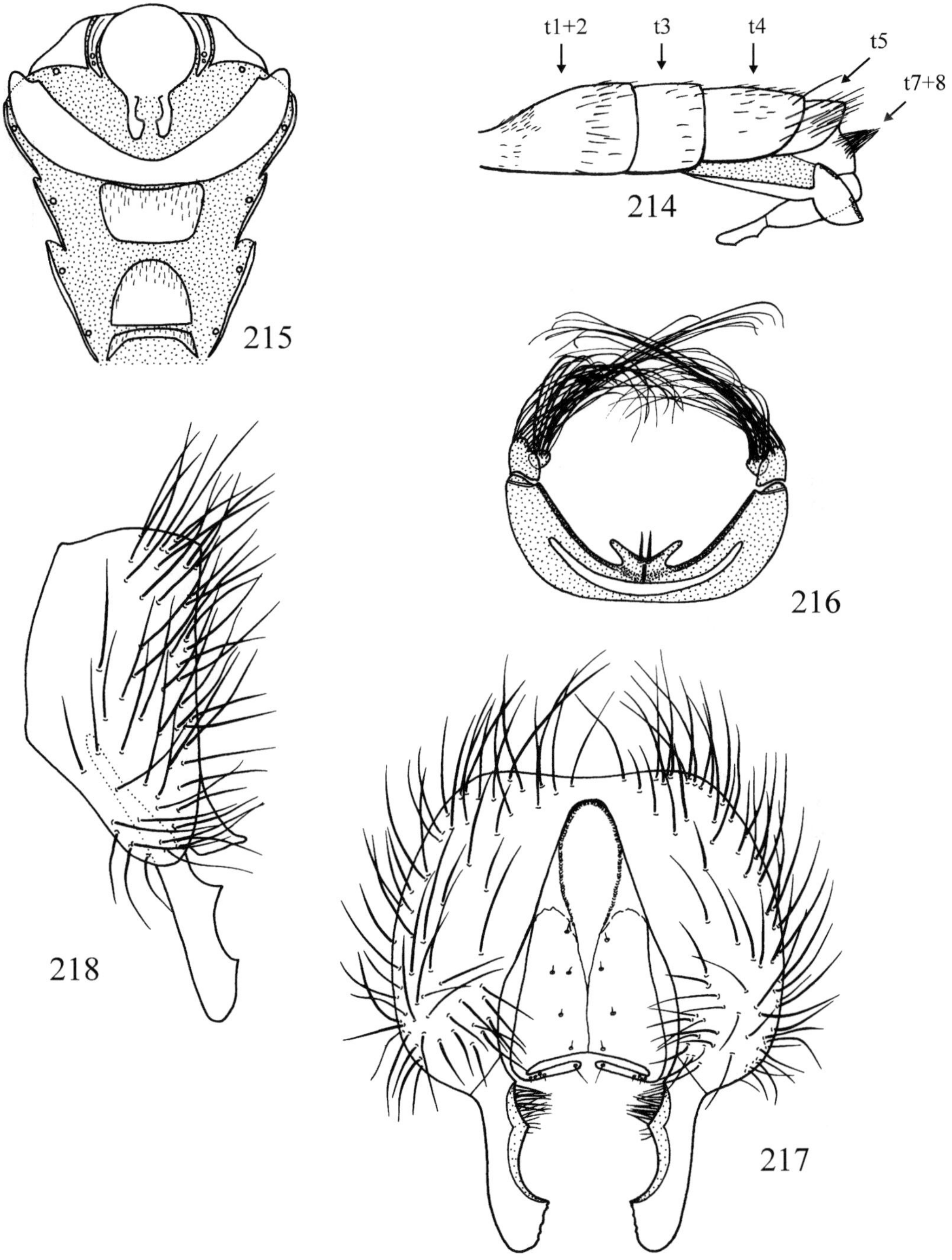

Figs 214-218. *Themira malformans* Melander & Spuler, ♂. 214: abdomen, lateral view; 215: abdomen, ventral view, position and shape of hypopygium and sternite 4 indicated; 216: sternites 4 and 5; 217: hypopygium, dorsal view; 218: hypopygium, lateral view. (t – tergite.)

E Siberia, Dudinka (UZMH). Nearctic: Canada, Hudson Bay (Steyskal, 1965: 684; Ozerov, 1999d: 207).

Collected from mid June to mid August.

Biology. – Subarctic. Adults have been found in wet meadows and on lake shores. Life-cycle and immature stages not known.

12. *Themira minor* (Haliday, 1833)

(Figs 26, 47, 94, 96, 98, 100, 134, 138, 219-228)

Sepsis halidayi Curtis, 1831: 230. Nomen nudum.
Sepsis minor Haliday, 1833: 170. Lectotype ♂, locality not stated [from title: IRELAND, Holywood], in NMI [see designation below].
Themira halidayi Curtis, 1837: 268. Nomen nudum.
Themira dentimana Van der Wulp, 1864: 135, plate 8 fig. 6. Holotype ♂, NETHERLANDS: near Arnhem, not in ZMAN [see below].
Themira consobrina Van der Wulp, 1871: 192. Syntypes ♂♀, NETHERLANDS: den Haag and Zutphen, in ZMAN [see below].

Diagnosis. – Adult recognised by the absence of the postpronotal seta, pruinose katepisternum, and ornamentation of the ♂ fore legs. The short but expanded posterior seta on tarsomere 4 of ♂ mid leg (Fig. 100) is diagnostic for the species.

Larval pseudocephalon with combs and ridges (Fig. 228); cephaloskeleton as in *leachi*; anterior spiracles with 11-15 lobes; anal division with a medium-sized subanal tubercle and a somewhat angular anal pad with a diagonal fold (Fig. 134); with six abdominal creeping welts; with small spines and spinules on dorsum not arranged in any order.

Adult. – Colour: Head and antenna black, sometimes browner on parafacial and gena. Thorax, scutellum and abdomen entirely black. Legs black, only fore coxa in front and most of trochanters yellow-brown. Wing clear; calypter white, margin and fringe-hairs creamy; haltere creamy.

Head (Fig. 94): Frons mostly shining, occiput weakly brownish-grey pruinose. Postocellar about half length of inner vertical; orbital short, subequal to postocellar. Gena at middle half length of antennal flagellomere; subvibrissal setae numerous and short.

Thorax: Scutum with a variable area shining, but at least antero-lateral area and postpronotal lobes shining, often much more; the remainder and scutellum brownish-grey matt. Setae: postpronotal absent; anterior notopleural very short, 1/3 to 1/2 length of posterior one; acrostichal setulae minute. Pleural pruinosity: anepisternum glossy; katepisternum entirely whitish-grey pruinose; anepimeron thinly pruinose in upper anterior corner and on posterior third, otherwise glossy; meron pruinose except for a large oval glossy patch, posterior margin pruinose (Fig. 96). Setae reduced; no katepisternal setula. Scutellum short, twice as wide as long, usually without a basal setula.

Legs: Fore coxa only glossy on lower 3/4 of outer surface, otherwise pruinose, mid and hind coxae mostly pruinose; trochanters and femora shining, tibiae subshining. Fore femur without or with 1-2 short posterodorsal seta; ♂ fore legs in Figs 219-220; ♂ fore tarsal ratio 10-5-4.3-2.5-2.5, i.e. tarsomere 2 longer than tarsomere 3; ♀ fore legs simple, femur without any anteroventral spinules. ♂ mid legs straight; tibia with the apical setae normal, without anteroventral seta in either sex; ♂ tarsi (Fig. 100) little modified, tarsomeres 4 and 5 rather enlarged and dorsoventrally flattened, at tip of tarsomere 4 with a short but expanded posterior seta. ♂ hind coxa and trochanter without modifications. Hind femur without anterodorsal setae. Hind tibia without setae; ♂ osmeterium in Fig. 221, the tibia with some longer and more erect setulae on anterodorsal surface.

Wings: Veins bare except for a few setulae in basal part of the stem-vein on ventral surface. Entirely covered with microtichiae.

Abdomen: Mostly shining black, with a few traces of shagreening. ♂ tergites 1+2 and 3 weakly, tergite 4 hardly, expanded behind (Figs 222-223). Without setae except for a few short ones at sides on hind-margin of tergite 5 in ♂; syntergite 1+2 without any erect setulae at sides. ♂ sternites with weak setulae (Fig. 223); sternite 4 and genitalia in Figs 224-227.

Wing-length: 2.4-3.1 mm (♂), 2.6-3.5 mm (♀).

Remarks. – *T. minor* was fully described and illustrated by Ozerov (1999d: 192-193, fig. 22); the ♂ genitalia were also illustrated by Ozerov (1999e: fig. 348.11).

This species is frequently confused with *lucida*, but can easily be distinguished in the ♂ sex by the presence of a short but expanded posterior seta on tarsomere 4 of mid leg (Fig. 100).

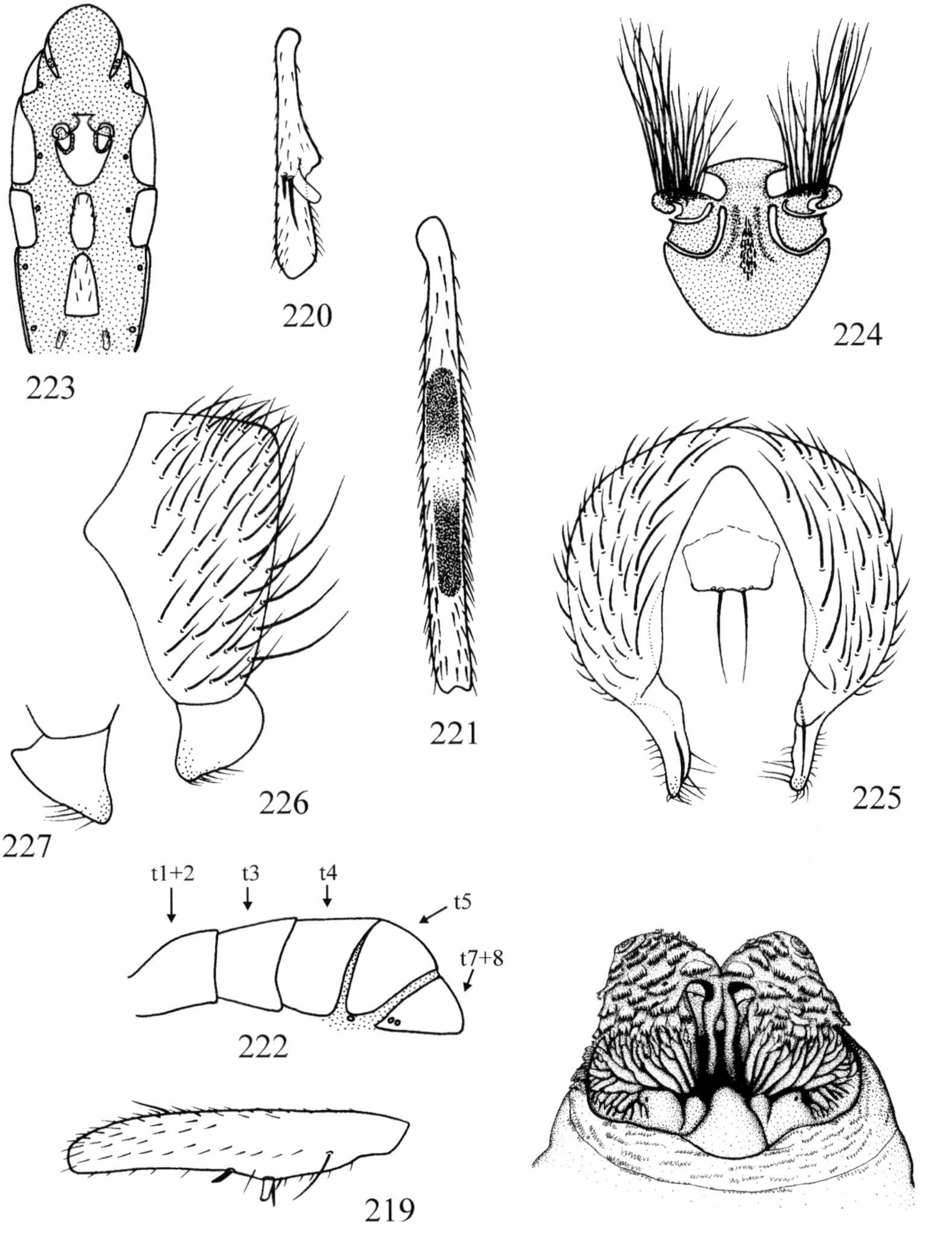

Figs 219-228. *Themira minor* (Haliday), ♂. 219: fore femur, anterior view; 220: fore tibia, anterior view; 221: hind tibia, dorsal view; 222: abdomen, lateral view; 223: abdomen, ventral view, hypopygium removed, position and shape of sternites 4 and 5 indicated; 224: sternites 4 and 5; 225: hypopygium, dorsal view; 226: hypopygium, lateral view, left side; 227: surstylus of the right side; 228: larval pseudocephalon, ventral. (t – tergite.)

Egg. – The egg has been described by Meier (1995b: 427-432, fig 10).

Larva. – The larva has been included in a key to larvae by Mangan (1977: 339). A full description of the 3rd instar larva has been given by Meier (1996: 113-116, figs 409-417).

Distribution. – A more southern species, but nowhere common. Denmark: most provinces (e.g. Lyneborg, 1965: 231). Norway: only Ø and AK. Sweden: SK to DR (e.g. Ringdahl, 1934: 4). Finland: southern provinces, and Om and ObS (e.g. Krogerus, 1932: 116); only rarely (W Finland) extending north of 61°N.

Widespread through the Palaearctic, from Iceland and Madeira eastwards, North Africa, Middle East, throughout Russia to the Far East province (Ozerov, 1989: 842; 1999d: 193), and Mongolia (Zuska & Pont, 1984: 165); Japan (Iwasa, 1981: 46; 1995: 790). The record from the Faroe Is should be deleted (see discussion above under *roseni*, p. 81). Nearctic: Alaska and British Columbia eastwards to Quebec, south to California, Utah, Illinois, Virginia (Steyskal, 1965: 685; Mangan, 1977: 338; Ozerov, 1999d: 207-208). Oriental: North Pakistan (Iwasa, 1989: 52).

Records are from April to mid September.

Biology. – *T. minor* is usually found in damp habitats and meadows, although there are also some records from drier habitats such as sand dune systems (Krogerus, 1932: 116; Ardö, 1957: 121). The adults have been recorded from cattle feeding pens (Mangan, 1977: 339), cow dung (Coffey, 1966: 214; Iwasa, 1981: 48), horse droppings (Coffey, 1966: 214), pig dung (Coffey, 1966: 214; Papp, 1974: 140), and a sewage treatment plant (U. Draeger, pers. comm.). Mangan (1976: 1026) observed territorial behaviour by males and also a courtship ritual in which the male displayed the tarsomeres of the mid leg before making attempts at genital contact. Occasionally, adults can also be found visiting flowers (Van der Goot, 1986a: 3).

Rearing records are from cattle dung (Pont, 1979: 14), pig dung (Papp, 1974: 139), and high-water flood debris (ZMUC; Boness, 1975: 390). The occurrence of larvae in cow dung has been confirmed by Iwasa (1981: 46) and Papp (1975: table 1), and larvae were also found in garbage pits, where they overwintered as puparia (Lobanov, 1962a: 37). In the laboratory, the flies can easily be reared on cow dung, and the development time from egg to adult is very short (8 days at 24°C).

Type-material. – *Sepsis minor*: Haliday described *minor* from an unspecified number of males, with the comment 'occurs in company with *S. putris*, and is often almost as common'. ACP has seen the 2♂ and 2♀ under *minor* that are in his collection in NMI. 1♂ from Ireland is pinned; it has been labelled and is herewith designated as lectotype. The second ♂, also from Ireland, is gummed on card; it has been labelled as paralectotype. The 2♀, one from England and one from Ireland, are not syntypes. The ♂ lectotype and both ♀♀ belong to *minor* as currently understood. The ♂ paralectotype is *Themira lucida*. Curtis' collection (NMV) contains 1♂ named as *minor*, without data, which is actually *lucida* (seen by ACP); there is a collection label 'Halidayi Curt.' standing over *minor*, but no specimens.

Themira dentimana and *consobrina*: Van der Wulp gave no details of his type-series of *consobrina*, but stated that it was the *minor* that he thought he had misidentified in his earlier paper (1864: 134). The 1864 material of *minor* is thus the syntype series of *consobrina*. Duda (1926a: 79) discussed Van der Wulp's confused treatment of the species *minor* (and of *leachi*) and his problems in differentiating males from females. According to H. de Jong (pers. comm.), there are several specimens in ZMAN collected by Van der Wulp at 's Gravenhage [= Den Haag] in June (4♂ 4♀), some of which may be Van der Wulp's syntypes of *consobrina*. No specimen from Zutphen was found. The single ♂ (holotype) of *dentimana* has not been found in ZMAN (see also Van der Goot, 1985a: 139).

13. ***Themira nigricornis*** (Meigen, 1826)

(Figs 44, 120, 229-238)

Sepsis nigricornis Meigen, 1826: 291. Holotype/syntypes ♀, locality not stated [GERMANY, probably Stolberg], not in MNHNP and presumed destroyed [see below].

Sepsis falleni Stæger *in* Schiødte, 1844: 32. Lectotype ♂, DENMARK: Charlottenlund, in ZMUC [see designation below].

Diagnosis. – Adult *nigricornis* is a large species, recognised by the striking orange colour of the gena and the wholly pruinose meron.

Larval pseudocephalon with combs and ridges (as in Fig. 228); cephaloskeleton as in *leachi*; anterior spiracles with 17-21 lobes; anal division with large subanal tubercles and a very small anal pad lacking a diagonal fold (Fig. 238); with three double rows of spines on dorsum of anal division; without hairs between the rows (Fig. 44).

Adult. – Colour: Vertex and occiput black, frons and face brown, parafacial and entire genal area (below eye) orange-yellow, with a clear line of demarcation between lower occiput and gena. Antennal pedicel and scape mostly orange to yellow, flagellomere black but orange to yellow on inner surface. Thorax, scutellum and abdomen entirely black. Coxae brown, yellow-tipped, fore coxa more extensively yellow; trochanters yellow; femora dark, narrowly brown at base and tip, ♀ mid and hind femora broadly yellow at base; fore tibia yellow, except part of apical half in ♂, and fore tarsus brown; mid and hind tibiae yellow at base and tip, more conspicuously in ♀, tarsi brown except for mid and sometimes hind tarsomere 1. Wing clear; calypter white, with creamy margin and white fringe-hairs; haltere creamy.

Head: Frons subshining, occiput thinly but conspicuously brownish-grey pruinose. Postocellar almost equal to inner vertical; orbital short, hardly 1/3 of inner vertical. Gena broad, at middle almost equal to length of antennal flagellomere; many short subvibrissal setae present.

Thorax: Scutum, postpronotal lobes and scutellum brownish-grey matt, not shining from any point-of-view. Setae: postpronotal present; anterior notopleural strong and hardly shorter than posterior one; acrostichal setulae minute; no auxiliary dorsocentrals. Pleural pruinosity: anepisternum mainly glossy, but pruinose around edges, especially in upper posterior corner; katepisternum entirely whitish-grey pruinose; anepimeron pruinose, with a small central glossy oval patch; meron entirely pruinose (Fig. 120). Setae reduced; proepisternal setula relatively long and conspicuous; anepisternal setula very indistinct; no katepisternal setula. Scutellum short, twice as wide as long, without basal setulae.

Legs: Fore coxa shining except basally and in front, mid and hind coxae pruinose; trochanters shining, femora subshining, tibiae pruinose. Fore femur with several posterodorsal setae in apical half; ♂ fore legs in Figs 229-230; ♂ fore tarsal ratio 10-4-3.5-2-2.7, i.e. tarsomere 1 of normal length and much longer than tarsomere 2; ♀ fore legs simple, femur with 4-5 anteroventral spinules. ♂ mid femur straight; tibia with the apical setae, and tarsi, normal; tibia with 1 anteroventral in both sexes. ♂ hind coxa and trochanter without modifications. Hind femur with a few short anterodorsal setae. Hind tibia without setae; ♂ osmeterium in Fig. 231, occupying 1/3 of tibial length, the tibia without erect fine anterodorsal setulae.

Wings: Veins bare except for a few inconspicuous setulae in basal part of stem-vein on ventral surface. Entirely covered with microtrichiae. Section of costa between humeral cross-vein and subcosta, as usual, much longer than that between subcosta and vein R_1 (10-7) (as in Fig. 125).

Abdomen: Glossy black; tergites 1+2, 3 and 4 hardly expanded behind (Figs 232-233). Without setae, except at sides of hind-margin of tergite 5 in ♂; syntergite 1+2 with erect setulae at sides. ♂ sternites with weak setulae (Fig. 233); sternite 4 and genitalia in Figs 234-237.

Wing-length: 2.9-3.8 mm (♂), 3.1-4.3 mm (♀).

Remarks. – *T. nigricornis* was fully described and illustrated by Ozerov (1999d: 182-183, fig. 12); the ♂ genitalia were also illustrated by Ozerov (1999e: figs 349.4, 8).

Larva. – A full description of the 3rd instar larvae was given by Meier (1996: 116-118, figs 418-426).

Distribution. – Widespread, occasionally abundant. Denmark: only EJ and NEZ (Stæger, 1844, type-locality of *falleni*). Norway: northwards to NT (e.g. Siebke, 1877: 153, as *falleni*; Hennig, 1949: 43). Sweden: SK to TO (e.g. Wahlgren, 1919: 284). Finland: northwards to Ks (Frey, 1908: 587).

A widespread Palaearctic species, through the central and northern zones of Europe eastwards to Russia (Far East province) (Zuska & Pont, 1984: 165), South Korea (Suh & Kwon, 1994: 62). Japanese records (Iwasa, 1981: 48) actually refer to *T. lutulenta* Ozerov, 1986 (Iwasa, 1995: 788). Nearctic: Illinois to New Jersey, Alaska (Steyskal, 1965: 684; Mangan, 1976: 1026; Ozerov, 1999d: 208). Oriental: North Pakistan (Iwasa, 1989: 52).

Records are from end March to August or, in the south of the area, to early October. Principally a spring species, as noted by several authors (e.g. Steyskal, 1946: 93; Minder, 1963: 388; Gregor, 1966: 82).

Biology. – Adults have a relatively high synanthropic index for a sepsid, being characterised

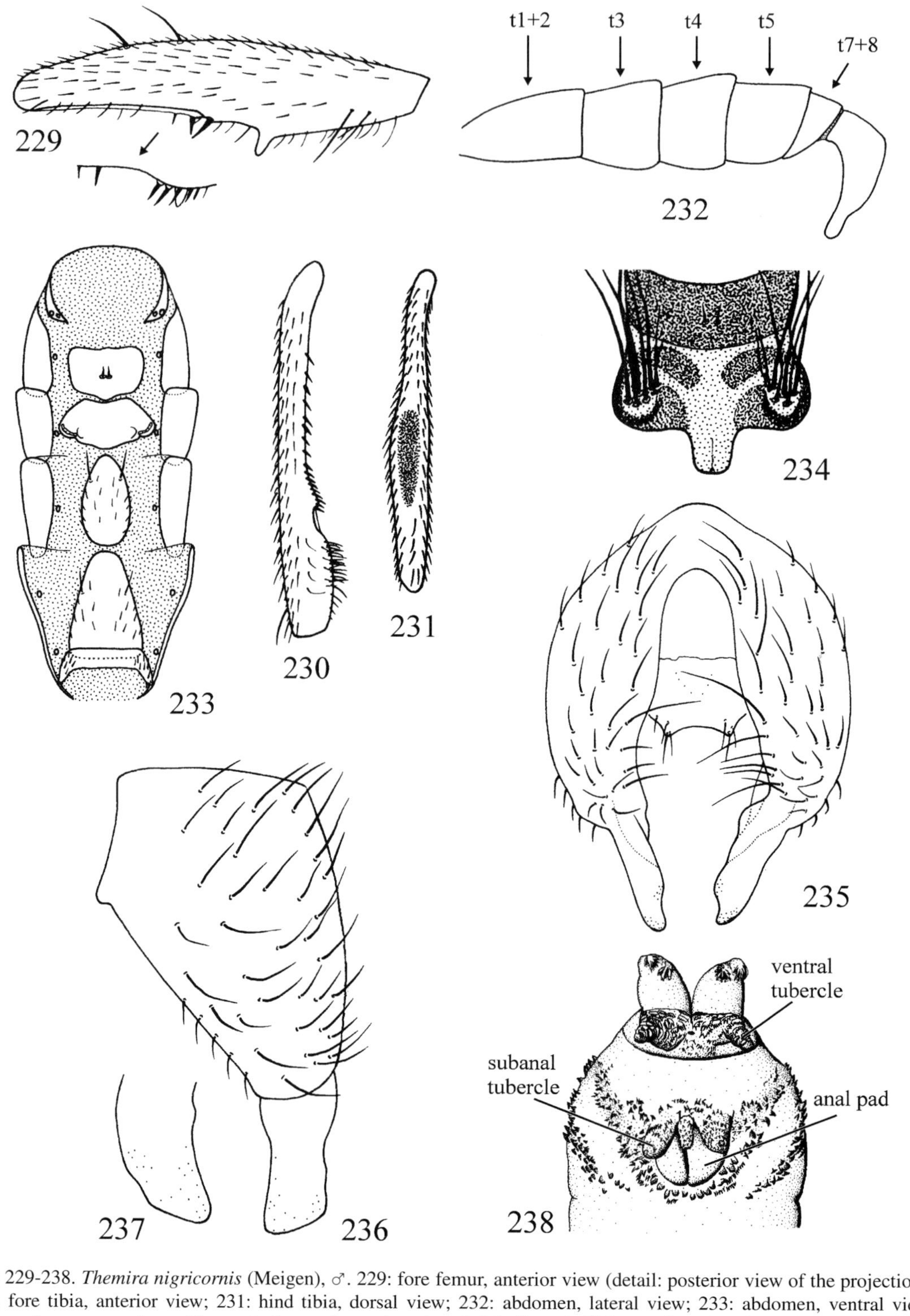

Figs 229-238. *Themira nigricornis* (Meigen), ♂. 229: fore femur, anterior view (detail: posterior view of the projection); 230: fore tibia, anterior view; 231: hind tibia, dorsal view; 232: abdomen, lateral view; 233: abdomen, ventral view, hypopygium removed, position and shape of sternites 4 and 5 indicated; 234: sternites 4 and 5; 235: hypopygium, dorsal view; 236: hypopygium, lateral view, left side; 237: surstylus of the right side; 238: larval anal division, ventral. (t – tergite.)

as communicative hemisynanthropes by Gregor (1966: 83). They are found near dung, and can be attracted to traps baited with fish and/or meat (Gregor, 1966: 82). Their frequency in food-processing plants, especially those where meat-based products are handled, is relatively high (Zuska & Laštovka, 1969: 209). They are also reported to occur indoors and in animal sheds (Petrova, 1968: 98). Adults have also been found on cow dung mixed with straw (Iwasa, 1981: 48), rotting fungi (Gregor, 1966: 83), and compost (RM, pers. obs.). In spring they are attracted to birch sap (Minder, 1963: 388). Territorial behaviour has been observed by Mangan (1976: 1026), who also noted that there is no courtship display but that the copulating pair is active.

Larvae have been found in cow manure (Iwasa, 1981: 48), hen manure (ZMUC), and puparia have been floated from marsh material in the subarctic (Mangan, 1976: 1026). Rearing records also come from garden soil and human excrement (Pont, 1979: 16). Lobanov (1962a: 37) reported that the puparia hibernate in cesspits, cow dung and chicken dung. This species was bred once in spring 1993 in the laboratory (U. Draeger, pers. comm.). The larvae developed through to the pupal stage, but no flies hatched for several weeks. Given that the species is mostly found in early spring, further studies should investigate whether it may have a summer diapause.

Type-material. – *Sepsis nigricornis*: There is no illustration of this species in the Meigen plates (Morge, 1976a, b). It was described by Meigen without citing a locality, 'nur einigemal das Weibchen'. The inference is that he collected the species himself, and the syntypes should therefore be in his collection in MNHNP. However, ACP has found no material in MNHNP: the name does not occur, nor are there any specimens misplaced under other names (see also Becker, 1902: 229). In NMW, Duda (1926a: 96) saw 1♂ 1♀ from 'Lüneburg' (which are *nigricornis*) and 1♀ from 'Paris' (which is *stercorarius*, i.e. *Meroplius minutus*). Becker (l.c.) had already defined the species as it is still currently understood on the basis of the ♂♀ on a single pin in NMW, and Duda followed this interpretation. ACP has seen the NMW material too, and concluded that none of the Winthem material (1♂ 3♀) can be syntypic. There seems no good reason for not continuing with the conventional interpretation of this name.

Sepsis falleni: In ZMUC, ACP found no specimens labelled from the type-locality, Charlottenlund. The type-collection includes 2♂ 1♀ from Enghave, which cannot be syntypes (which are *nigricornis*), and 4♂ without locality. He concludes that these 4♂ are part or all of Stæger's syntype series, and herewith designates one in excellent condition as lectotype and the others as paralectotypes; all are *nigricornis*. ACP has seen several Stæger specimens in ZML, in Zetterstedt's Diptera Scandinaviae collection, drawer 38, but they too are without locality.

14. *Themira paludosa* Elberg
(Figs 239-246)

Themira paludosa Elberg, 1963: 909, figs 1-3. Holotype ♂, W ESTONIA: Tukhu Bog, in ASSP [see below].

Diagnosis. – Adult *paludosa* is recognised by the absence of the postpronotal seta, yellow base to mid and hind femora, and, in the ♂, the fore leg ornamentation and genitalia. It is most similar to *germanica* Duda.

Adult. – Colour: Head mostly black; lunule, parafacial, lower facial margin and vibrissal area often brown. Antenna dark brown, articulation of pedicel and flagellomere reddish-brown on inner surface. Thorax, scutellum and abdomen entirely black. Fore coxa, except dorsal tip and in ♀ on anterior surface, and trochanters yellow, mid and hind coxae mostly dark brown; fore legs black, extreme base of femur usually yellow in ♂; mid and hind legs dark, but femora broadly yellow at base, often less so in ♀ than in ♂. Wing almost clear; calypter white, margin and fringe-hairs creamy; haltere creamy.

Head: Frons mostly shining, occiput weakly brownish-grey pruinose. Postocellar about 2/3 length of inner vertical; orbital very short, only half as long as postocellar. Gena at middle much less than half length of antennal flagellomere; subvibrissal setae numerous and short.

Thorax: Scutum and scutellum brownish-grey matt and nowhere shining, but postpronotal lobes shining especially on posterior part. Setae: postpronotal absent; anterior notopleural minute and hair-like, hardly 1/3 length of posterior one; acrostichal setulae minute. Pleural pruinosity: anepisternum glossy; katepisternum entirely whitish-

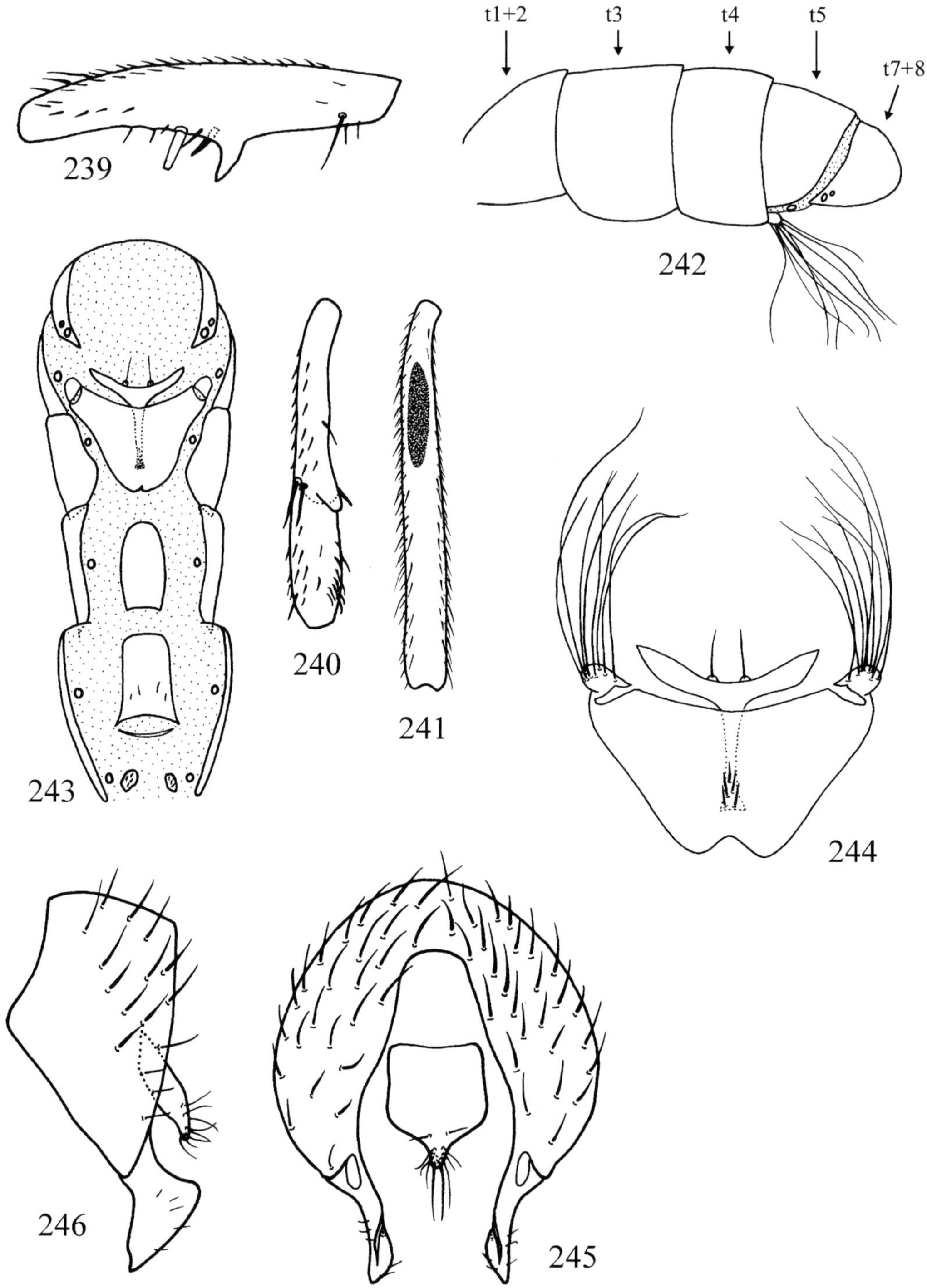

Figs 239-246. *Themira paludosa* Elberg, ♂. 239: fore femur, anterior view; 240: fore tibia, anterior view; 241: hind tibia, dorsal view; 242: abdomen, lateral view; 243: abdomen, ventral view, hypopygium removed, position and shape of sternites 4 and 5 indicated; 244: sternites 4 and 5; 245: hypopygium, dorsal view; 246: hypopygium, lateral view. (t – tergite.)

grey pruinose; anepimeron thinly pruinose on upper anterior corner and on posterior fifth, otherwise glossy; meron pruinose except for a large oval glossy patch, posterior margin pruinose (as in minor, Fig. 96). Setae reduced; no katepisternal setula. Scutellum short, twice as wide as long, with a pair of tiny basal setulae.

Legs: Fore coxa apparently mostly shining, except basally, mid and hind coxae mostly pruinose; trochanters shining, femora and tibiae subshining. Fore femur with 1 weak posterodorsal seta before tip; ♂ fore legs in Figs 239-240; ♂ fore tarsal ratio 10-4-3-2-2.4, i.e. tarsomere 2 longer than tarsomere 3. ♀ fore legs simple, femur without any anteroventral spinules. ♂ mid legs straight; tibia with the apical setae normal, without anteroventral setae; ♂ tarsi not modified. ♂ hind coxa and trochanter without modifications. Hind femur without anterodorsal setae. Hind tibia without setae; ♂ osmeterium in Fig. 241, the tibia without erect fine anterodorsal setulae.

Wings: Veins bare except for a few setulae in basal part of stem-vein on ventral surface. Entirely covered with microtrichiae.

Abdomen: Mostly shining black, with a few traces of shagreening. ♂ tergites 1+2 and 3 weakly, tergite 4 hardly, expanded behind (Figs 242-243). Without setae except for a few short ones at sides of hind-margin of tergite 5 in ♂; syntergite 1+2 without any erect setulae at sides. ♂ sternites with weak setulae (Fig. 243); sternite 4 and genitalia in Figs 244-246.

Wing-length: 1.9-2.3 mm (♂), 2.2 mm (♀).

Remarks. – *T. paludosa* was described and illustrated from Elberg's original description by Ozerov (1999d: 193-195, fig. 23).

Egg and larva. – Not known.

Distribution. – Apart from the type-locality in Estonia, known only from a few localities in Finland (Ab, N, St, Sb, Ks) and from the adjacent part of Russia (Kr, Valamo) (Pont, 1990: 154). The Valamo ♂ in UZMH was the basis for Frey's (1908: 587) record of *gracilis*. The species recorded from Japan as *paludosa* by Iwasa (1985: 634) is actually *T. kanoi* Iwasa, 1995.

Collected only in August.

Biology. – The type-series was collected in a bog, but otherwise nothing is known of the habits and biology of this species.

Type-material. – ACP has studied the holotype ♂, which is in ASSP. It is labelled: 'Zap. Estoniya [= West Estonia] / Boloto [= bog] Tukhu 12.viii / K. Elberg 1961'. Its condition is good; left fore tarsal segments 2-5 missing; genitalia exposed.

15. ***Themira pusilla*** (Zetterstedt, 1847)
(Figs 127-129, 247-254)

Sepsis pusilla Zetterstedt, 1847: 2295. Lectotype ♂, SWEDEN: Åsele Lpm, in ZML [see designation below].

Themira spinosa Verrall, 1886: 179. Nomen nudum.

Themira spinosa Verrall, 1886: 233. Lectotype ♂, SCOTLAND: islands on Loch Maree, in UMO [see designation below].

Diagnosis. – Adult *pusilla* is recognised among the species with postpronotal seta and pruinose katepisternum by the ornamentation of the ♂ legs, especially fore legs and hind tibia (Figs 247-249).

Adult. – Colour: Head black, sometimes browner on parafacial, face and gena. Antenna black. Thorax, scutellum and abdomen entirely black. Legs black. Wing clear; calypter and fringe-hairs creamy, margin yellow; haltere yellow.

Head: Frons mostly shining, occiput thinly brownish-grey pruinose. Postocellar subequal to inner vertical; orbital 1/3 to 1/2 length of inner vertical. Gena at middle half width of antennal flagellomere; several short subvibrissal setae present.

Thorax: Scutum and postpronotal lobes entirely though thinly brownish-grey pruinose, subshining from some points-of-view but nowhere glossy; scutellum more densely brown pruinose. Setae: postpronotal present; anterior notopleural weak, 1/3 to 1/2 length of posterior one; acrostichal setulae minute. Pleural pruinosity: anepisternum entirely glossy; katepisternum whitish-grey pruinose; anepimeron thinly pruinose on upper anterior part and narrowly along hind-margin; meron pruinose, with a central glossy oval patch. Setae reduced; no katepisternal setula. Scutellum short, twice as wide as long, with a pair of tiny lateral setulae.

Legs: Fore coxa shining except at extreme base, mid and hind coxae pruinose; trochanters and femora shining, tibiae shining to subshining. Fore femur with 1 or more posterodorsal setae in apical

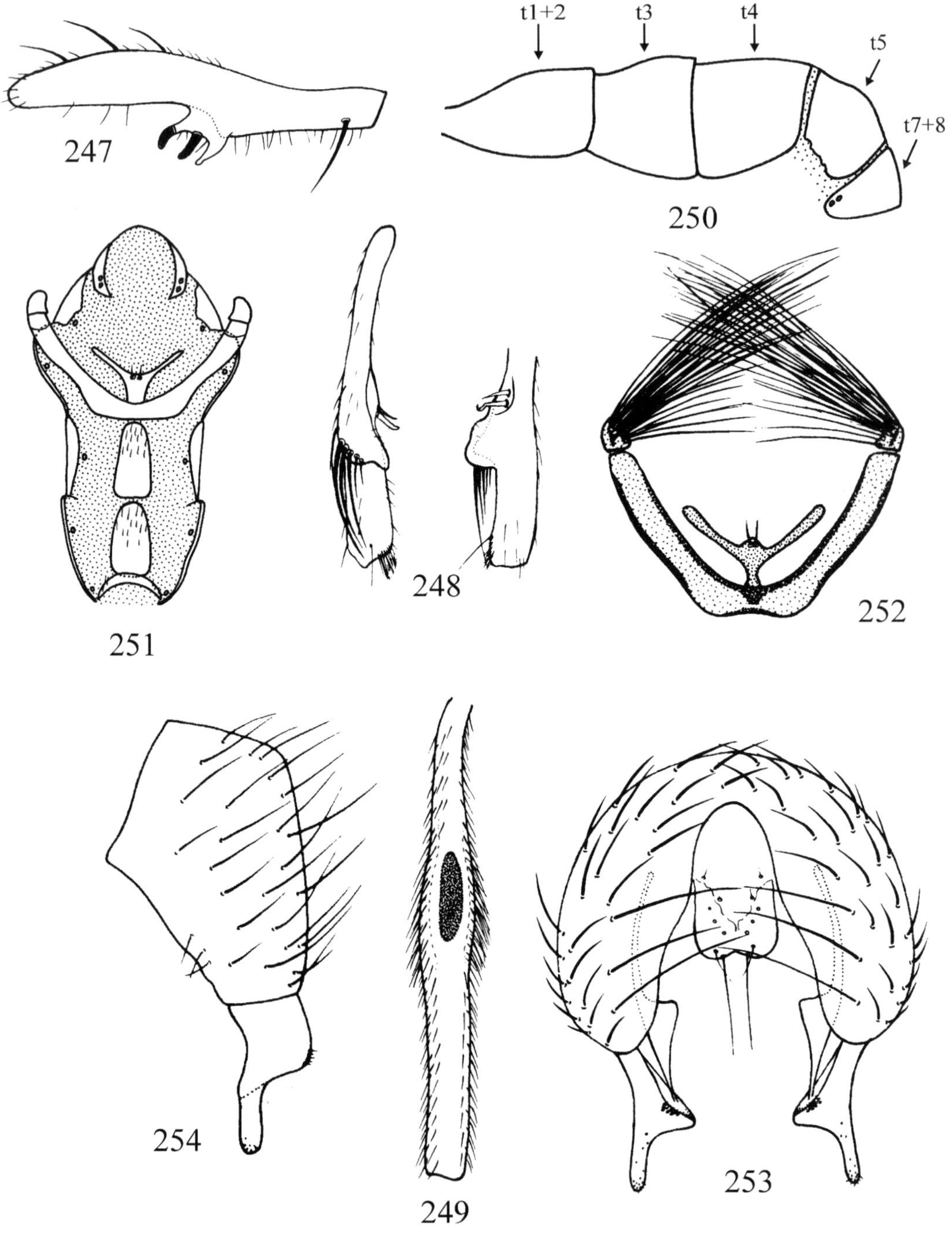

Figs 247-254. *Themira pusilla* (Zetterstedt), ♂. 247: fore femur, anterior view; 248: fore tibia, anterior view (detail: ventral view of apical half); 249: hind tibia, dorsal view; 250: abdomen, lateral view; 251: abdomen, ventral view, hypopygium removed, position and shape of sternites 4 and 5 indicated; 252: sternites 4 and 5; 253: hypopygium, dorsal view; 254: hypopygium, lateral view. (t – tergite.)

half; ♂ fore legs in Figs 247-248; ♂ fore tarsal ratio 10-4-4.8-2.8-3, i.e. tarsomere 3 slightly longer than tarsomere 2; ♀ fore legs simple, femur with 2-3 anteroventral spinules. ♂ mid legs straight; tibia with several fine posterior apical setae, often coalescing to form a fasciculus, and a row of erect posterior to posteroventral setulae in apical third (Fig. 128). ♂ hind coxa and trochanter without modifications. Hind femur with several short anterodorsal setae; in ♂ (Fig. 127) with 3-4 erect posteroventrals at middle and a row of short curled anteroventrals on basal half. Hind tibia without setae; in ♂ (Fig. 129) with a ventral excision near middle, and the tibia bent at this point; ♂ osmeterium in Fig. 249, short and narrow, the tibia without any erect anterodorsal setulae in apical half.

Wings: Veins bare except for a few setulae on basal part of stem-vein on ventral surface. Entirely covered with microtrichiae.

Abdomen: Glossy black. ♂ tergites 1+2 and 3 hardly, tergite 4 strongly, expanded behind (Figs 250-251). Without setae, except at sides of hind-margin of tergite 5 in ♂; syntergite 1+2 with erect setulae at sides. ♂ sternites with weak setulae (Fig. 251); sternite 4 and genitalia in Figs 252-254.

Wing-length: 2.6-3.1 mm (♂), 3.0-3.6 mm (♀).

Remarks. – *T. pusilla* was fully described and illustrated by Ozerov (1999d: 202-203, fig. 32); the ♂ genitalia were also illustrated by Ozerov (1999e: fig. 350.1).

Egg and larva. – Not known.

Distribution. – An uncommon species, with a marked northern distribution. Denmark: absent. Norway: O. Sweden: SK, and UP to TO (e.g. Ringdahl, 1934: 5; 1950: 115); its occurrence in south Swedish mosses is a relict occurrence. Finland: Sb to Li (e.g. Frey, 1908: 586). Duda (1926a: 82) and Hennig (1949: 47) refer to a specimen from Ab, Karislojo, but it must have been misidentified as the species does not occur so far south.

Outside Fennoscandia known from Ireland and Great Britain (Zuska & Pont, 1984: 166; see also Pont, 1987b: 17), Iceland (Andersson, 1967: 107) including Surtsey (Lindroth et al., 1973: 35), and Russia: Kr and Lr, including the Kola Peninsula (UZMH), Leningrad district (Shtakel'berg, 1958: 133), Moscow (Minder, 1963: 389), and Kamchatka (UZMH). Mongolia (Soós, 1972: 356). Nearctic: Alaska (Mangan, 1976: 1027), Washington State (Steyskal, 1965: 684) and New York (Ozerov, 1999d: 208).

Records are from June to early September.

Biology. – Adults are associated with dung of the muskrat (*Ondatra zibethicus*), and are reported from river banks and lake margins, on sand and decaying algae (Ozerov, 1999d: 203). Otherwise nothing is known of its habits and biology.

Type-material. – *Sepsis pusilla*: This was described by Zetterstedt 'ad ripam fluminis Ångermanniae ad Almsele et Åsele'. ACP has studied 1♂ in ZML, with head and right fore leg missing, in Zetterstedt's Insecta Lapponica collection, Diptera box 17. It is labelled by Zetterstedt with a black tag and 'S. pusilla / ♂. putris. b. / Zett. Ins. Lapp. / Åsele Lapp.'. ACP has labelled it and herewith designates it as lectotype. It belongs to *pusilla* as currently understood, and is probably the specimen referred to as 'type' by Collin (1910: 176). In the Diptera Scandinaviae collection, drawer 38, there is a second ♂, with a black tag and labelled 'S. pusilla. ♂. / putris. b. Zett. / ... Lap. Åsele'; it is very badly damaged. ACP has labelled it as paralectotype; it is conspecific with the lectotype.

Themira spinosa: Verrall described this from 'several specimens ... on the islands on Loch Maree, on June 9th, 1884, and one male at Tongue on June 18th, 1884'. ACP located 4♂ 3♀ syntypes from the Loch Maree islands in UMO and 2♂ 2♀ syntypes in BMNH (see also Pont, 1995: 23); the ♂ from Tongue is in UMO. ACP has labelled and herewith designates a ♂ in UMO from the Loch Maree islands as lectotype; it lacks the left fore leg, but is otherwise in good condition. The remaining 6♂ 5♀ have been labelled as paralectotypes. All 12 specimens are *pusilla* as currently understood.

16. *Themira putris* (Linnaeus, 1758)

(Figs 121-122, 124-125, 139, 255-262)

Musca putris Linnaeus, 1758: 597. Holotype/syntypes, locality not stated ['habitat α in Fimetis, β in Caseo', probably SWEDEN], not in LSL and presumed destroyed [see below].

Musca fimeti Linnaeus, 1761: 456. Holotype/syntypes, locality not stated ['habitat in fimetis copiosissime & instar grandinum', from title: SWEDEN], not in LSL and presumed destroyed [see below].

Musca conssencis Harris, 1780b: 123, plate 36 fig. 47. Syntypes, locality not stated [from title: ENGLAND], not found and presumed destroyed [see below].

Themira pilosa Robineau-Desvoidy, 1830: 746. Holotype/syntypes, FRANCE: Saint-Sauveur, not in MNHNP and presumed destroyed.

Themira 'setosa' Robineau-Desvoidy [erroneous subsequent spelling of *Themira pilosa* Robineau-Desvoidy, 1830, by Rondani, 1856: 117].

Diagnosis. – Adult *putris* is a relatively large and very common species with broad gena, similar to *biloba* and best distinguished by the characters given in the key (p. 59).

Larval morphology very similar to that of *minor*, except for the presence of three double-rows of spines on the dorsal side of the anal division and anterior spiracles with 15-23 lobes.

Adult. – Colour: Head mostly black, parafacial, gena and face brown. Antenna black. Thorax, scutellum and abdomen entirely black. Legs wholly black. Wing clear; calypter white, with creamy margin and white fringe-hairs; haltere creamy.

Head: Frons mostly shining, occiput weakly brownish-grey pruinose. Postocellar almost equal to inner vertical; orbital half as long as inner vertical. Gena broad, at middle almost equal to length of antennal flagellomere (Fig. 124); many short subvibrissal setae present.

Thorax: Scutum and postpronotal lobes thinly brownish-grey matt, subshining from most points-of-view; scutellum more densely matt. Setae: postpronotal present; anterior notopleural short, 1/4 (♂) to 1/2 (♀) length of posterior one; acrostichal setulae minute; no auxiliary dorsocentrals. Pleural pruinosity: anepisternum almost wholly glossy, thinly pruinose along hind-margin; katepisternum entirely whitish-grey pruinose; anepimeron thinly pruinose in upper anterior corner and on posterior 1/4, otherwise glossy; meron pruinose, with a ventral glossy oval patch (Fig. 121). Setae reduced; anepisternal setula distinct; no katepisternal setula. Scutellum short, twice as wide as long, with a pair of short basal setulae.

Legs: Fore coxa shining except basally and in front, mid and hind coxae pruinose; trochanters and femora shining, tibiae pruinose. Fore femur with several posterodorsal setae in apical half; ♂ fore legs in Figs 255-256; ♂ fore tarsal ratio 10-10-5.8-3.3-4, i.e. tarsomere 1 short and subequal to tarsomere 2; ♀ fore legs simple, femur with 7-8 anteroventral spinules. ♂ mid femur weakly curved; tibia with the apical setae normal; tarsomeres 4 and 5 rather flattened and expanded; tibia with 1 anteroventral seta in both sexes. ♂ hind coxa and trochanter without modifications. Hind femur with a few short anterodorsal setae. Hind tibia without setae; ♂ osmeterium in Fig. 257, occupying almost 1/3 of tibial length, the tibia with fine anterodorsal setulae, especially before apex.

Wings: Veins bare except for a few setulae in basal part of stem-vein on ventral surface. Entirely covered with microtrichiae. Section of costa between humeral cross-vein and subcosta much longer than that between subcosta and vein R_1 (10-7 or 7.5) (Fig. 125).

Abdomen: Glossy black. ♂ tergites 1+2, 3 and 4 moderately expanded behind (Figs 258-259). Without setae; syntergite 1+2 with erect setulae at sides. ♂ sternites with short setulae (Fig. 259); sternite 4 and genitalia in Figs 260-262.

Wing-length: 3.3-4.1 mm (♂), 3.8-4.3 mm (♀).

Remarks. – *T. putris* was fully described and illustrated by Ozerov (1999d: 183-185, fig. 13); the ♂ genitalia were also illustrated by Ozerov (1999e: fig. 350.7, 8).

Egg. – The egg has been described by Meier (1995b: 427-432, fig. 4).

Larva. – The puparium was first described by Zetterstedt (1847: 2290) and more recently by Hennig (1949: 19, figs 23, 24, 33, 41, 45-47). The larva has been included in a key to larvae by Mangan (1977: 341). A full description of the 3rd instar larva is given by Meier (1996: 118-121, figs 427-435).

Distribution. – The most abundant Fennoscandian *Themira*, widespread and common. Denmark: almost all provinces (Stæger, 1844: 29). Norway: Ø and AK to F (Siebke, 1877: 153). Sweden: SK to TO (Zetterstedt, 1847: 2290). Finland: almost all provinces (Frey, 1908: 586).

Throughout Europe, from Ireland eastwards to Mongolia, China, Japan (Iwasa, 1981: 48), the Far East province of Russia (Ozerov, 1999e: 563), Sakhalin Island, but absent from North Africa and the Middle East (Zuska & Pont, 1984: 166). Probably absent from the Faroes (Lyneborg, 1968: 25) though once recorded from there. Nearctic: Alaska and British Columbia to Quebec, south to California, Nebraska, Illinois, Ohio and Pennsylvania (Steyskal, 1965: 684; Ozerov, 1999d: 208).

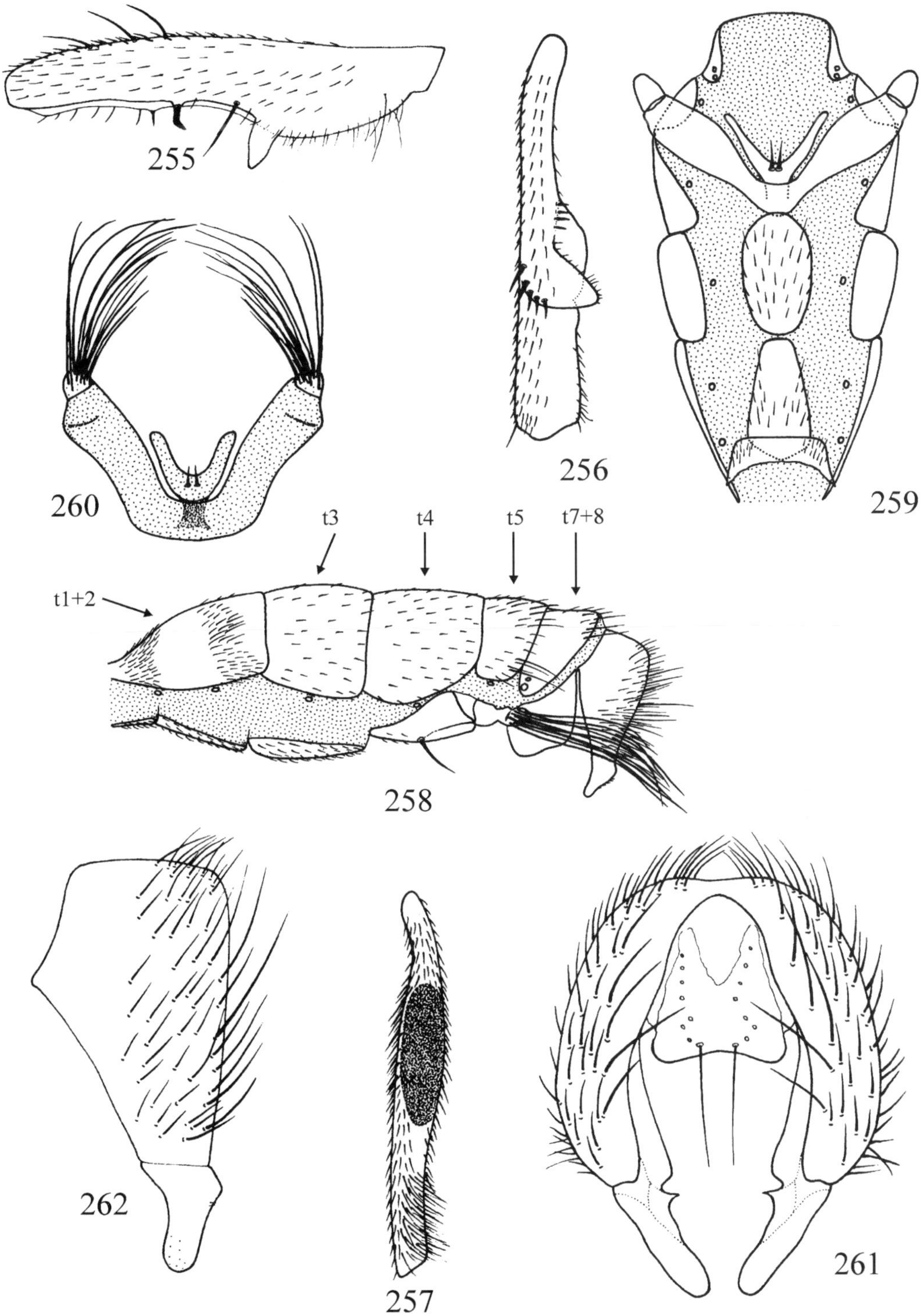

Figs 255-262. *Themira putris* (Linnaeus), ♂. 255: fore femur, anterior view; 256: fore tibia, anterior view; 257: hind tibia, dorsal view; 258: abdomen, lateral view; 259: abdomen, ventral view, hypopygium removed, position and shape of sternites 4 and 5 indicated; 260: sternites 4 and 5; 261: hypopygium, dorsal view; 262: hypopygium, lateral view. (t – tergite.)

Records are from early April to early September in the south, early May to August in the north.

Biology. – *T. putris* has been known from the earliest days as a filth fly, though sometimes confused as an adult with the cheese skipper, *Piophila casei* (Linnaeus). Wilcke (1789: 27) reported that females lay their eggs 'in volutabris' [places where pigs roll]. The habits and life-cycle have been succinctly described by Minder (1963: 387-388).

Adults are most usually found around slurry or any liquid manure or garbage, steaming manure heaps, or occasionally cattle dung. They have also been reported on human excrement (Pont, 1979: 15), cow dung (Duda, 1926a: 20; Coffey, 1966: 214), cattle manure heaps (Andersson, 1975: 60; Iwasa, 1981: 49), liquid manure (Duda, 1926a: 94; Hennig, 1949: 15, 19, 42; Minder, 1963: 387; Coffey, 1966: 214; numerous pers. obs.), sewage (Melander & Spuler, 1917: 44; Duda, 1926a: 19; Cole, 1969: 366; Draeger & Scheurer, 1999: 67) pig dung (Papp, 1974b: 140), cattle feeding pens, rotting vegetation (Andersson, 1975: 60; Mangan, 1977: 339), and green algae on the Baltic Sea coast (Remmert *in* Moeller, 1965: 574). The species is notorious for large outbreaks in sewage works (Pont, 1979: 15; Draeger & Scheurer, 1999: 67), and can be caught on fish bait but also visits fruits and faeces (Gregor, 1966: 82). *T. putris* is obligatorily dependent on a source of carbohydrates (Draeger & Scheurer, 1999: 68). It is thus commonly found on flowers (e.g. Alfken, 1924: passim; *Cornus mas*; Van der Goot, 1986a: 3; Bährmann, 1993b: 105; Draeger & Scheurer, 1999: table 1, with a list of plants), and also utilizes honeydew excreted by aphids or the exudates of ergot-infested grass (Draeger & Scheurer, 1999: 69, 71). Adults are very tolerant, being reported from sand dune systems (Ardö, 1957: 121), in mountains and the far north (Ringdahl, 1951: 143). They have a moderately high synanthropic index, being recorded from meat and dairy processing plants (Zuska & Laštovka, 1969: 209), and have been classified as communicative hemisynanthropes (Gregor, 1966: 83). The courtship display has been briefly described by Mangan (1976: 1026): males do not display their mid tarsomeres, but the tufts of setae on segment 4 are extended laterally and vibrated; receptive females respond by remaining still and vibrating their wings; the male circles to the side of the female and attempts to mount, making genital contact directly after mounting.

Larvae are found in various kinds of dung or manure, the older and more liquid the better. They have been found at the edges of patches of liquid manure (Minder, 1963: 387), cow dung and cow manure (Hennig, 1949: 17; Papp, 1975: table 2; Iwasa, 1981: 48), liquid pig manure (Coffey, 1966: 214; Papp, 1974b: 140, 144), settlement tanks containing either regular household sewage (Draeger & Scheurer, 1999: 67) or soil and fine root waste produced by washing sugar beet (Dinter & Paarmann, 1989: 770), high-water flood debris (ZMUC; Boness, 1975: 390), decaying algal masses on beaches (Remmert, 1955: 8, 46), and rotting carrots (Werner, 1997: 97). The species has also been reared from such unusual substrates as a bird's nest (Pont, 1979: 15) and sea wrack that had been exposed to fresh water (Moeller, 1965: 564). The cycle from egg to adult takes 20-30 days. At least some puparia hibernate in the Moscow and Ivanovo areas of Russia (Minder, 1963: 388; Lobanov, 1962a: 37), while Lavčiev & Tsankova (1982: 61) report overwintering as larvae. The species produces four generations per year in the Berlin area (Draeger & Scheurer, 1999: 67). Under laboratory conditions it breeds readily in cow dung.

The following puparial parasitoids have been recorded (Ursu & Tudor, 1975; Fabritius, 1981): *Spalangia endius* Walker, *Spalangia subpunctata* Forster, *Muscidifurax raptor* Girault & Sanders, *Urolepis maritima* Walker (Pteromalidae), and *Kleidotoma truncata* Cameron (Eucoilidae).

Sukhacheva (1963: 600) has discussed the beneficial activity of the larvae in destroying ascarid eggs, or at least ruining their viability.

Type-material. – *Musca putris* and *fimeti*: Linnaeus (1758) described his *Musca* no. 68 *putris* with two unnamed forms, 'α in Fimetis, β in Caseo'. Subsequently (Linnaeus, 1761: 456), he again listed *putris* and divided it into two named species, α *fimeti* and β *casei*. The form α *putris*, has always been interpreted as the common large *Themira*, whilst form β, *casei*, is now *Piophila casei* (Piophilidae), the well-known cheese skipper. Study of Linnaeus' collection (LSL) by ACP revealed, as is so often the case, an unpalatable situation. There are 2 specimens under *putris*, which is *Musca* no. 89 in Diptera box 13. One is labelled by Linnaeus 'putris 68.' and is *Lonchaea vaginalis* Meigen, possibly ♂ (Lonchaeidae) (det. B. H. Cogan), as already pointed out by Haliday (1851: 144) and Zetterstedt (1852: 4332).

The second is unlabelled, and is *Themira nigricornis* Mg. ♀ (erroneously listed by Haliday (l.c.) as a ♂ of *Nemopoda cylindrica* F., i.e. *nitidula* Fall.). Neither of these can be a syntype: they do not agree with Linnaeus' description, particularly 'alarum costa nigra', expanded in 1761 to 'alae albide margine crassiore nigra linea'; *T. nigricornis* does not agree with the statement 'pedes toti atri', and *L. vaginalis* does not live as a larva 'in fimetis'. The conclusion must be that the syntypes of *putris* are lost, and so we continue to use the name in the sense in which it has been consistently used since 1758.

No material exists of *Musca fimeti*, which was in any case an unnecessary name and should be treated as an objective junior synonym of *putris*.

A little further on in the Linnaean collection, under *Musca* no. 95 in Diptera box 14, there are 4 specimens of *Musca pubera* Linnaeus (1758: 598, no. 74), which ACP has also studied: 1♂, labelled by Linnaeus 'pubera 74.'; 1♂ 1♀ on one pin, unlabelled; 1♂, unlabelled. All four specimens are in fact the species *Themira putris* as currently understood. These specimens fit Linnaeus' sparse description extraordinarily well, especially 'abdominis ultimo segmento incurvato: pube utrinque reflexa': the pregenital tuft of setae is clearly visible in the ♂ bearing Linnaeus' label. Later Linnaeus (1761: 457) added that the species was smaller than the preceding one in his list, which was the small green stratiomyid *Microchrysa polita* Linnaeus. Unfortunately, the name *pubera* has always been used for a species of *Cordilura* Fallén (Scathophagidae), but in fact this large dung-fly does not agree with Linnaeus' descriptions: it has no tuft of setae near the tip of the ♂ abdomen, the fore and mid tibiae are yellow, the body is not shining black, and in body-length it is almost twice as large as *Themira putris* and *Microchrysa polita*. The future usage of the name *pubera* needs to be discussed with workers on the Scathophagidae, and a ruling by the *International Commission for Zoological Nomenclature* is required so that it can be retained in the Scathophagidae.

M. conssensis: Harris (1780b: 123) wrote quaintly of his species: 'The author found a great number of these in a hole of a wall, where they seemed very busy, as they appeared to be all in motion; but what business they were met upon, he could not discover. Found in June.' Note that *Musca putris* of Harris (1780b: 73, plate 21 figs 1-2) is a *Scathophaga* and probably *stercoraria* Linnaeus. No material from Harris' collection is known to exist (Pont & Michelsen, 1982: 25-27).

17. ***Themira ringdahli*** Pont, **sp. n.**
(Figs 115, 119, 263-271)

Themira arctica Becker, sensu Ringdahl, 1934: 3, 5, fig. 4 [misidentification].

Note. – Ringdahl (1934) distinguished two very similar species from North Sweden to which he attached the names *dampfi* Becker and *arctica* Becker. However, the types of these two names have proved to be conspecific (see discussion under *arctica*, p. 64), and *arctica* sensu Ringdahl (1934) is an undescribed species.

Diagnosis. – Adult *ringdahli* is most similar to *arctica* Becker, but distinguished in the ♂ sex by the very short anterior notopleural seta, longer fore tarsomere 1, long fine anteroventral and ventral apical setae on mid tibia (Fig. 115), and details of the fore leg ornamentation and genitalia. ♀ not known.

Adult. – Colour: Head mostly black, gena, facial carina and lunule brown. Antenna black. Thorax, scutellum and abdomen entirely black. Legs black, trochanters and knees brownish. Wing clear; calypter creamy, margin and fringe-hairs yellow; haltere yellow.

Head: Frons mostly shining, the rest brownish-grey pruinose. Postocellar subequal to inner vertical; orbital only slightly shorter than this. Gena equal to length of antennal flagellomere; subvibrissal setae numerous and short.

Thorax: Scutum, postpronotal lobes and scutellum thinly brownish-grey pruinose. Setae: postpronotal present; anterior notopleural fine but distinct, almost half as long as posterior one; acrostichal setulae minute. Pleural pruinosity: anepisternum glossy, except for a tiny pruinose proepimeral area; katepisternum as in *arctica* (Fig. 105) but with rather more pruinosity on posterior margin; anepimeron and meron pruinose, each with a central undusted oval patch. Setae reduced; no katepisternal setula. Scutellum short, twice as wide as long, without basal setulae.

Legs: Fore coxa shining except above, mid and hind coxae pruinose; trochanters and femora shining. Fore femur with some short posterodorsal setae before tip; ♂ fore legs in Figs 263-265. ♂ mid legs straight; tibia without an anteroventral seta, apical setae fine and elongate (Fig. 115). ♂ hind

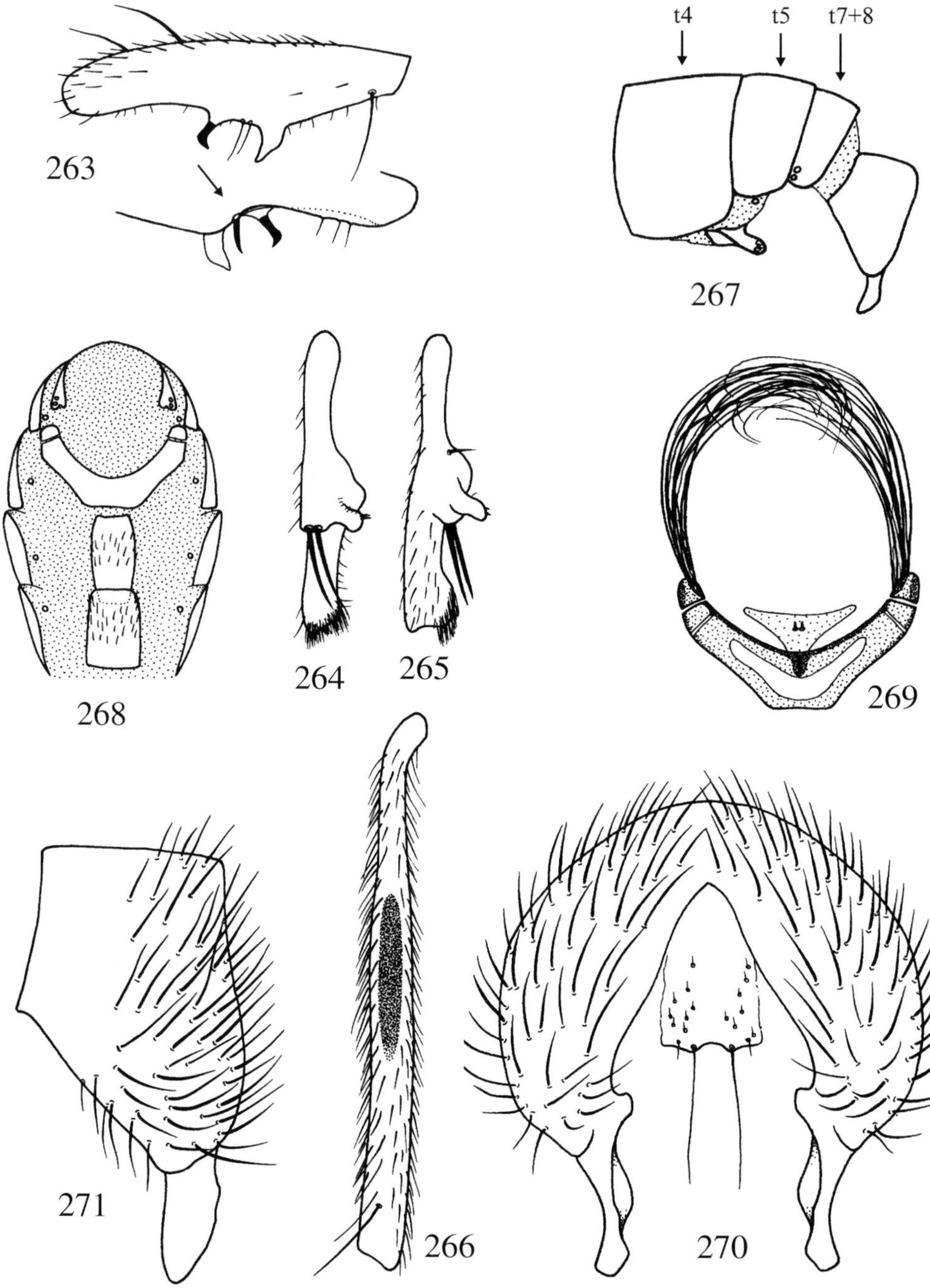

Figs 263-271. *Themira ringdahli* Pont, sp. n., ♂. 263: fore femur, anterior view (detail: posterior view of protuberance); 264: fore tibia, anterior view; 265, fore tibia, dorsal view; 266: hind tibia, dorsal view; 267: apical part of abdomen, lateral view; 268: abdomen, ventral view, hypopygium removed, position and shape of sternite 4 indicated; 269: sternites 4 and 5; 270: hypopygium, dorsal view; 271: hypopygium, lateral view. (t – tergite.)

coxa and hind trochanter without modifications. Hind femur with the anterodorsal setae very fine. Hind tibia without setae; ♂ with the osmeterium in Fig. 266, with the dorsal ground-setulae fine and erect just before apex.

Wings: Veins bare except for several setulae in basal part of stem-vein on ventral surface. Entirely covered with microtrichiae.

Abdomen: Glossy black. ♂ tergites 1+2, 3 and 4 moderately expanded behind (Figs 267-268). Without setae except at sides of tergites 3 and 4; syntergite 1+2 with erect setulae at sides. ♂ sternites with weak setulae (Fig. 268); sternite 4 and genitalia in Figs 269-271.

Wing-length: 2.9 mm (♂).

Type-material. – Holotype ♂, NORTH SWEDEN, TO: Abisko, regio subarctica, 29.vii.1918 (O. Ringdahl), in ZML.

Egg and larva. – Not known.

Distribution. – Known only from the type-locality.

Biology. – Nothing is known of the habits or biology. Ringdahl (1934: 5, as *arctica*) recorded it from the regio subarctica, i.e. the birch forest zone, but searches by ACP for further material around the many tarns and pools at Abisko, as well as along the shores of Lake Torneträsk and the banks of the Abiskojokk, have not been successful.

18. *Themira simplicipes* (Duda, 1926)
(Figs 90, 92, 136, 272-278)

Enicita simplicipes Duda, 1926a: 33, 72. Lectotype ♂, ROMANIA, Berzáska, in ZMHU [by designation of Hennig, 1949: 38, figs 11, 25, 89, 146; by inference, see discussion below].

Diagnosis. – Adult recognised by the absence of the apical scutellar seta, and also by the absence of a postpronotal seta and the wholly pruinose katepisternum and anepimeron (Fig. 90).

The larva is indistinguishable from that of *annulipes* except for a somewhat smaller number of lobes on the anterior spiracles (10 instead of 12-14 in *annulipes*). However, only 5 larvae of *simplicipes* have been studied.

Adult. – Colour: Most of frons, face, parafacial and gena yellow; upper third of fronto-orbital plate, vertex and occiput dark. Antenna brown, tip of pedicel and base of flagellomere yellow. Thorax, scutellum and abdomen dark, only postpronotal lobe and proepisternal area yellowish. Fore legs yellow, femur rather darkened outside, and tarsi dark except at base of tarsomere 1; mid and hind legs mostly dark, apical part of coxae, trochanters, basal quarter of femora, knees and tips of tibiae yellow, mid tarsus with tarsomere 1 mainly white and tarsomere 2 basally white (Fig. 274). Wing clear; calypter dirty white, margin and fringe-hairs brown; haltere yellow. Abdomen black; hypopygium pale brown, paler than preceding abdominal segments.

Head: Frons subshining, the dark areas and the occiput brownish-grey pruinose. Postocellar short, half as long as inner vertical; orbital even shorter, hair-like and inconspicuous (Fig. 92). Gena less than half length of flagellomere; subvibrissal setae short and numerous.

Thorax: Scutum, postpronotal lobes and scutellum brown dusted and nowhere shining. Setae: postpronotal absent; anterior notopleural very short, at most 1/3 length of posterior one; acrostichal setulae minute. Pleural pruinosity: anepisternum glossy, with weak and scarcely perceptible dusting in upper posterior corner; katepisternum entirely white pruinose; anepimeron grey pruinose; meron pruinose, with a central oval undusted patch (Fig. 90). Setae reduced; no katepisternal setula. Scutellum short, twice as wide as long, without basal setula or apical seta.

Legs: Fore coxa apparently shining except on anterior edge, mid and hind coxae pruinose; trochanters and femora matt, dusted. Fore femur with a short posterodorsal seta at apical third (only in lectotype); ♂ fore legs in Figs 272-273. ♀ fore femur without setae. ♂ fore tarsal ratio 3-1-1-3/4-4/5 (= 10-3.3-3.3-2.5-2.5), i.e. tarsomeres 2 and 3 subequal. ♂ mid legs straight; tibia usually with an anteroventral seta, the apical setae normal; tarsi not modified (Fig 274). Hind coxa and hind trochanter without modifications. Hind femur with a few anterodorsal setae. Hind tibia without setae; ♂ with the osmeterium in Fig. 275, the tibia without any erect dorsal setulae.

Wings: Veins bare except for several setulae in basal part of stem-vein on ventral surface. Entirely covered with microtrichiae.

Abdomen: Shining black, but not glossy, the posterior segments in particular appearing rather shagreened. ♂ tergites slightly expanded behind (Fig. 276). Without setae, except in posterior lateral corners of tergites 4 and 5; syntergite 1+2

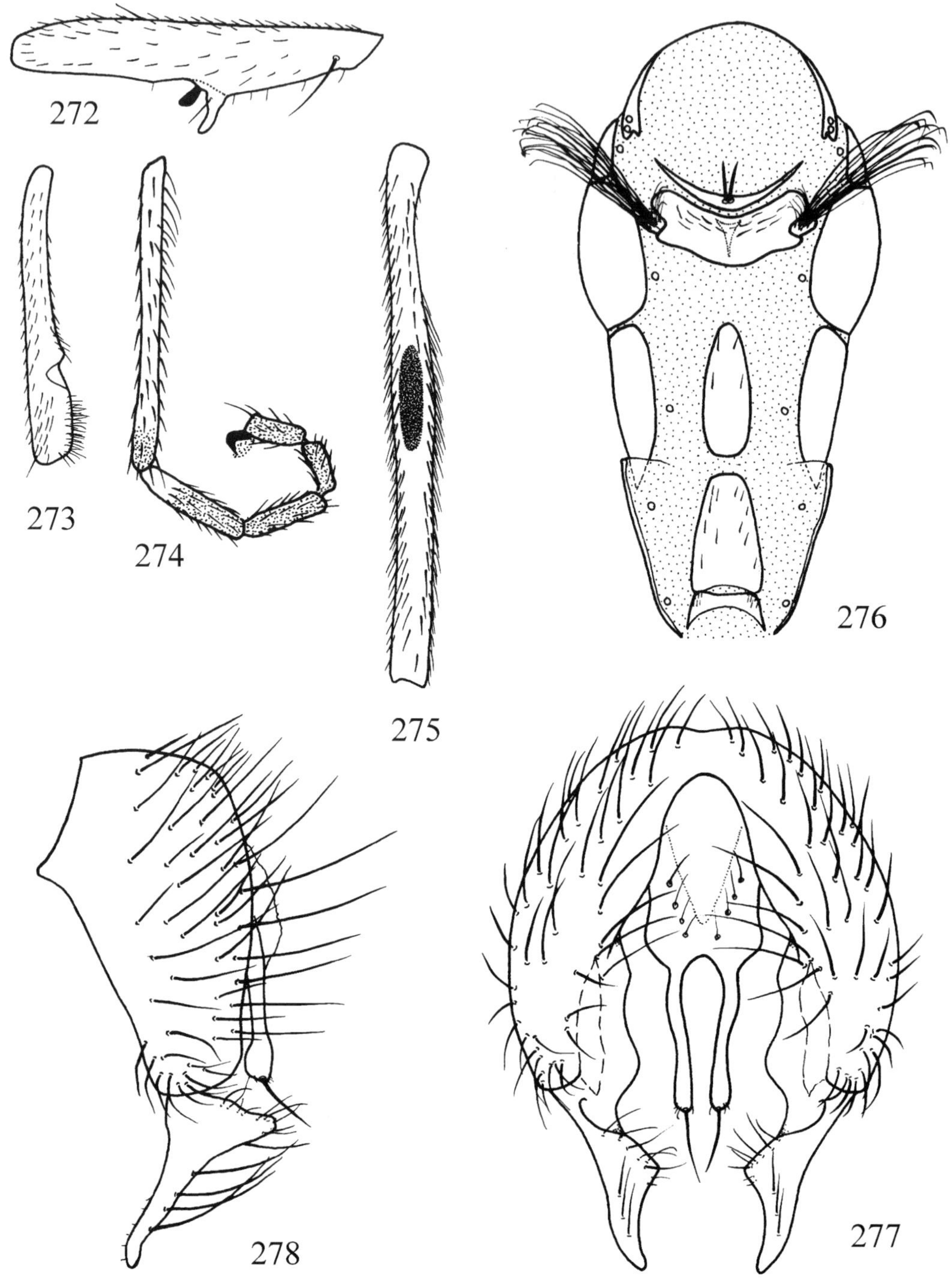

Figs 272-278. *Themira simplicipes* (Duda), ♂. 272: fore femur, anterior view; 273: fore tibia, anterior view; 274, mid tarsus, anterior view; 275: hind tibia, dorsal view; 276: abdomen, ventral view, hypopygium removed; 277: hypopygium, dorsal view; 278: hypopygium, lateral view.

without any erect setulae at sides. ♂ sternites 1-3 with weak setulae (Fig. 276); sternite 4 and genitalia in Figs 276-278.

Wing-length: 2.76-3.2 mm (Ozerov, 1999d: 191).

Remarks. – *T. simplicipes* was fully described and illustrated by Ozerov (1999d: 190-191, fig. 20).

Larva. – A description of the 3rd instar larva has been given by Meier (1996: 121, figs 436-444).

Distribution. – Known only from the type-series, from Hungary and Romania (Zuska & Pont, 1984: 166), and from Russia, North Ossetia (Ozerov, 1999d: 191).

Biology. – Ozerov (1999d: 191) collected adults among grasses, and found larvae in the decaying stem of a chervil, *Chaerophyllum rubellum*.

Type-material. – Duda described this species from 2♂ from Hungary (Berzaska and Ugod), both in ZMHU. ACP has seen both of these. The ♂ labelled 'Berzáska / Pável' and, on the reverse, 'vii.98', which is actually from Romania, was discussed and illustrated by Hennig (1949: 38, figs 11, 25, 89, 146) as the 'Typus', which is a lectotype designation by inference. It is very greasy; the abdomen after segment 1+2, wings, right mid leg and left fore leg are missing; unfortunately, no slide with these parts has been found, either in DEI or ZMHU, and Hennig's dissections of the lectotype must now be presumed lost (H. Wendt and J. Ziegler, pers. comm.). The second ♂, labelled as paralectotype, is in excellent condition, from Hungary, Ugod, 4.vi.1906 (Kertész).

19. *Themira superba* (Haliday, 1833)

(Figs 103, 106, 108-110, 116, 279-286)

Sepsis superba Haliday, 1833: 170. Lectotype ♂, IRELAND: Holywood, in NMI [see designation below].

Sepsis ciliata Stæger *in* Schiødte, 1844: 30. Lectotype ♂, DENMARK: near Leersøen, in ZMUC [see designation below].

Enicita nigra Duda, 1926a: 72, 88. Unavailable name; citation of a Meigen MS name, in synonymy with *Enicita annulipes* and *Themira superba*.

Themira fasciculata Duda, 1926a: 88. Unavailable name; citation of a Meigen MS name, in synonymy with *Themira superba*.

Diagnosis. – The adult is recognised by the glossy katepisternum and presence of a postpronotal seta, together with the highly ornamented ♂ fore and hind legs (Figs 108-109, 279-281).

Larval morphology very similar to that of *lucida*, except for regularly distributed spines on the dorsal side of the anal division; anterior spiracles with 15-17 lobes.

Adult. – Colour: Head and antenna black. Thorax, scutellum and abdomen entirely black. Legs wholly black. Wing clear; calypter creamy, margin and fringe-hairs yellow; haltere yellow.

Head: Frons mostly shining, the rest dark brown pruinose. Postocellar subequal to inner vertical; orbital half as long as this in ♂, even shorter in ♀. Gena just over half length of antennal flagellomere; subvibrissal setae numerous and very short.

Thorax: Scutum glossy, narrowly brown pruinose along notopleural, prealar, postalar and prescutellar areas. Postpronotal lobes glossy, pruinose below. Scutellum pruinose. Setae: postpronotal present; anterior notopleural minute, hair-like; acrostichal setulae minute. Pleural pruinosity in Fig. 103: anepisternum and katepisternum with tiny pruinose areas, anepimeron narrowly pruinose behind, meron pruinose except for an oval central area. Setae reduced; no katepisternal setula. Scutellum short, twice as wide as long, without basal setulae.

Legs: Fore coxa wholly shining, mid and hind coxae shining outside; trochanters, femora and tibiae mostly shining, but hind tibia pigmented all along anterodorsal surface beyond osmeterium. Fore femur with some short posterodorsal setae before tip; ♂ fore legs in Figs 279-280; ♀ fore legs simple, femur with 4-5 strong anteroventral spinules around middle. ♂ mid femur curved and slightly swollen apically, with a series of fine curled posteroventral setae in apical half (Fig. 110); normal and bare in ♀. Mid tibia with a short anteroventral seta; in ♂ slightly curved and with the ground-setulae rather fine and erect, apical setae normal. ♂ hind coxa without modifications (Fig. 106). ♂ hind trochanter with a downwardly-directed thorn-like process on inner surface (Figs 108-109). Hind femur with some anterodorsal setae around middle; in ♂ with long fine posterior setae on all but apical third, and in ♀ with short fine posterior setae in basal third. Hind tibia without setae; ♂ with the osmeterium in Fig. 281, the

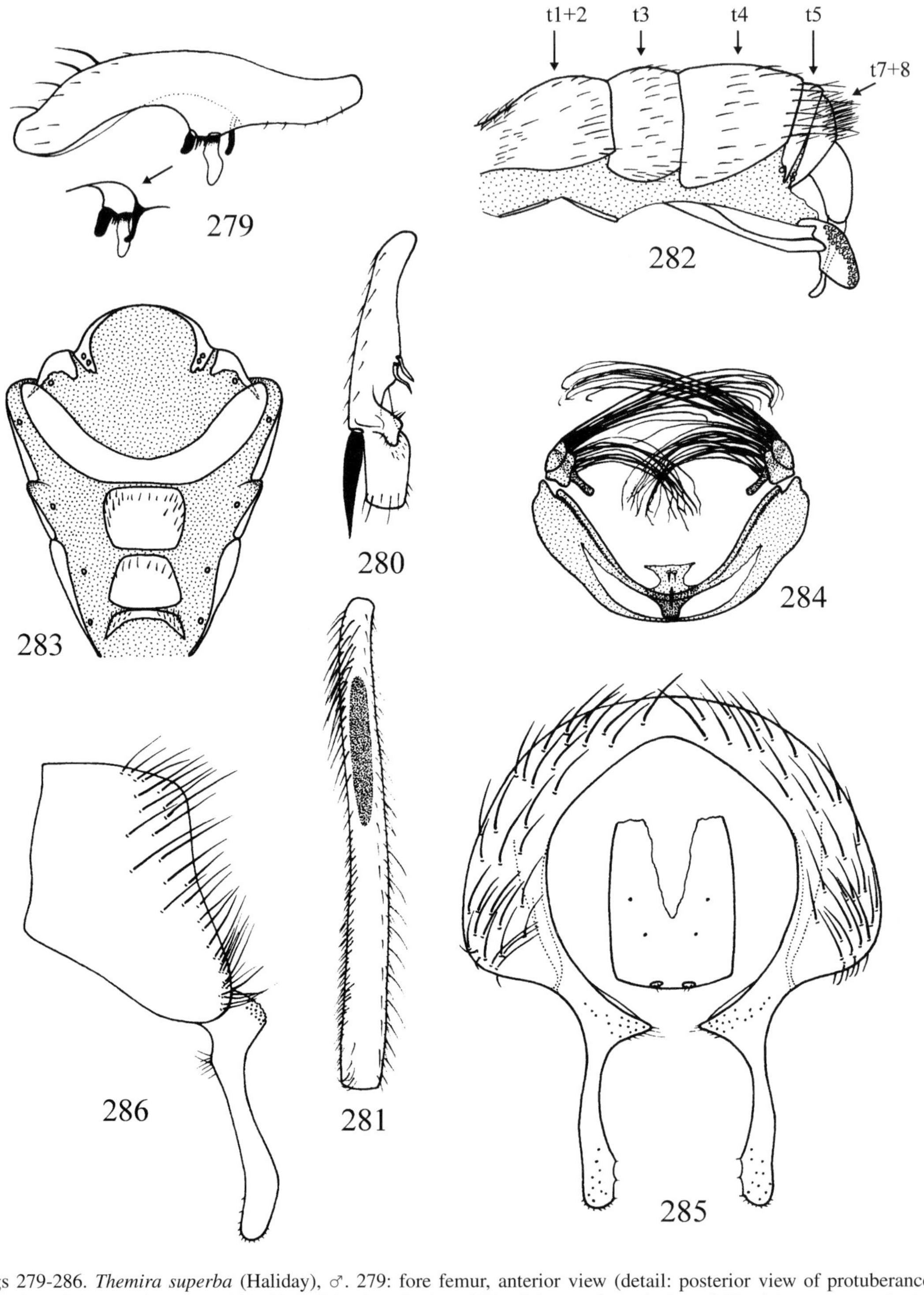

Figs 279-286. *Themira superba* (Haliday), ♂. 279: fore femur, anterior view (detail: posterior view of protuberance); 280: fore tibia, anterior view; 281: hind tibia, dorsal view; 282: abdomen, lateral view; 283: abdomen, ventral view, hypopygium removed, position and shape of sternite 4 indicated; 284: sternites 4 and 5; 285: hypopygium, dorsal view; 286: hypopygium, lateral view. (t – tergite.)

tibia with some fine erect dorsal setulae on apical 2/5.

Wings: Veins bare except for several setulae in basal part of stem-vein on ventral surface. Covered with microtrichiae except for a narrow band in posterior cubital cell and along subcosta just beyond humeral cross-vein.

Abdomen: Glossy black. ♂ tergites 1+2, 3 and 4 greatly expanded behind, tergite 4 very broad and concealing the following tergites (Figs 282-283). Without setae except at sides of tergite 4 in ♂; syntergite 1+2 with erect setulae at sides; ♂ tergite 5 reduced, tergite 7+8 setulose (Fig. 282); ♂ sternites with weak setulae (Fig. 283); sternite 4 and genitalia in Figs 284-286.

Wing-length: 2.9-3.4 mm (♂), 3.0-3.8 mm (♀).

Remarks. – *T. superba* was fully described and illustrated by Ozerov (1999d: 203-204, fig. 33); the ♂ genitalia were also illustrated by Ozerov (1999e: fig. 350.6).

Larva. – A description of the 3rd instar larva has been given by Meier (1996: 121, figs 436-444).

Distribution. – Widespread but seldom common. Denmark: most provinces (e.g. Lyneborg, 1965: 257). Norway: O and R (e.g. Duda, 1926a: 88). Sweden: SK to ÖG, and LY (e.g. Ringdahl, 1934: 5; 1951: 143); 1♀ from Sarek National Park, LU (Poppius, Lundström & Frey, 1916: 695) has proved to be *arctica* Becker. Finland: northwards to LkW (first correctly recognised, as *ciliata*, by Frey, 1906: 67).

Through the central and northern zones of Europe, from Ireland eastwards to Ukraine (Zuska & Pont, 1984: 166), and in Russia eastwards to Krasnoyarsk province and Yakutiya (Ozerov, 1999d: 204).

Records are from April to early September.

Biology. – Adults are usually found near creeks, pools and tarns (e.g. Duda, 1926a: 88), especially on waterfowl droppings where they can be common. A mass occurrence was reported by Naswaty (1960: 91-92). The species has also been found in cattle sheds (Minder, 1963: 388) and sand dune systems (Krogerus, 1932: 116; Ardö, 1957: 121), sometimes near accumulations of manure. It has been reared from high-water flood debris (ZMUC). Ozerov (1999d: 204) found larvae in a dirty pool. In the laboratory, *superba* breeds readily on goose and duck droppings.

Type-material. – *Sepsis superba*: Haliday described this species from the ♂ only, 'near Holywood, and again in Wicklow'. His collection in NMI contains 4♂ 2♀ under this name. and these have been studied by ACP. 1♂ is labelled by Haliday 'H' [= Holywood], and ACP has labelled this and herewith designates it as lectotype; it is pinned, and in good condition. A second ♂ labelled 'Wic' [= Wicklow] is gummed to card; it has been labelled as paralectotype. Both specimens belong to *superba* as currently understood. The other 2♂, labelled 'British' and 'Ireland', and the 2♀ are not syntypes; 2♂ 1♀ are *superba*, 1♀ is *lucida*. Curtis' collection (NMV) contains 3♂ named as *superba*, without data, and all are *superba* (seen by ACP).

Sepsis ciliata: Stæger described his *ciliata* from 'Leersøen'. In ZMUC ACP saw 2♀ of *ciliata* in the Type collection (which are *superba*) and 1♀ in the General collection (which is actually *lucida*, and evidently the ♀ seen by Duda (1926a: 89)), but these were all collected by Stæger at Vedbæk and cannot be syntypes. 1♂ from the General collection is labelled by Stæger '♂ / Lersø' / St.', and ACP has labelled it and designates it herewith as lectotype; left fore leg missing, but condition otherwise good. It is conspecific with *superba* as currently understood. ACP has seen 1♂ from Stæger (Leersøen, 11.v.1842) in ZML in Zetterstedt's Göteborg collection. This is a syntype, and has been labelled as paralectotype. There are also 2♀ in ZML in Zetterstedt's Diptera Scandinaviae collection (drawer 38), mentioned by Zetterstedt (1847: 2292) as given him by Stæger: neither has any locality data; 1 is *lucida*, 1 is *superba*.

Genus *Susanomira* Pont, 1987

Susanomira Pont, 1987c: 265. Type-species: *Susanomira caucasica* Pont, 1987, by original designation. Gender: feminine.

Adult – *Susanomira* can be distinguished from the other genera with a full complement of head setae (inner vertical, outer vertical, postocellar, orbital) by the bare prosternum, absence of basal scutellar seta, and presence of well-developed dorsocentral setae (1+3). It has an unusual and striking appearance which is unique among European Sepsidae.

Adult. – Head rather oval (Fig. 19), facial carina well-developed (Fig. 20), gena broad; face sloping forwards and lower margin well in front of profrons. A full complement of head setae: orbital, inner vertical, outer vertical, postocellar. Arista bare. Scutum and scutellum matt. Postpronotal seta present. 1+3 dorsocentrals; anterior two pairs short but strong, 3-4 times as long as a ground-setula, posterior two pairs long and strong. 0 intraalar, 1 supraalar, 1 strong postalar. 1 short proepisternal present, proepimeral area bare. Anepisternal present, strong. Katepisternal absent. Anepisternum and anepimeron subshining, katepisternum and meron mainly thinly brownish-grey pruinose. Meral bridge absent. Prosternum bare, but with some obvious pile. 1 pair of scutellars (apical). Fore femur without posterodorsal setae. Tibiae without anterodorsal or posterodorsal setae. ♂ fore femur and tibia modified. ♂ hind tibia without osmeterium; the tibia rather flattened along whole length in both sexes. Wing without a dark spot, but clouded along costa and over apical part of veins R_{2+3} and R_{4+5} (Fig. 294). Abdomen without dorsal setae except for 2-3 pairs at apical lateral corners of tergites 4 and 5 in ♂.

♂ abdomen. No striking constriction between syntergite 1+2 and tergite 3. Tergite 6 absent; tergite 7+8 present (Fig. 296). Spiracles 1-5 in the intersegmental membrane, well-separated from the tergal margin; spiracles 7 and 8 at the ventral edge of tergite 7+8, a clear suture visible at the fusion of these two tergites and the spiracles in the 'tergite 7' component (Fig. 295). Sternite 1 band-shaped, with a few setulae at sides; sternite 2 subquadrate, sparsely setulose; sternite 3 more elongate, with the medio-posterior area weakly sclerotised, basally with tufts of long dense setae, apically with several sparse strong setae; sternite 4 mostly covered with fine setulae, and with 3 distinct groups of tufted setae, lateral, latero-apical and medio-apical (Fig. 295); sternite 5 with 3 pairs of short stout setae; sternite 7+8 band-shaped, connected to the 'tergite 7' component of tergite 7+8 (Fig. 295). Epandrium without an outstanding spinose seta on upper part. Cercal plate (Fig. 297) bilobed, bearing some very long setae. Surstylus (Figs 297-298) weakly attached to epandrium. Sperm pump large, 'dumb-bell' in shape.

♀ abdomen. Sternites 2-5 normal in shape, plate-like. Spiracles 1-5 in the intersegmental membrane, spiracles 6 and 7 in tergites 6 and 7. Postabdominal segments 6 and 7 with the tergites and sternites separated on sides of segments, the sternites (like the tergites) rather broad and short, sternite 7 (and tergite 7) longer than sternite 6, tergite 6 with a long anterior extension, tergite 7 and sternites 6 and 7 without extensions; segment 8, when not invaginated, long, extended posteriorly and ventrally, with a ventral element (sternite 8) on each side that remains separated at tip and a dorsal element (tergite 8) that forms the usual pair of ring-like bars that do not quite touch apically. Cerci small, round. Epiproct present. Hypoproct present, weakly sclerotised. 2 spermathecae, one much larger than the other.

Egg. – Egg of *S. caucasica* with a single respiratory filament. In all the specimens studied, filament twisted around its axis (Fig. 30). In cross section, filament with a central canal with irregularly arranged struts, as in *Themira* (as in Fig. 21). Surface of filament not communicating with the surrounding air, but instead with a regular pattern of elevations and depressions, the elevations apparently indicating underlying struts (see Meier, 1995b: fig. 12).

Larva. – Pseudocephalon bilobed, cephalic lobes with few combs; facial mask consisting of a small number of combs.

Anal division (Fig. 300) short; lateral tubercles present; stigmatophores with a dorsal tubercle; ventral tubercles present, simple; subanal tubercle small; anal pad large, broad, with a deep diagonal furrow, rounded laterally; anal tuft and preanal protuberance present, but very small and hardly visible; anal division between anterior margin of anal pad and origin of stigmatophores about as wide as long; with few spines on lateral aspect of anal division but three double rows of spines present on the dorsum; integument between rows bare (as in Fig. 44).

Cephaloskeleton: labial and dental sclerites absent; bridge of intermediate sclerite approximately at half length of sclerite (as in Fig. 138), anteriorly with a flat plate; epistomal sclerite longer than wide, without posterior bars; parastomal bars long, only slightly shorter than intermediate sclerite (as in Fig. 48); basal sclerite 2.5 times as long as high, dorsal cornua shorter and three times narrower than ventral cornua.

Anterior spiracles bush-like, with more than 30 long, thin lobes; lobes 5-8 times as long as wide (Fig. 40).

Posterior spiracles with almost equal, narrow, weakly arched spiracular openings on oval plate; two spiracular openings parallel or almost parallel, third one perpendicular to the other two (Fig. 301).

Discussion. – Plesiomorphies (i.e. ground-plan characters of the Sepsinae) include the following: outer vertical setae present; postocellar setae present; orbital setae present; supraalar seta present; postalar seta present; anepisternal seta present; scutellum short; wing with the basal radial and median cells distinct; fore femur without posterodorsal setae; hind tibia without anterodorsal or posterodorsal setae; ♂ hind tibia without an osmeterium; ♂ sternite 7+8 present as a band. *Susanomira* is apomorphous relative to the groundplan of the Sepsinae with regard to the following charaacters: strong facial carina (shared with *Saltella*), several pairs (1+3) of dorsocentral setae present, ♂ surstylus not fused with epandrium.

Autapomorphies of the genus are:

1. ♂ tergite 3 densely setose
2. ♂ tergite 4 with multiple tufts of setae
3. ♂ tergite 7+8 with a distinct suture, indicating the original two elements

Susanomira is a very striking sepsid, quite unlike any other Palaearctic species in size and general appearance. Characters of the immature stages suggest that *Susanomira* and *Saltella* are sister-groups, but this is difficult to justify on adult characters (see discussion under *Saltella*). *Susanomira* also differs from *Saltella* in the elaborate development of the ♂ secondary sexual characters on the fore leg and on sternites 3-5. In these respects *Susanomira* resembles *Themira* and *Zuskamira*, and *Saltella* resembles *Nemopoda*.

One species, from the Caucasus Mts. Larva in cow dung.

20. *Susanomira caucasica* Pont, 1987

(Figs 19-20, 27, 30, 40, 46, 287-301)

Susanomira caucasica Pont, 1987c: 266, figs 1-11. Holotype ♂, GEORGIA: Caucasus Mts, Kazbegi, Mt Sameba, in ASSP.

Diagnosis. – As for the genus.

Adult. – Colour: Frons, occiput, gena and lower margin of head black; lunula and facial carina yellow to orange; face, parafacial and vibrissal area brown. Antennal scape and pedicel yellow, darkened on outer surface; flagellomere yellow at base, then orange, mostly dark after aristal insertion. Thorax, scutellum and abdomen black. Legs black. Wing smoky along costal margin, with a conspicuous dark suffusion across apical part of costa, R_{2+3} and R_{4+5}, slightly less extensive in female than in male; basal costal and costal cells dark brown, basal area otherwise yellow. Calypter creamy, with yellow margin and fringe. Haltere yellow.

Head (Figs 19-20): Fronto-orbital plate shining from some angles, frontal vitta matt, ocellar triangle and occiput weakly grey pruinose. Gena broad, subequal to length of antennal flagellomere; with a row of short genal setae.

Thorax: Scutum, postpronotal lobe, notopleuron and scutellum matt, with a slight burnished tinge. Scutum with the setae (except dorsocentrals) rather short; 2 strong subequal notopleurals; acrostichal setulae conspicuous in front, in 1-2 rows. Lower part of postpronotal lobe, proepisternal depression, anepisternum and anepimeron subshining, the anepisternum more matt in female; katatergite and anatergite weakly grey pruinose; katepisternum and meron weakly brownish-grey pruinose, more thinly so and subshining on a diagonal band across katepisternum (from mid upper margin to behind fore coxa) and on anterior margin of meron. Prosternum with short pile, but no setae or setulae. Anterior margin of katepisternum bare. Scutellum short, twice as wide as long; without trace of basal lateral setulae.

Legs: Fore coxa matt, mid and hind coxae brownish-grey dusted; trochanters and basal part of male hind femora shining; rest of femora and tibiae matt. ♂ fore legs modified in Figs 287-289; fore tarsomeres in the ratio 10 : 4.8 : 3.3 : 2.6 : 2.8. ♀ fore femur with short stout spinules on apical two-fifths, 3 posteroventral and 3-4 anteroventral; tibia not modified. Mid coxa with a strong seta on outer surface (Fig. 293). Mid femur in ♂ with 2 strong anteroventral setae at middle, in ♀ without setae. ♂ mid tibia (Figs 290-291) with dense soft hairs in apical half, between ventral and posteroventral surfaces; around middle with 3 erect strong ventral setae and with a comb-like row of short setae on apical third of posterior surface; apical setae: short anterior and dorsal, long anteroventral, posterior and posteroventral. ♀ mid tibia bare except for 2 stout anteroventral setae around middle, each subequal to tibial depth. ♂ mid tarsomere 1 rather attenuated; on basal half with fine, rather tufted, curled hairs on ventral to anterior surfaces; at tip of tarsomere 1, along tarsomere 2 and at base of tarsomere 3 with fine curled anterodorsal to anteroventral hairs

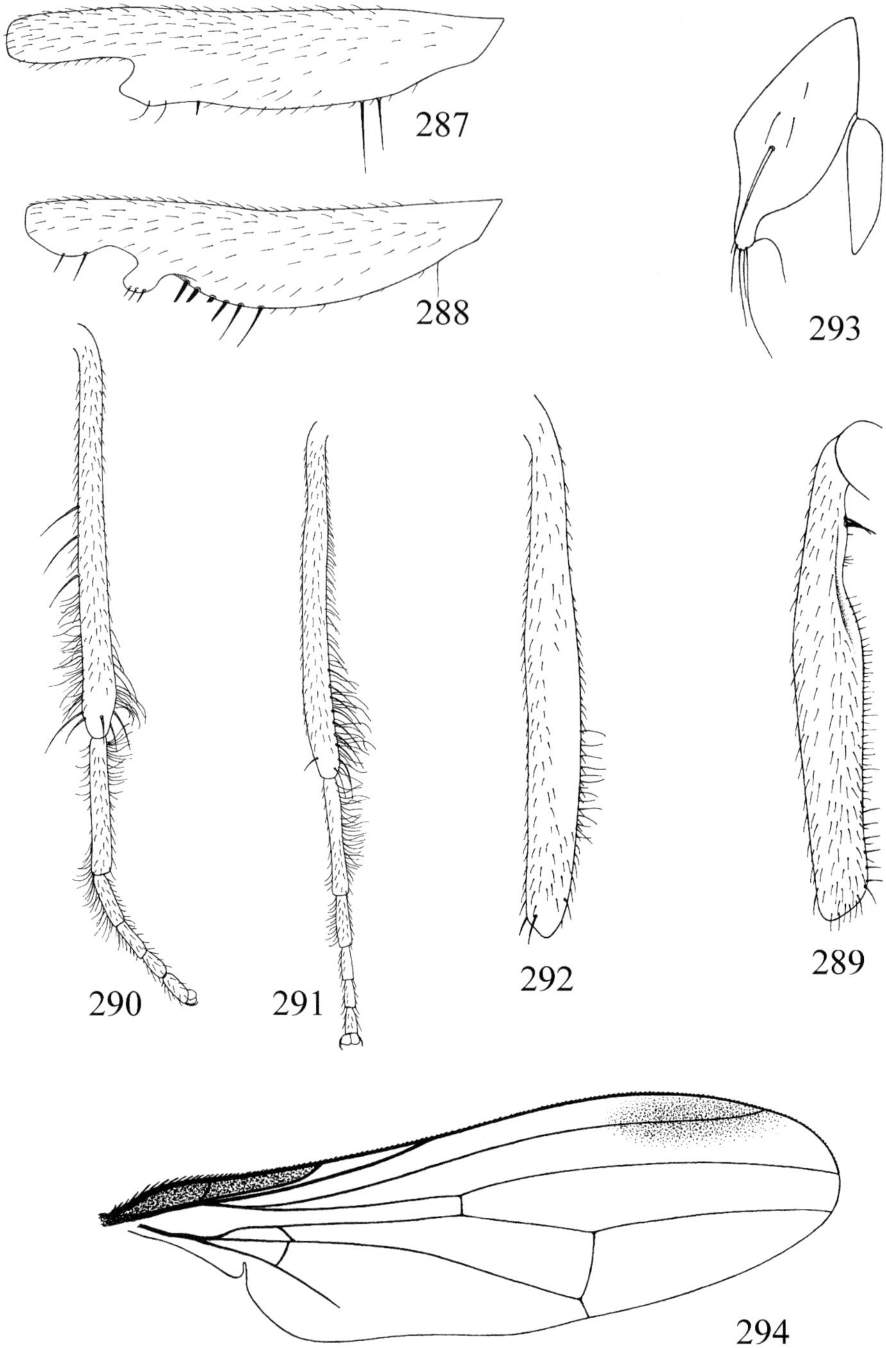

Figs 287-294. *Susanomira caucasica* Pont, ♂. 287: fore femur, anterior view; 288: fore femur, posterior view; 289: fore tibia, anterior view; 290: mid tibia and tarsus, anterodorsal view; 291: mid tibia and tarsus, posterodorsal view; 292: hind tibia, anterodorsal view; 293: mid coxa, lateral view; 294: wing.

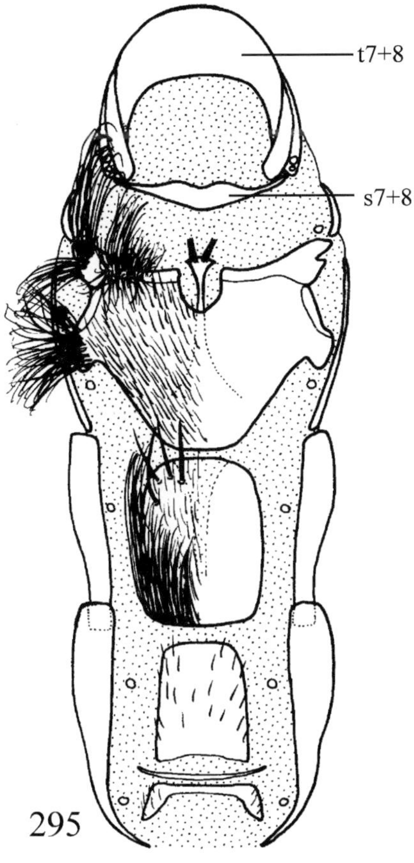

Fig. 295. *Susanomira caucasica* Pont, ♂. Abdomen, ventral view, hypopygium removed. (s – sternite, t – tergite.)

(Figs 290-291). ♂ hind coxa and trochanter without modifications. Hind femur with 2-4 short stout anterodorsal setae around middle. Hind tibia (Fig. 292) rather compressed; without setae except at apex; in ♂ with some sparse fine erect anterodorsal hairs on apical third; ♂ without osmeterium but with a pigmented and sparsely setulose area on 3rd and 4th fifths of anterior to anterodorsal surface.

Wing: (Fig. 294) Veins bare except for a few setulae in basal part of stem-vein on ventral surface. Entirely covered with microtrichiae.

Abdomen: Matt, with a slightly burnished tinge, the genitalia subshining. In ♂ slightly constricted after synsegment 1+2 and hardly expanded apically; synsegment 1+2 slightly wider than long, segments 3 and 4 only slightly longer than wide. Without any setae on dorsal surface except for 2-3 pairs at apical lateral corners of tergites 4 and 5 in ♂. ♂: ventral margins of tergites 4 and 5 with rather tufted groups of pale setae (Fig. 296). ♂ sternite 4 and genitalia in Figs 295-299. ♀ sternites 1-6 matt; sternites 3-5 each with a stout seta at each posterior angle.

Wing-length: 4.1 mm (♂), 5.2 mm (♀).

Body-length: 5.7 mm (♂), 6.8 mm (♀).

Egg. – See under the generic description. A full description of the egg was given by Meier (1995b: 427-431, figs 8, 12, 20).

Larva. – See under the generic description. A full description of the 3rd instar larva was given by Meier (1996: 102-104, figs 355-363).

Distribution. – Known only from the Caucasus Mts (Georgia and North Ossetia).

Adults have been collected from late May to early July.

Biology. – The holotype was collected on dry horse dung in a high pasture, above the tree-line, but since then the species has been found to be more closely associated with cow dung. The adult and larval habitats are discussed by Ozerov (1997d: 90).

Genus ***Saltella*** Robineau-Desvoidy, 1830

Saltella Robineau-Desvoidy, 1830: 746. Type-species: *Saltella nigripes* Robineau-Desvoidy, 1830, by designation of Westwood, 1840: 147. Gender: feminine.

Brachygaster Meigen, 1826: 244, 245. [Preocc. by Leach, 1815.] Type-species: *Brachygaster analis* Meigen, 1826 [= *Trupanea sphondylii* Schrank, 1803], by designation of Zuska & Pont, 1984: 157. Gender: feminine.

Pandora Haliday, 1833: 169. [Preocc. by Bruguière, 1797.] Type-species: *Piophila scutellaris* Fallén, 1820 [= *Trupanea sphondylii* Schrank, 1803], by designation of Hennig, 1949: 27. Gender: feminine.

Anisophysa Macquart, 1835: 543. Type-species: *Piophila scutellaris* Fallén, 1820 [= *Trupanea sphondylii* Schrank, 1803], by designation of Hennig, 1949: 27. Gender: feminine.

Pseudopandora Rapp, 1946: 500. Replacement name for *Pandora* Haliday, 1833. Type-species:

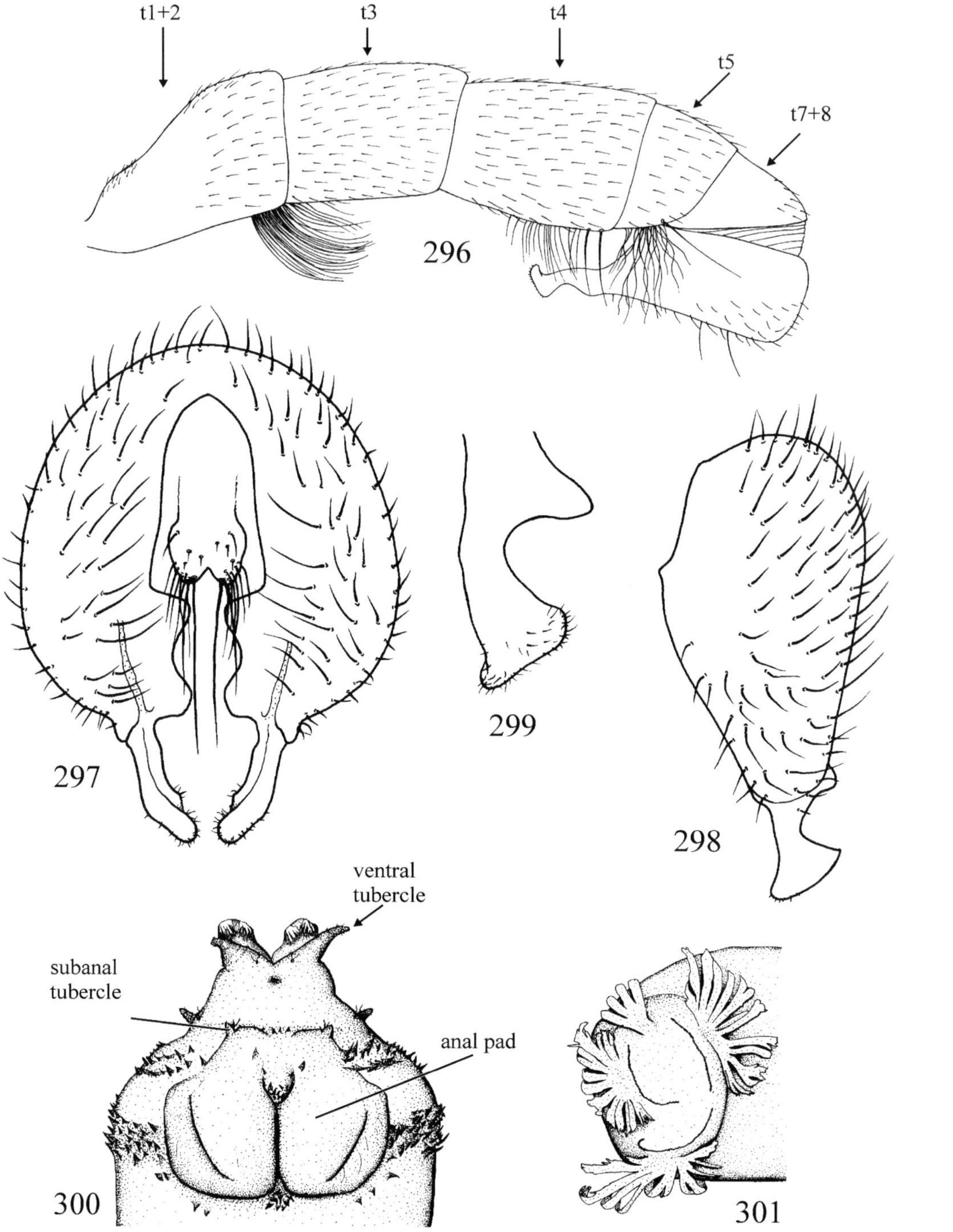

Figs 296-301. *Susanomira caucasica* Pont, ♂. 296: abdomen, lateral view; 297: hypopygium, dorsal view; 298: hypopygium, lateral view; 299: surstylus, enlarged view; 300: larval anal division, ventral; 301: larval posterior spiracles. (t – tergite.)

Trupanea sphondylii Schrank, 1803, automatic. Gender: feminine.

Adult *Saltella* can be distinguished from all other Palaearctic genera by the fused basal radial and medial wing-cells (Fig. 13), together with the enlarged and flattened scutellum (Fig. 2).

Adult. – Head round to oval, facial carina well-developed, gena broad; face receding and lower facial margin behind level of profrons (*nigripes*) or face sloping forwards and lower facial margin well in front of profrons (*sphondylii*) (Figs 302-303). A full complement of head setae (Fig. 1): inner vertical, outer vertical, postocellar and orbital. Scutum matt to subshining, scutellum velvety. Postpronotal present. 1-2 dorsocentrals, the rows of dorsocentral setulae well-developed. 0 intraalar, 1 supraalar, 1 postalar. Proepisternal strong. Proepimeral strong (*sphondylii*) or weak to absent (*nigripes*). Anepisternal present. Katepisternal absent. Anepisternum, anepimeron and meron chagreened, katepisternum partly (*nigripes*) or wholly (*sphondylii*) white pruinose. Prosternum setulose. Scutellum flattened and enlarged, almost as long as broad, with 2 pairs of strong setae. Fore femur with several posterodorsal setae. Tibiae without anterodorsal or posterodorsal setae. ♂ fore femur with several short stout anteroventral and posteroventral spines around middle. ♂ hind tibia with a small or large but inconspicuous osmeterium. Wing without a dark spot; basal radial and medial wing-cells fused. Abdominal setae absent except for 1-2 lateral marginals on tergites 3-5.

♂ abdomen. Tergites 1+2 to 5 normal; tergite 6 absent; tergite 7+8 with 2 spiracles (Fig. 312). Spiracle 1 in syntergite 1+2, spiracles 2-5 in the intersegmental membrane (*sphondylii*, Fig. 319) or at ventral edge of the tergites (*nigripes*, Figs 312-313). Sternite 1 narrow, strap-like, bare; sternites 2-5 simple, plate-like, not divided, sternite 5 deeply incised on posterior margin; sternite 7+8 present as a very narrow band connecting tergite 7+8 across the ventral part of the abdomen. Cercal plate simple, without projections. Surstylus simple, heavily spinose in *nigripes* (Fig. 314); asymmetrical in *sphondylii* and only the left surstylus with spines (Fig. 321). Sperm pump large, 'dumbbell' in shape.

♀ abdomen. Sternites 2-5 large, broad, plate-like, sternites 3-5 each with 3 stout setae along each outer edge. Spiracles 1-7 in the tergites. Postabdominal segments 6 and 7 with the tergites and sternites hardly separated, both well-developed and strongly sclerotised, the tergites each with an anterior extension, that on tergite 7 long and much longer than that on tergite 6; segment 8, when not invaginated, long, extended posteriorly and ventrally, with a ventral element (sternite 8) on each side that remains separated at tip and a dorsal element (tergite 8) that forms the usual pair of ring-like bars that do not quite touch apically. Cerci small, round. Epiproct absent. Hypoproct present. 2 spermathecae, one slightly larger than the other.

Egg. – Description based on *sphondylii*. Eggs with a single respiratory filament, and morphology of the micropyle as in Fig. 29. In cross section, filament round with a ventral central canal surrounded by an air space. Both surrounded by a loose but highly organized and distinctive meshwork of struts (Fig. 27). Usually two struts fused into small, more or less polygonal plates which form the exochorion (Fig. 27).

Larva. – Pseudocephalon bilobed, much longer than wide, cephalic lobes with very few combs; facial mask consisting of a small number of ridges (Figs 304-305).

Anal division short (Fig. 306) or slightly elongated (Fig. 34); lateral tubercles present (Figs 34, 306); stigmatophores with a dorsal tubercle (Fig. 306); ventral tubercles present, clearly bifurcated (Figs 34, 306); subanal tubercles small; anal pad large, very broad and swollen, with a deep diagonal furrow, rounded laterally; anal tuft present and very small (Figs 34, 306); anal division between margin of anal pad and origin of stigmatophores 1-1.5 times as long as wide (Figs 34, 306); with small spines on lateral and dorsal aspect of anal division (Fig. 306).

Cephaloskeleton: labial sclerites present, fused into one V-shaped structure (Fig. 35); dental sclerites absent; bridge of intermediate sclerite approximately at half length of sclerite (as in Fig. 138), wide, anteriorly with a flat plate; epistomal sclerite longer than wide, without posterior bars; parastomal bars long, only slightly shorter than intermediate sclerite (as in Fig. 47); basal sclerite 1.5-3 times as long as high, dorsal cornua shorter and 3-4 times narrower than ventral cornua.

Anterior spiracles tree-like, rather short (Fig. 307) to very long (Fig. 308), basally or in basal third strongly widened, with 14-21 lobes; with basal lobes that are longer (5-7 times as long as wide) than apical ones (2.5-3 times as long as wide) (Figs 307-308).

Posterior spiracles with almost equal, narrow, weakly arched to straight spiracular openings on oval plate; two spiracular openings parallel, third one perpendicular to the other two (as in Fig. 301).

Discussion. – Plesiomorphies (i.e. ground-plan characters of the Sepsinae) include the following: outer vertical setae present; postocellar setae present; orbital setae present; supraalar seta present; postalar seta present; anepisternal seta present; hind tibia without anterodorsal or posterodorsal setae; ♂ sternite 7+8 present as a narrow band; ♀ with postabdominal tergites 6 and 7 each with an anterior prolongation. *Saltella* is apomorphous relative to the ground-plan of the Sepsinae with regard to the following characters: presence of a facial keel (shared with *Susanomira*), wing with the basal medial and radial cells fused (shared with *Sepsis niveipennis*), fore femur with posterodorsal setae (shared with some *Themira*), mid tibia with 1 anterodorsal and 2 or more posterodorsal setae (shared with some *Sepsis*), ♂ hind tibia with an osmeterium, ♂ sperm-pump shaped like a 'dumb-bell.'

Autapomorphies of the genus are:

1. Scutellum elongated, as long as broad, and strikingly flat
2. ♂ spiracle 1 in syntergite 1+2, not in the intersegmental membrane
3. ♀ with abdominal spiracles 1-5 in the tergites

Saltella is another well-founded monophyletic genus, based on a suite of characters from the adult and immature stages. More problematic is the systematic position of the genus and the identity of its sister-group. *Saltella* has always been assigned a rather basal position in the family (e.g. Hennig, 1949), but characters of the immature stages indicate that this is probably not correct and that the genus is considerably more derived than *Ortalischema*, *Zuskamira* and even *Themira*. The same character suite also suggests that the sister-group of *Saltella* is *Susanomira*. This is more difficult to justify on adult characters. The two genera share two possible synapomorphies, neither of which is particularly convincing: the presence of a well-developed facial keel and, perhaps, the presence of a weak proepisternal seta.

Six species: 1 Holarctic, 1 West Palaearctic, 1 East Palaearctic, 2 Oriental, 1 Afrotropical. Larvae in cow dung; Palaearctic larvae described by Ozerov (1986c).

Key to European species of *Saltella* Robineau-Desvoidy

Adults

1 Katepisternum matt, with a patch of white pruinosity confined to upper posterior corner. Wing with very sparse microtrichiae, the basal cells and more basal part of wing more or less devoid of microtrichiae, the wing having a rather milky appearance. Proepimeral seta shorter, at most subequal to the proepisternal or virtually absent. Lower facial margin slightly behind level of profrons (Fig. 302). ♂: ground-setulae on scutum very stout and spinulose outside the dorsocentrals, especially on postpronotal lobes; hind tibia with the osmeterium longer and broader, occupying all of basal half of posterodorsal surface; sternites 2-5 each with a pair of stout setae in posterior corner (Fig. 313). ♀: fore femur with a series of 5-7 short but stout anteroventral spinules occupying all but basal quarter .... 21. *nigripes* Robineau-Desvoidy

– Katepisternum entirely white pruinose (this is more difficult to see in specimens with yellow pleura than in those with dark). Wing entirely covered with rather dense and close-set microtrichiae, producing an opaque appearance. Proepimeral seta strong, longer than the proepisternal seta. Lower facial margin projecting far beyond level of profrons (Fig. 303). ♂: ground-setulae on scutum normal, weak; hind tibia with the osmeterium small and narrow, placed just after middle, on 3rd quarter of anterodorsal surface; sternites 2-5 with fine setae in posterior corners (Fig. 320). ♀: fore femur without anteroventral spinules .... 22. *sphondylii* (Schrank)

Larvae

1 Anal division short (Fig. 306); spines on anal division dark brown; anterior spiracles very broad and not slender, with 8-13 lobes (Fig.

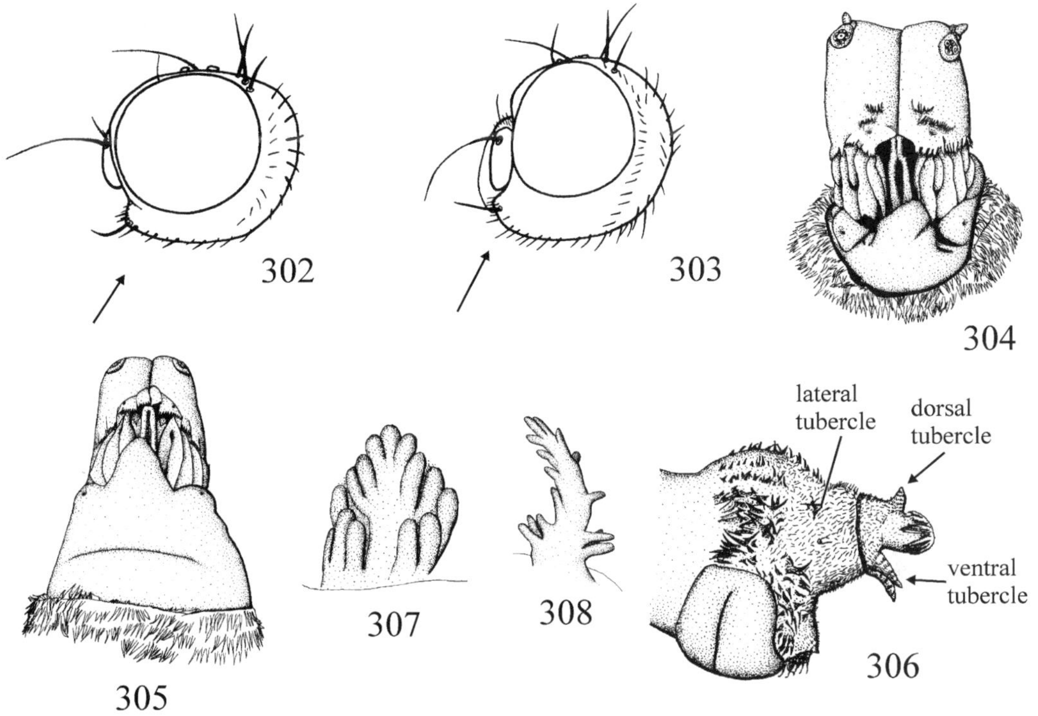

Figs 302-308. Heads, lateral view. 302: *Saltella nigripes* Robineau-Desvoidy; 303: *Saltella sphondylii* (Schrank). 304-308. Larval morphology. 304: *Saltella nigripes* Robineau-Desvoidy, pseudocephalon, ventral; 305: *Saltella sphondylii* (Schrank), pseudocephalon, ventral; 306: *Saltella nigripes* Robineau-Desvoidy, anal division, lateral; 307: *Saltella nigripes* Robineau-Desvoidy, anterior spiracles; 308: *Saltella sphondylii* (Schrank), anterior spiracles.

307); facial mask in posterior half of pseudocephalon (Fig. 304)........................ ....21. *nigripes* Robineau-Desvoidy

– Anal division elongated (Fig. 34); spines on anal division of the same colour as the cuticle; anterior spiracles very long and slender, with 14-17 lobes (Fig. 308); facial mask positioned far anterior on pseudocephalon (Fig. 305)... ....22. *sphondylii* (Schrank)

21. ***Saltella nigripes*** Robineau-Desvoidy, 1830

(Figs 302, 304, 306-307, 309-315)

Saltella nigripes Robineau-Desvoidy, 1830: 747. Holotype/syntypes, sex not stated, FRANCE, Saint-Sauveur, not in MNHNP and presumed destroyed.

Nemopoda ferruginea Robineau-Desvoidy, 1830: 744. Holotype/syntypes, sex not stated, FRANCE, locality not stated, not in MNHNP and presumed destroyed.

Nemopoda gagatea Robineau-Desvoidy, 1830: 745. Syntypes ♂♀, FRANCE, Saint-Sauveur, not in MNHNP and presumed destroyed. **Syn. n.**

Diagnosis. – Adult recognised by the receding face, sparsely haired wing-membrane, and restricted white katepisternal pruinosity.

Larval pseudocephalon with facial mask in posterior half (Fig. 304); anterior spiracles not very long, with 8-13 lobes (Fig. 307); spines on anal division dark-pigmented; anal division shorter than in *sphondylii* (compare Figs 34 & 306).

Adult. – Colour: Frontal plate, face, parafacial, gena and most of occiput reddish, ocellar triangle and vertex dark. Antennal scape, pedicel and flagellomere yellowish. Thorax, scutellum and abdomen entirely dark, at most postpronotal lobe and proepisternal area reddish-yellow. Coxae and

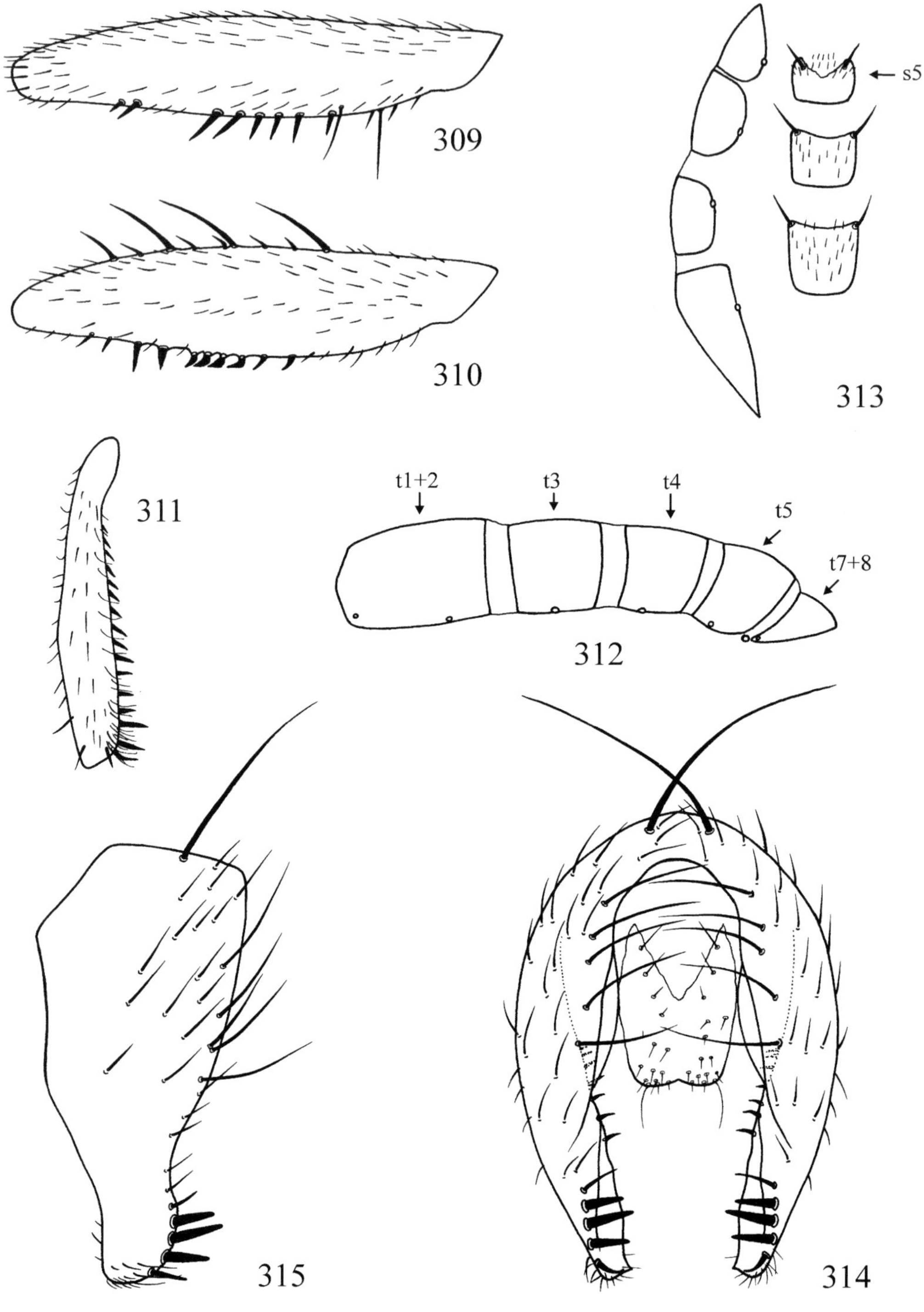

Figs 309-315. *Saltella nigripes* Robineau-Desvoidy, ♂. 309: fore femur, anterior view; 310: fore femur, posterior view; 311: fore tibia, posterior view; 312: abdomen, lateral view; 313: part of abdomen, ventral view; 314: hypopygium, dorsal view; 315: hypopygium, lateral view. (s – sternite, t – tergite.)

trochanters yellow, ♂ mid and hind coxae partly dark; femora narrowly to broadly yellow at base, and knees reddish, otherwise femora, tibiae and tarsi dark. Wings clear. Calypters and fringe-hairs creamy. Knob of halteres black.

Head (Fig. 302): A full complement of setae present. Vibrissae and subvibrissal setae stout in ♂, finer in ♀.

Thorax: Scutum mostly matt, subshining towards sides and behind. Acrostichal setulae distinct, in 3 rows, and a complete row of dorsocentral setulae present from neck to prescutellar setae, stout in ♂ but fine in ♀. Setulae outside dorsocentrals stout and spinulose, especially on postpronotal lobes. Proepimeral seta at most subequal to proepisternal, or virtually absent. Pleura chagreened except for a small band of white pruinosity in upper posterior corner of katepisternum. Scutellum bare except for the basal and apical setae.

Legs: ♂ fore legs in Figs 309-311; fore femur with the anteroventral and posteroventral spinules more numerous and generally stouter than in *sphondylii*. ♀ fore femur with 5-7 anteroventral spinules, without posteroventrals. Mid femur with some short strong anteroventral and posteroventral spinules in ♂, these virtually absent in ♀. Mid tibia with 2-4 each anteroventral, posteroventral and posterior setae in ♂, these reduced to 2 weak anteroventrals and posteroventrals in ♀. Hind femur with a row of anterodorsal setae in apical half and, in ♂, some short anteroventral spinules in apical third. Hind tibia without setae except for a short anterodorsal preapical; ♂ with a depressed bare area occupying basal half of posterodorsal surface (an osmeterium?).

Wings: Veins bare except for a few setulae on basal part of stem-vein on ventral surface. Microtrichiae very sparse, giving the wing a milky appearance; the basal cells and more basal part of the wing more or less devoid of microtrichiae.

Abdomen: Subshining. Without setae except for some lateral marginals on tergites 3-5; syntergite 1+2 with setulae at sides. ♂ sternites 2-5 each with a pair of stout setae in posterior corners (Fig. 313). ♂ spiracles at lower edge of tergites 1+2 to 5 (Figs 312-313). ♂ genitalia in Figs 314-315; surstylus heavily spinose.

Wing-length: 2.9-3.1 mm (♂), 2.2-3.1 mm (♀).

Larva. – The immature stages were described by Ursu (1973: 79-80). More recently, the 3rd instar larva has been described by Ozerov (1986c: 45), whose work includes a revision of the larvae of all three Palaearctic *Saltella*. A full description based on an SEM study is included in Meier (1996: 51-53, figs 158-166).

Distribution. – Not in Fennoscandia.

A central and south European species, extending from France eastwards to European Russia and Afghanistan, and south to Israel (Zuska & Pont, 1984: 158).

Biology. – Adults can be collected on cow dung (Papp, 1971a: 94), where the larvae also breed (Papp & Garzó, 1985: 156). The species overwinters as a puparium (Lobanov, 1962a: 37). Two hymenopterous parasitoids (Eucoilidae) have been reared from puparia: *Cothonaspis pusilla* Giraud and *Cothonaspis gracilis* Hartig.

22. *Saltella sphondylii* (Schrank, 1803)

(Figs 1-3, 13, 27, 34-35, 303, 305, 308, 316-323)

Trupanea sphondylii Schrank, 1803: 149. Holotype/syntypes, sex not stated, locality not stated [from title: GERMANY, Bavaria], destroyed [see below].

Piophila scutellaris Fallén, 1820a: 10. Syntypes, SWEDEN, 'in vicinitate Lundae' [Skåne, near Lund], not in NRS or ZML and presumed destroyed [see below].

Ochtera melanocephala Drapiez, 1820: 354, plate 68 fig. 7. Holotype, sex not stated, ITALY, Piedmont, location not known and presumed destroyed. **Syn. n.**

Cordylura analis Meigen, 1826: 244; Meigen plate 145 fig. 12 [Morge, 1976a]. Holotype/syntypes, AUSTRIA, not in NMW and presumed destroyed [see below].

Cordylura varia Meigen, 1826: 245; Meigen plate 145 fig. 13 [Morge, 1976a]. Holotype/syntypes, AUSTRIA, not in NMW and presumed destroyed [see below].

Brachygaster hemorrhoidalis Meigen, 1826: 245; Meigen plate 133 fig. 7 [Morge, 1976a]. Unavailable name, proposed in synonymy with *Cordylura varia* Meigen.

Saltella longipes Robineau-Desvoidy, 1830: 747. Holotype/syntypes, sex not stated, FRANCE, Saint-Sauveur, not in MNHNP and presumed destroyed. **Syn. n.**

Saltella nana Robineau-Desvoidy, 1830: 747. Holotype/syntypes, sex not stated, FRANCE,

Saint-Sauveur, not in MNHNP and presumed destroyed. **Syn. n.**

Pandora basalis Haliday, 1833: 169. Holotype/syntypes, locality not stated [from title: IRELAND, Holywood], not in NMI and presumed destroyed [see below].

Nemopoda scutellata Macquart, 1835: 481. Holotype/syntypes ♀, FRANCE: Bordeaux, not in MHNL and presumed destroyed [see below].

Nemopoda ruficoxa Macquart, 1835: 481. Holotype/syntypes ♂, 'du nord de la France' [FRANCE, presumably Lille district], not in MHNL and presumed destroyed [see below].

Anisophysa albipennis Macquart, 1835: 545. Holotype ♀ [not ♂], FRANCE, in MHNL [see below].

Musca spondylii Schrank in Gistel, 1837: 18. Holotype/syntypes, sex not stated, 'in umbellis Heraclei spondylii' [probably GERMANY, Bavaria], destroyed [see below].

Saltella sellata Curtis, 1837: 268. Nomen nudum [see below].

Saltella sellata Haliday, 1838: 186. Holotype ♀, IRELAND: Holywood, not in NMI and presumed destroyed [see below].

Piophila distincta Meigen, 1838: 360; Meigen plate 279 fig. 4 [Morge, 1976c]. Holotype ♀, 'Lüttich' [BELGIUM, Liège], in MNHNP [see below].

Piophila analis Roser, 1840: 61. [Junior secondary homonym, preocc. in *Saltella* by *Cordylura analis* Meigen, 1826.] Holotype/syntypes, locality not stated [presumably GERMANY, Württemberg], not in SMN and presumed destroyed [Becker, 1903a: 56].

Piophila humeralis Roser, 1840: 61. Holotype ♂, locality not stated [presumably GERMANY, Württemberg], in SMN [Becker, 1903a: 56; Duda, 1926a: 65].

Saltella parmensis Rondani, 1874: 179. Syntypes ♂♀, ITALY, Parma and Piedmont, probably in MZLS; 1♂ in UMO [seen by ACP].

Saltella nigerrima Rondani, 1874: 180. Holotype/syntypes ♀, ITALY, Parma, probably in MZLS.

Pandora nigripes Duda, 1926a: 65. Unavailable name; citation of a Meigen MS name, in synonymy with *Pandora scutellaris* Fallén.

Pandora fumipennis Duda, 1926a: 65. Unavailable name; citation of a Meigen MS name, in synonymy with *Pandora scutellaris* Fallén.

Chlorops floreus Duda, 1926a: 65. Unavailable name; citation of a Bouché MS name, in synonymy with *Pandora scutellaris* Fallén.

Saltella 'chaerophylli' Schrank; error or emendation (e.g. Grzegorzek, 1873: 33; Kowarz, 1873: 463) [see below].

Diagnosis. – Adult recognised by the projecting face and lower facial margin (Fig. 303), densely haired wing-membrane, and wholly white pruinose katepisternum. The only member of the genus in Fennoscandia.

Larval pseudocephalon with facial mask in anterior half (Fig. 305); anterior spiracles very long, with 8-14 lobes (Fig. 308); spines on anal division of the same colour as cuticle; anal division longer than in *nigripes* (compare Figs 34 & 306).

Adult. – Colour: Extremely variable in colour: males may be almost wholly yellow through to almost black, whilst females tend to be mainly to wholly black. Ocellar triangle and vertex black, rest of head yellow to dark brown. Antennal scape and pedicel yellow, flagellomere orange to brown. Scutum dark, sometimes with postpronotal lobe and notopleuron yellow. Pleura almost wholly yellow, with katepisternum or meron partly dark, to wholly dark; anatergite black. Scutellum wholly yellow to wholly black. Coxae wholly yellow or mostly brown; trochanters usually yellow; femora entirely yellow, to entirely brown with yellow tips; tibiae entirely yellow with fore tibia usually partly brown, to entirely brown; tarsi brown. Wing clear. Calypter and fringe-hairs creamy. Knob of haltere brown to black. Abdomen usually entirely dark with just the ♂ genital capsule yellow, sometimes broadly yellow basally.

Head (Fig. 1). A full complement of setae present. Subvibrissal setae short and weak.

Thorax: Scutum entirely matt. Acrostichals distinct, in 3 irregular rows, and a complete row of strong dorsocentral setulae present from neck to prescutellar setae. Setulae outside dorsocentrals distinct but not spinulose. Proepimeral seta longer than proepisternal. Pleura shagreened except for katepisternum which is entirely white pruinose, a small band of this pruinosity extending to proepimeral area. Scutellum bare except for the basal and apical setae.

Legs: ♂ fore legs in Figs 316-318; fore femur with anteroventral and posteroventral spinules less numerous and stout than in nigripes. ♀ fore femur without anteroventral or posteroventral spinules.

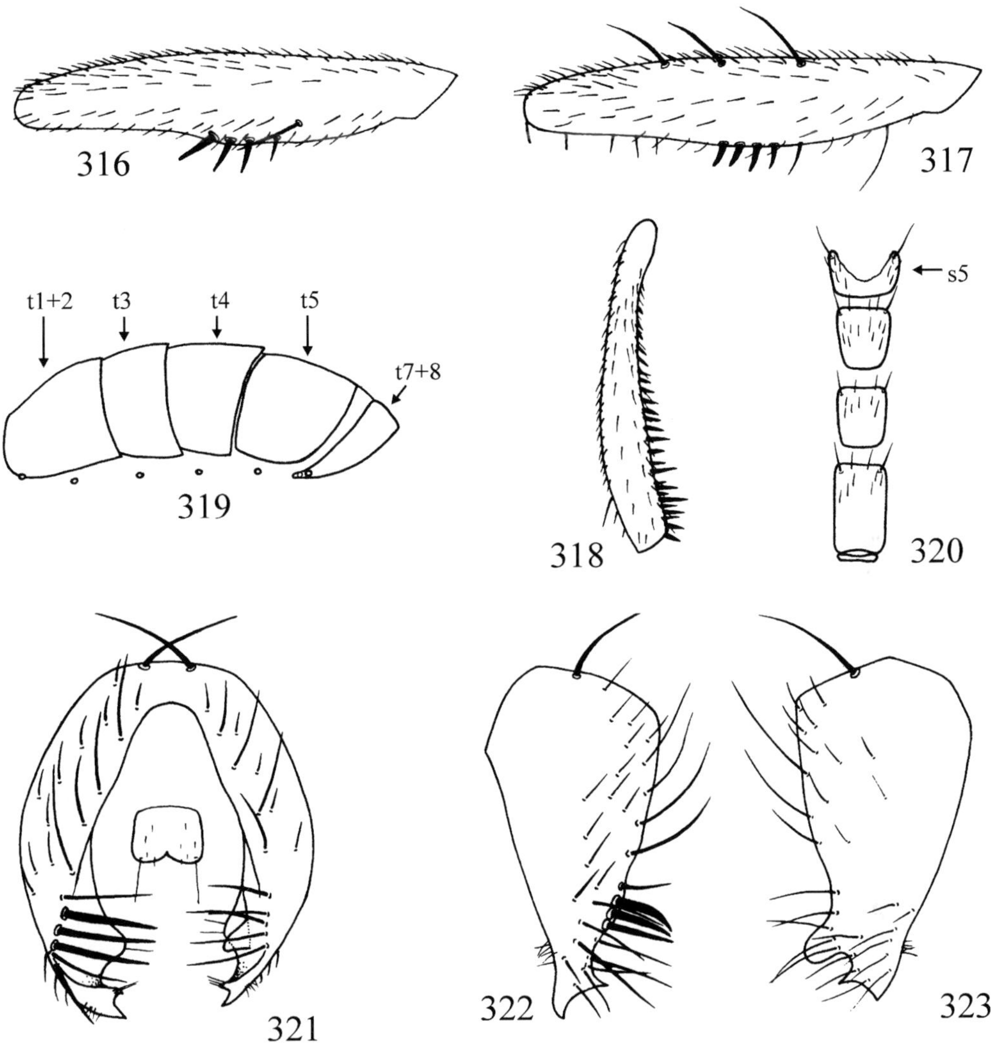

Figs 316-323. *Saltella sphondylii* (Schrank), ♂. 316: fore femur, anterior view; 317: fore femur, posterior view; 318: fore tibia, posterior view; 319: abdomen, lateral view; 320: abdominal sternites 1-5; 321: hypopygium, dorsal view; 322: hypopygium, left lateral view; 323: hypopygium, right lateral view. (s – sternite, t – tergite.)

Mid femur without anteroventral or posteroventral spinules, around middle with 1 (♀) or several (♂) short setae. Mid tibia with 2 strong anteroventral and 6-9 irregular weak posterior and posteroventral setae in ♂, all these reduced to 1-2 anteroventrals in ♀. Hind femur with several anterodorsal setae in apical half, bare ventrally. Hind tibia without setae except for a short anterodorsal preapical; ♂ with a narrow osmeterium occupying 3rd quarter of anterodorsal surface.

Wings: Veins bare except for a few fine setulae on basal part of stem-vein on ventral surface. Entirely covered with microtrichiae which are dense and close-set, producing a very opaque appearance.

Abdomen: Subshining. Without setae except for some lateral marginals on tergites 3-5; syntergite 1+2 with setulae at sides. ♂ sternites 2-5 each with a pair of weak setae in posterior corners (Fig. 320). ♂ spiracle 1 at edge of syntergite 1+2, spiracles 2-5 in the intersegmental membrane (Fig. 319). ♂ genitalia in Figs 321-323; surstyli asymmetrical (first pointed out by Minder, 1963), only the left-hand surstylus (in dorsal view) with heavy spines; genitalia also illustrated by Ozerov (1999e: figs 347.8-9).

Wing-length: 1.9-2.8 mm (♂), 2.0-2.7 mm (♀).

Variation. – The colour variation in this species is very striking, and has given rise to many new 'species' and 'varieties' that are no more than elements in a single cline of variation. This was first recognised by Loew (1841: 189-190), who gave a very detailed and thorough account. More recently it has been briefly discussed by Minder (1963: 384) and Pont (1979: 8), and reference is made to it in the description given above. The larger and more orange forms are almost always males, whereas the small dark forms may be males or females.

Egg. – A description of the egg and some functional considerations are included in Hinton (1960: 51-53, fig. 2; 1961: 145). A full description of the egg has been given by Meier (1995b: 427-431, figs 4, 11, 22)

Larva. – The immature stages have been described by Hammer (1941: 112), Hennig (1949: 19, figs 21, 28, 31, 43, 62, 63), Schumann (1962: 429) and Ozerov (1986c: 45), whose work includes a revision of the larvae of all three Palaearctic *Saltella*. The species was also included in a key to sepsid larvae by Mangan (1977: 341). A full description of the 3rd instar larvae was given by Meier (1996: 55-57, figs 167-175).

Distribution. – Widespread but nowhere abundant. Denmark: all provinces. Norway: only from Ry (L. G. Jensen, pers. comm.). Sweden: a few records in the south (e.g. Ringdahl, 1948: 2; 1959b: 45), and DR. Finland: Ta, Kb. Longitude 62°N seems to mark its northern limit.

Mainland Europe and British Isles, though less common around the Mediterranean (Zuska & Pont, 1984: 158), western Russia (e.g. Lobanov, 1962a: 37; Minder, 1963: 388; Tamarina et al., 1982: 87), eastwards through Turkmenistan, Altai, Transbaikalia (Ozerov, 1999e: 559), to Mongolia (Papp, 1976: 130), the Russian Far East province (Ozerov, 1989: 840; 1999e: 559) and Hokkaido (Iwasa, 1984: 296; 1995: 785). Nearctic: from Quebec to Washington, south to California (Steyskal, 1965: 683; Mangan, 1977: 338).

Records are from May to August.

Biology. – The species is very closely associated with cow dung, usually in open fields, and adults are seldom found far away from cow pads, preferring older crusted pads (Hammer, 1941; Schumann, 1962: 429; Pont, 1979: 12; Schweiger, 1988). When disturbed, the flies seek shelter in the tunnels and cavities created by dung beetles. The occurrence of *sphondylii* in the vicinity of cattle or yak dung is well documented in the literature (Zetterstedt, 1847: 2520; Valiela, 1969: 213; Papp, 1976: table 6; Mangan, 1977: 339; Iwasa, 1984b: 297; Van der Goot, 1986a: 2). In addition, oviposition on or rearing from cow dung has been described by Coffey (1966: 214), Foster (1967: 1035-1036), Hinton (1960: 52), Hammer (1941), Laurence (1954: 250), Ozerov (1986c: 46), Pont (1979: 12), Schumann (1962: 429-430), Schweiger (1988), Valiela (1974: 374), Papp (1976: table 1), and Wharton & Moon (1979: 88). Records not involving cow dung are rare, but there are a few reports from liquid manure (Hennig, 1949: 15), stable manure (Duda, 1926a: 19; Hennig, 1949: 15), manure heaps (Minder, 1963: 389), and horse dung (Coffey, 1966: 214).

Adults are also commonly found on flowers, especially umbellifers (Macquart in Duda, 1926a: 65; Wahlgren, 1919: 284; Ringdahl, 1948 2; Minder, 1963: 389; Van der Goot, 1986a: 2; Ozerov, 1989: 840; RM, pers. obs.). Occasionally, copulating pairs have been found on flowers (Schulz, 1999: 69). A courtship display or dance has been described by Hammer (1941: 83) for a '*Pandora*' species which is clearly the present species. He referred to it as 'like a little elegant minuet'. Ovipositing couples have been observed by Schumann (1962: 429), and Foster (1967: 1035-1036) observed that males protect their mates by remaining in a mounted position after copulation and during oviposition. Such guarding has recently been confirmed (Schulz, 1999: 68-69). What is remarkable about male guarding in *sphondylii* is that it is postcopulatory guarding, while all other sepsid species with male guarding have precopulatory guarding (Schulz, 1999: 70).

During oviposition, 10-20 eggs are usually placed at the base of open crevices on the pad (0.5-1.5 cm away from entrance), often in beetle tunnels (Hammer, 1941). In midsummer, development takes 29-35 days (Hammer, 1941: 273). However, Schumann (1962: 429) reported a much more rapid development (6 days from egg to puparium). The larvae pupariate in or below the cow pad and in Denmark the species is either uni- or bivoltine (Hammer, 1941: 186). Hibernation takes place as a puparium (Hammer, 1941: 212; Lobanov, 1962a: 37).

Type-material. – *Trupanea sphondylii*: No type-material is known to exist. Loew (1870: 15) introduced this name, replacing the better-known *scutellaris* Fallén, and several of his contemporaries used an apparent misspelling or emendation '*chaerophylli*' Schrank (e.g. Grzegorzek, 1873: 33; Kowarz, 1873: 463). The name *spondylii*, listed by Thompson & Pont (1994: 162), was introduced by Gistel from Schrank's posthumous manuscripts.

Piophila scutellaris: Fallén described this species from two specimens collected by Zetterstedt ('In vicinitate Lundae a D. Zetterstedt, mense majo capta ... Duo tantum specimina a nobis observata'). In drawer 24 of Fallén's collection in NRS, under *Piophila scutellaris*, there is a label in Fallén's hand 'Piophila / scutellaris / Zetterst. / Collectio'. However, no specimens labelled by Fallén have been located by ACP in NRS or in the Zetterstedt collection in ZML, and the syntypes must be presumed destroyed.

Cordylura analis was described from material sent to Meigen by Megerle. Megerle's collection is known to have been destroyed (Pont, 1986b: 204). ACP has seen 1♂ in NMW labelled as *analis*, but it is from Winthem's collection and not from Megerle (see also Duda, 1926a: 65). This ♂ is *sphondylii*.

Cordylura varia was also described from material sent to Meigen by Megerle, labelled as '*Brachygaster hemorrhoidalis*'. ACP has seen 3♂ in NMW labelled as *varia*, but they are from Winthem's collection not from Megerle (see also Duda, 1926a: 65). All 3 are *sphondylii*. There is no material in NMW labelled as *hemorrhoidalis*.

Pandora basalis and *Saltella sellata*: Haliday's collection in NMI contains no material labelled as *basalis* or *sellata*. Under the genus *Pandora* (i.e. *Saltella*), ACP has seen: named as *nigripes*, 1♀ (no locality) and 1♀ ('Ireland'); named as *scutellaris*, 1♂ (Orsova, vi.1871, Kowarz) and 2♂ ('Holywood'). All 5 specimens are *Saltella sphondylii*, and all but one bear determination labels by J. E. Collin so that Haliday's original identifications are not clear. Type-material of *basalis* and *sellata* may well be present here: the ♀ labelled 'Ireland' may be the holotype of *sellata*, and the 2♂ labelled 'Holywood' may be syntypes of *basalis*. At all events, since *sphondylii* is the only species of the genus to occur in the British Isles, Haliday's names can be safely synonymised with *sphondylii*. The Curtis collection (NMV) contains 1♀ without data labelled as *basalis* (which is actually *sphondylii*) (seen by ACP), and no material of *sellata*.

Nemopoda scutellata and *Nemopoda ruficoxa*: the type-material of these two Macquart species has not been found by ACP in MHNL or MNHNP and is presumed destroyed, but the holotype of *Anisophysa albipennis* has been studied by ACP in MHNL, in box T.11 of the Macquart collection. It is a ♀, not ♂ as stated by Macquart.

Piophila distincta: Meigen described this simply from 'Lüttich', without stating the sex or number of specimens. ACP has studied the holotype ♀ in MNHNP, Meigen collection no. 2359. It is labelled 'Piophila distincta' by Meigen. It was also seen by Becker (1902: 249). It is *Saltella sphondylii*. Duda (1926a: 65) mentioned 4♀ of *distincta* in NMW, but ACP was unable to locate these.

Genus *Nemopoda* Robineau-Desvoidy, 1830

Nemopoda Robineau-Desvoidy, 1830: 743. Type-species: *Nemopoda putris* Robineau-Desvoidy, 1830 [= *Musca nitidula* Fallén, 1820], by designation of d'Orbigny, 1846: 611 [see Evenhuis & Thompson, 1990: 237]. Gender: feminine.

Pseudonemopoda Duda, 1926a: 30. Type-species: *Pseudonemopoda speiseri* Duda, 1926a, by monotypy. Gender: feminine.

Adult *Nemopoda* can be recognised in both sexes by the characteristic fore femoral armature: both sexes have a close-set row of of short anteroventral spinules on apical half and the males also have a loose row of setae or spinose setae on posteroventral surface.

Adult. – Head roundish, facial carina small, face receding, gena narrow. Outer vertical short or absent (Figs 324-325); postocellar present; orbital very short or absent. Scutum and scutellum matt. Postpronotal present or absent. 1 dorsocentral. 0 intraalar, 1 supraalar, 1 strong postalar. Proepisternal hairs present, proepimeral area bare. Anepisternal present. Katepisternal short or absent. Anepisternum, anepimeron and meron shining; katepisternal pruinosity confined to a band along upper or upper posterior part (Figs 327,

329). Prosternum setulose. 1 scutellar (apical). Fore femur without posterodorsal setae. Tibiae without anterodorsal or posterodorsal setae. ♂ fore tibia with a ventral row of tiny spinules (Fig. 335). ♂ hind tibia with the osmeterium small, in apical half, or absent. Wing without a dark spot. Abdominal setae absent except on tergite 5.

♂ abdomen. Tergites 1+2 to 5 normal; tergite 6 absent; tergite 7+8 narrow, with a small strip of chitin present between it and tergite 5 that appears to be a vestigial sternite 7+8. Spiracles 1-5 in the intersegmental membrane close to the edge of the tergites (Fig. 337). Sternite 1 divided into two plates; sternites 2-4 complete but 2-3 often with a median line of weakness, all three fully divided longitudinally in *speiseri* (Fig. 353); each apical lobe of sternite 4 with a brush of bushy setae at tip; sternite 5 partly concealed beneath sternite 4. Cercal plate consisting of a pair of long setiferous teat-like projections (Figs 339, 346, 355). Surstylus completely fused with epandrium, simple, not spinose, in *speiseri* bearing a bifurcate internal process near tip. Sperm pump large, 'dumb-bell' in shape.

♀ abdomen. Sternites 2-5 normal in shape, plate-like. Spiracles 1-5 in the intersegmental membrane, spiracles 6 and 7 in tergites 6 and 7. Postabdominal segments 6 and 7 with the tergites and sternites separated on sides of segments, the sternites (like the tergites) thus very broad and short, the tergites but not the sternites each with an anterior extension (but tergite 6 without an anterior extension in *speiseri*); segment 8, when not invaginated, long, extended posteriorly and ventrally, with a ventral element (sternite 8) on each side that remains separated at tip and a dorsal element (tergite 8) that forms the usual pair of ring-like bars that do not quite touch apically. Cerci small, round. Epiproct absent. Hypoproct present, weakly sclerotised. 2 spermathecae, one much larger than the other.

Egg. – Description based on *nitidula*. Egg with a single respiratory filament. Micropyle fused to dorsal side of filament base (Fig. 29). Filament dorso-ventrally compressed and ribbon-like (Fig. 23). Its exochorion open to the surrounding atmosphere and broken up into polygonal plates. Filaments with a single central canal; the struts that are connected to the plates v-shaped and regularly arranged (Fig. 23). The two layers of struts from opposite sides of the filament irregularly connected to either side of the central canal (Fig. 23).

Larva. – Pseudocephalon bilobed, cephalic lobes partly covered by combs; facial mask consisting of numerous ridges. All body segments very hairy (as in Fig. 45).

Anal division (Figs. 331-332) short; lateral tubercles present; stigmatophores with a dorsal tubercle; ventral tubercles present, simple to large; subanal tubercle large and hairy; anal pad moderately large, with a lateral lobe ('wing-shaped'), anal tuft present; preanal protuberance absent; anal division between anterior margin of anal pad and origin of stigmatophores about as wide as long or slightly longer; with two double rows of spines on lateral aspect of anal division; dorsally with three double rows of spines and an additional anterior transverse row (Fig. 43).

Cephaloskeleton: labial sclerites absent; dental sclerites present (Fig. 330) or absent; bridge of intermediate sclerite at half length of sclerite (as in Fig. 138), anteriorly with or without a flat plate; epistomal sclerite longer than wide, with or without posterior bars; parastomal bars long, only slightly shorter than intermediate sclerite (Fig. 330); basal sclerite 2.3-2.5 times as long as high, dorsal cornua shorter and 2-3 times narrower than ventral cornua.

Anterior spiracles tree-like, long, with 11-15 lobes; lobes 2.5-4 times as long as wide (as in Fig. 42).

Posterior spiracles with almost equal, straight or slightly arched spiracular openings on an oval plate; two spiracular openings parallel, third one perpendicular to the other two (as in Fig. 167), or all three parallel to each other (*speiseri*).

Discussion. – Plesiomorphies (in terms of the ground-plan of the Sepsinae) include the following: postocellar seta present; supraalar seta present; postalar seta present; anepisternal seta present; scutellum reduced in size; wing with basal radial and medial cells distinct; abdominal spiracles 1-5 in the intersegmental membrane; ♂ hind tibia with the osmeterium inconspicuous (absent in *speiseri*); ♀ with postabdominal tergites 6 and 7 each with an anterior prolongation. *Nemopoda* is apomorphous relative to the ground-plan of the Sepsinae with regard to the following characters: outer vertical seta short, or absent (*speiseri*), orbital seta short, or absent (*speiseri*), katepisternum with a short seta in upper posterior corner (absent in *speiseri*), ♂ sternite 4 with hair tufts, ♂ sperm-pump shaped like a 'dumb-bell'.

The only autapomorphy that can be given at present is the apparent reduction of the ♂ sternite 7+8. There is a small narrow strip present between tergite 5 and tergite 7+8 (Figs 336, 352), without an associated spiracle, and we interpret this as a sternite 7+8 that is reduced from a single band to two bands and displaced dorsally to a position between tergites 5 and 7+8.

Nemopoda appears to be a monophyletic genus with species of characteristic appearance, but it is only weakly founded on adult characters. There is considerable morphological diversity within the genus, and this has led to the separation of *speiseri* into the genus *Pseudonemopoda* (e.g. Ozerov, 1992a: 46-47).

Four Palaearctic species (key in Ozerov, 1997c), one of which is also Nearctic and Afrotropical; 1 Afrotropical species. The larvae of *nitidula*, *pectinulata* and *speiseri* were described by Ozerov (1987a). Larvae breed in various kinds of decaying organic matter (vegetation, fungi, carrion, excrement).

Key to European species of *Nemopoda* Robineau-Desvoidy

Adults

1 Postpronotal, orbital and outer vertical setae minute or absent (Fig. 325). Katepisternum bare (Fig. 329). Hind femur with 2-3 anterodorsal setae around middle, and in ♂ without any posteroventrals. Mid coxa without a seta on fore-margin (Fig. 329). Syntergite 1+2 virtually bare along sides. ♂: fore femur with 9-11 stout anteroventral spinules in apical half (Fig. 348), their length at least equal to basal tibial width; posteroventral surface (Fig. 349) with a comb-like row of 8-10 short spinules, the longest equal to about half femoral depth; hind tibia without dorsal setae but at apical sixth with 3-4 short stout close-set decumbent posteroventral setae (Fig. 351). ♀: fore femur with a comb-like row of 3-6 short stout posteroventral spinules .... 25. *speiseri* (Duda)

– Postpronotal seta strong [NOTE: this is easily broken off, so check for the presence of a scar], orbital and outer vertical setae short but well-developed (Fig. 324). Katepisternum with a short but distinct seta in upper posterior corner (Fig. 327). Hind femur without anterodorsal setae, and in ♂ with 2-4 posteroventrals in basal half. Mid coxa with a strong seta at middle of fore-margin (Fig. 327). Syntergite 1+2 with the usual setulae along sides. ♂: fore femur with the anteroventral spinules minute (Figs 333, 341), their length less than half basal tibial width; posteroventral surface with sparser spinules (Figs 334, 342), the longest ones over half femoral depth; hind tibia without setae on shaft except for a few fine erect dorsals just before tip. ♀: fore femur without any posteroventral spinules 2

2(1) Katepisternum, on upper part of fore-margin behind fore coxa, with a vertical row of 2-3 weak setulae (Fig. 327). Apical part of vein R_{2+3} weakly tinged with brownish. ♂: hind trochanter (Fig. 326) with 2, rarely 3, stout thorns on inner surface, directed downwards and best seen when the fly is viewed from behind and below; fore femur (Figs 333-334) with fewer, 8-10, short anteroventral spinules, and with only about 8 short posteroventral spinules; sternites 2 and 3 weakly concave or straight on posterior margins (Fig. 337). ♀: sternites 1-5 narrower, sternite 5 1.2 times as long as broad, sternite 6 twice as broad as sternite 4 (10 : 5) .... 23. *nitidula* (Fallén)

– Katepisternum, on upper part of fore-margin behind fore coxa, with a vertical row of 4-6 stronger and more spinulose setulae (Fig. 328). Apical part of vein R_{2+3} clear. ♂: hind trochanter bare on inner surface; fore femur (Figs 341-342) with more numerous, 11-12, short anteroventral spinules, and with more numerous, 10-13, and longer posteroventral spinules; sternites 2 and 3 each with a deep V-shaped incision on posterior margin (Fig. 344). ♀: sternites 1-5 broader, sternite 5 as long as broad, sternite 6 less than twice as long as broad (10 : 6) .... 24. *pectinulata* Loew

Larvae

1 Cephaloskeleton without dental sclerites .... 25. *speiseri* (Duda)

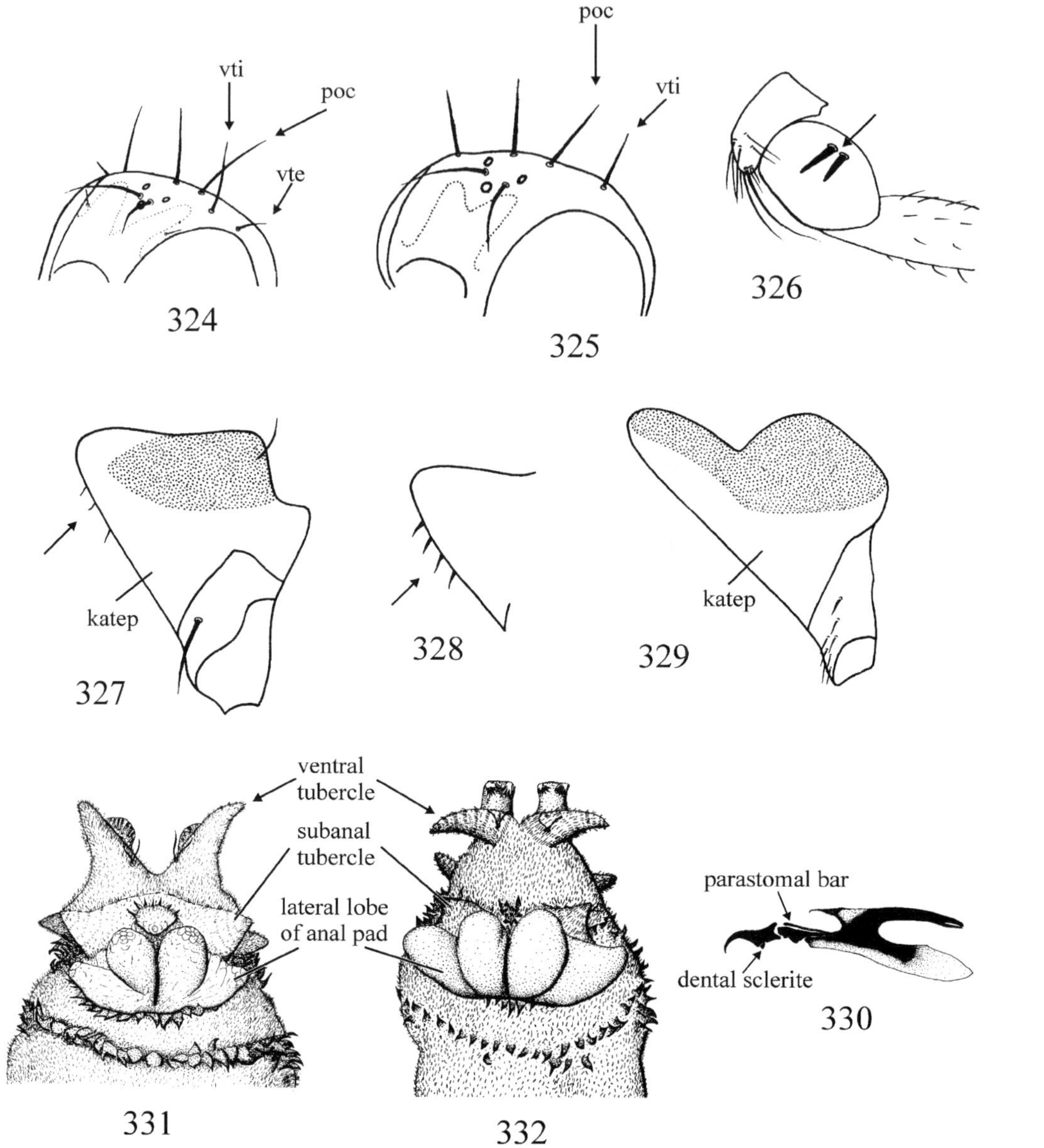

Figs 324-332. Frons, oblique view. 324: *Nemopoda nitidula* (Fallén); 325: *Nemopoda speiseri* Duda. 326: *Nemopoda nitidula* (Fallén), hind trochanter of ♂, inner surface. ♂ katepisternum. 327: *Nemopoda nitidula* (Fallén); 328: *Nemopoda pectinulata* Loew; 329: *Nemopoda speiseri* Duda. 330-332: larval morphology. 330: *Nemopoda nitidula* (Fallén), cephaloskeleton, lateral; 331: *Nemopoda nitidula* (Fallén), anal division, ventral; 332: *Nemopoda pectinulata* Loew, anal division, ventral. (katep – katepisternum, poc – postocellar, vte – outer vertical, vti – inner vertical.)

– Cephaloskeleton with dental sclerites (Fig. 330) 2

2(1) With a row of spines along entire anterior side of anal pad (Fig. 331) .... 23. *nitidula* (Fallén)

– With at most 6 spines on anterior side of anal pad, the spines restricted to central area of pad (Fig. 332) .... 24. *pectinulata Loew*

23. ***Nemopoda nitidula*** (Fallén, 1820)
(Figs 23, 29, 324, 326-327, 330-331, 333-340)

Musca cylindrica Fabricius, 1794: 336. [Junior primary homonym of *Musca cylindrica* De Geer, 1776.] Holotype ♀, 'Germania' [GERMANY], in ZMUC [see below].

Musca nitidula Fallén, 1820b: 21. Lectotype ♂, locality not stated [from title: SWEDEN], in NRS [see designation below].

Nemopoda putris Robineau-Desvoidy, 1830: 744. Syntypes ♂♀, locality not stated, 'cette espèce est excessivement commune' [FRANCE: probably Saint-Sauveur], not in MNHNP and presumed destroyed.

Nemopoda viridis Macquart, 1835: 481. Holotype ♀, FRANCE, Bordeaux, not in MHNL and presumed destroyed.

Diagnosis. – The adult is very similar to *pectinulata*, and is separated by the weaker setulae along fore-margin of katepisternum (Fig. 327) and, in the ♂, by the presence of 2 stout thorns on inner edge of hind trochanter (Fig. 326).

Larval cephaloskeleton with dental sclerites (Fig. 330); anterior spiracles with 11-16 lobes; anal pad with a long row of spines along anterior margin; lateral lobe smaller than in *pectinulata* (compare Figs 331 & 332).

Adult. – Colour: Most of frontal plate, face, parafacial and gena yellow; fronto-orbital plate, vertex and occiput dark. Antenna orange, flagellomere partly darkened. Thorax, scutellum and abdomen dark, just the proepisternal area (including also part of postpronotal lobe and anepisternum) yellow, sometimes also parts of anepimeron and even meron yellow. Fore coxa and fore trochanter yellow; fore femur mostly yellow, sometimes dark outside; fore tibia yellow; fore tarsus yellow at base, apical tarsomeres dark; mid and hind legs with coxae and trochanters yellow, femora broadly yellow at base and at tip, tibiae broadly yellow at tip, tarsi mostly brown. Wing clear, apical part of vein R_{2+3} conspicuously smoky, especially in ♂. Calypter and fringe-hairs smoky brown. Haltere yellow.

Head: Frons in Fig. 324. Inner vertical and postocellar strong, outer vertical and orbital weak. Fronto-orbital plate narrow. Subvibrissal setae numerous and short.

Thorax: Scutum, with postpronotal lobes and scutellum, matt. Setae: 1 dorsocentral, 1 postalar, 1 supraalar, 2 notopleurals, 1 postpronotal; acrostichal setulae minute. Pleura shining, except for a band of white pruinosity along upper margin of katepisternum that is usually obsolescent in front (Fig. 327). Setae reduced: anepisternal present, proepisternals reduced to setulae, 1 short katepisternal in upper posterior corner; anterior margin of katepisternum, behind fore coxa, with a vertical row of 2-3 weak setulae (Fig. 327). Scutellum short, twice as wide as long, with only a strong pair of apical setae.

Legs: ♂ fore legs in Figs 333-335; fore femur with the anteroventral spinules and posteroventral setae less numerous than in *pectinulata*. ♀ fore femur with 7-12 anteroventral spinules, without posteroventrals. Mid coxa with a strong seta at middle of fore margin (Fig. 327). Mid femur without setae, mid tibia with a short anteroventral at apical third. ♂ hind trochanter with 2 stout thorns on inner surface (Fig. 326), very rarely with a third thorn (e.g. Naswaty, 1960: 113). ♂ hind femur without anterodorsals, with 1 ventral at base and 2-4 posteroventrals in basal half; hind tibia without posteroventrals, with a few erect dorsal setulae just before tip, with a small osmeterium in apical half. ♀ hind femur and tibia without setae.

Wings: Veins bare except for several setulae in basal part of stem-vein on ventral surface. Wing-surface covered with microtrichiae.

Abdomen: Subshining black, with some violet reflections. Without setae except on margin and sides of ♂ tergite 5; syntergite 1+2 with setulae at sides. ♂ sternites 2-4 complete, sternites 2 and 3 straight on posterior margins (Fig. 337). ♂ sternites 4 and 5 and genitalia in figs 338-340; also figured by Gregor & Povolný (1964: figs 65, 68), Zuska (1965: figs 8-9, 12, 14) and Ozerov (1997c: figs 13, 15, but his fig. 13 actually appears to be of *pectinulata*).

Wing-length: 2.7-4.0 mm (♂), 2.7-4.3 mm (♀).

Remarks. – A reliable separation of *nitidula* and *pectinulata* is only possible in the male sex. In northern Europe both species are equally common, and occur together ecologically and seasonally; both occur equally frequently in traps baited with faeces and meat. Adults of *pectinulata* often appear smaller and less robust than *nitidula*, but this is by no means invariably so and there is a complete overlap in the size-range of these two species.

In addition to the absence of the two stout thorns on the inside of the hind trochanter and

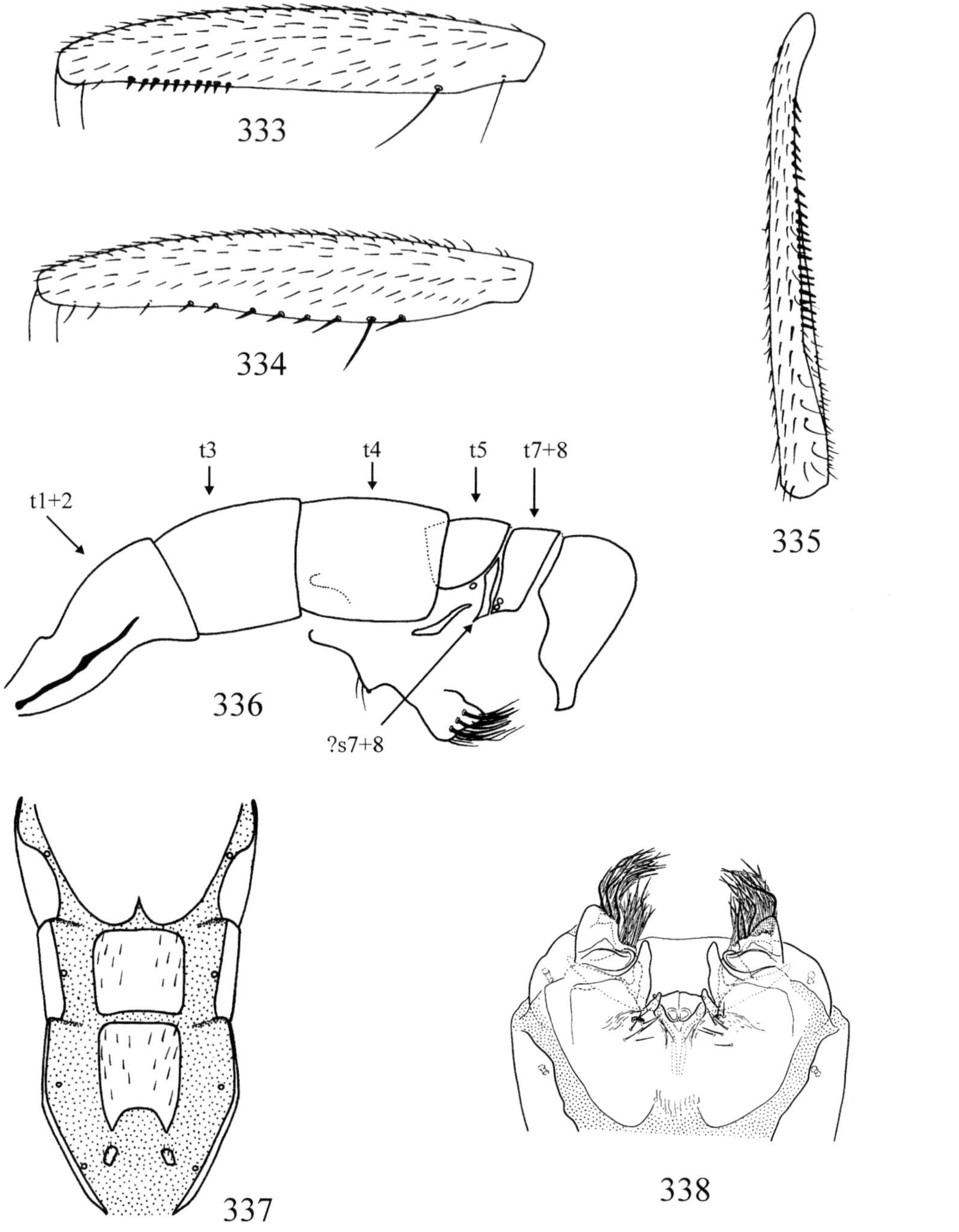

Figs 333-338. *Nemopoda nitidula* (Fallén), ♂. 333: fore femur, anterior view; 334: fore femur, posterior view; 335: fore tibia, posterior view; 336: abdomen, lateral view; 337: abdomen, ventral view, without genital segments and sternites 4 and 5; 338: sternites 4 and 5. (s – sternite, t – tergite.)

more numerous posteroventral setae on fore femur, males of *pectinulata* can be distinguished from those of *nitidula* by the clear wing (no conspicuously smoky area around apical part of vein R_{2+3}) and the presence of more numerous and more spinulose vertically-arranged setulae on the anterior margin of katepisternum (Fig. 328). These two features appear to hold good for females, par-

339

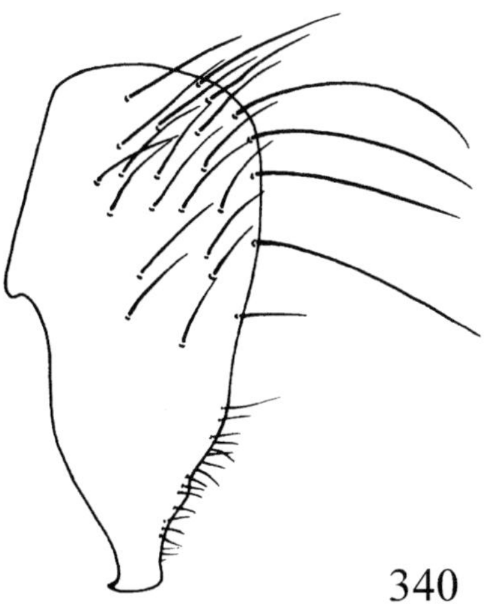

340

Figs 339-340. *Nemopoda nitidula* (Fallén). 339: hypopygium, dorsal view; 340: hypopygium, lateral view.

ticularly the katepisternal setulae, and will separate the majority of specimens, though there can be some ambiguity.

Egg. – The egg has been described by Meier (1995b: 427-431, figs 5, 13, 14, 19).

Larva. – The immature stages have been described by Bouché (1834: 95, plate 6 figs 9-11), Porchinskiy (1910: 87, fig. 91), Hennig (1949: 17, figs 20, 30, 37, 42, 48-49, 54-55), and Ozerov (1987a: 107-109, figs 2-5). The larva was included in a key by Mangan (1977: 341). Most recently, a full description of the 3rd instar larva has been given by Meier (1996: 27-29, figs 55-63).

Distribution. – Common and widespread. Denmark: all provinces (Stæger, 1844: 33). Norway: most provinces (e.g. Siebke, 1877: 153). Sweden: most provinces (Zetterstedt, 1838: 748; 1847: 2301; Wahlgren, 1919: 281; Ringdahl, 1951: 142). Finland: almost all provinces (Frey, 1908: 585).

Widely distributed throughout Europe, eastwards to Mongolia and the Far East province of Russia (Zuska & Pont, 1984: 156; Ozerov, 1989: 840; 1999e: 566); Japan (Iwasa, 1984: 303; 1995: 792). Nearctic: Alberta to Quebec, south to Indiana, Ohio & Virginia (Steyskal, 1965: 682; Mangan, 1977: 338). Afrotropical: Zaire (Zuska, 1980: 592).

Records are from May to early October in the south, mid May to August in the north.

Biology. – Adults are shade-loving, and are usually found in the ground-vegetation in or around woodlands, especially in damp situations (Bährmann, 1993a: 104-105). In Fennoscandia they are common in the deciduous, coniferous and birch forest zones. Their synanthropy has been discussed by Gregor & Povolný (1961: 27) and Mihályi (1967a: 375). They have been recorded as the prey of *Empis livida* L. (Tuomikoski, 1952: 171). Adults carry the mite *Anoetus sapromyzarum* Dufour (Ursu, 1969b: 537).

Larvae have been found in or adults reared from the following substrates: rotting fungus (Zetterstedt, 1847: 2301; Brindle, 1965: 139), slime moulds (Buxton, 1954: 169), high-water flood refuse (ZMUC), rotting log (Smith, 1957: 215), vertebrate carcasses (Reed, 1958: 238; Hanski, 1976: 114; Smith, 1975: 278; Van der Goot, 1986a: 3; Ozerov, 1989: 840), invertebrate carrion (Keilin, 1919: 451; Iwasa, 1984b: 303; Ozerov, 1989: 840), human excrement (Bouché, 1834: 95; Mihályi, 1967b: 332; Cole, 1969: 365; Sychevskaya, 1972a: 149; Iwasa, 1984b: 303; Ozerov, 1989: 840), and cow dung (RM, pers. obs.). Puparia are reported by Lobanov (1962a: 37) to overwinter in garbage pits, cesspits, pig dung and piles of rotting grass. Under laboratory conditions, females favour rotting flesh over cow dung (RM, pers. obs.). The larvae are able to complete their development in both substrates. Development times: 21 days at 20°C, 14 days at 24°C (RM, pers. obs.). Hennig (1952: 233) speculated that sepsids on carcasses may be visiting the gut contents of the dead animal, but this is certainly not the case in *nitidula* which definitely breeds in decaying flesh.

Two hymenopterous parasitoids have been reared from puparia (Ursu, 1969b: 537): *Alysia manducator* Panzer (Braconidae) and *Melittobia acasta* Walker (Eulophidae).

Type-material. – *Musca cylindrica*: Fabricius wrote of this species: 'Habitat in Germania Dom. Smidt'. There is 1♀, the presumed holotype, in his collection in ZMUC, which ACP has seen. It is fragmentary, consisting of thorax, two wings, and both fore legs; nevertheless, it can be recognised as being a ♀ of *nitidula* as defined in this paper. Alongside this ♀ are two other specimens, equally fragmentary, without labels, not syntypes and in fact not even Sepsidae.

Musca nitidula: Fallén gave no locality for this species, but described its habits as 'in nuptiis observata, cum praecedente [*putris*] frequenter cohabitans'. In the Fallén collection in NRS, drawer 22, ACP has studied 1♂ and 5♀ which are considered to be syntypes. The ♂, labelled by Fallén 'S. nitidu= / la ♂', is in good condition; ACP has labelled it and herewith designates it as lectotype; it is *nitidula* as currently understood. The 5♀ have been labelled as paralectotypes, and appear to be conspecific with the lectotype.

24. ***Nemopoda pectinulata*** Loew, 1873
(Figs 43, 45, 328, 332, 341-347)

Nemopoda pectinulata Loew, 1873: 305. Lectotype ♂, without locality [from title: 'die posener Gegend', = POLAND: Poznań], in ZMHU [by designation of Ozerov, 1997a: 483].

Diagnosis. – The adult is very similar to *nitidula*, and separated by the stronger setulae along fore-margin of katepisternum (Fig. 328) and, in the ♂, by the absence of stout thorns on the inner surface of hind trochanter. See also the notes under *nitidula* (p. 121-123).

Larval cephaloskeleton with dental sclerites (as in Fig. 330); anterior spiracles with 12-14 lobes; anal pad with a short row of spines along anterior margin (Fig. 332); lateral lobe larger than in *speiseri* and *nitidula* (compare Figs 331 & 332).

Adult. – Colour: Most of frontal plate, face, parafacial and gena yellow; fronto-orbital plate, vertex and occiput dark. Antenna orange, flagellomere partly darkened. Thorax, scutellum and abdomen dark, just the proepisternal area (including also part of postpronotal lobe and anepisternum) yellow, sometimes also parts of anepimeron and even meron yellow. Fore coxa and fore trochanter yellow; fore femur mostly yellow, sometimes dark outside; fore tibia yellow; fore tarsus yellow at base, apical tarsomeres dark; mid and hind legs with coxae and trochanters yellow, femora broadly yellow at base and at tip, tibiae broadly yellow at tip, tarsi mainly brown. Wing clear, apical part of vein R_{2+3} broadly infumated or clear. Calypter and fringe-hairs smoky brown. Haltere yellow.

Head: Inner vertical and postocellar strong, outer vertical and orbital weak. Fronto-orbital plate narrow. Subvibrissal setae numerous and short.

Thorax: Scutum, with postpronotal lobes and scutellum, matt. Setae: 1 dorsocentral, 1 postalar, 1 supraalar, 2 notopleurals, 1 postpronotal; acrostichal setulae minute. Pleura shining, except for a band of white pruinosity along upper margin of katepisternum that is usually obsolescent in front (as in Fig. 327). Setae reduced: anepisternal present, proepisternals reduced to setulae, 1 short katepisternal in upper posterior corner; anterior margin of katepisternum, behind fore coxa, with a vertical row of 4-6 stronger and more spinulose setulae (Fig. 328). Scutellum short, twice as wide as long, with only a strong pair of apical setae.

Legs: ♂ fore legs in Figs 341-343; fore femur with the anteroventral spinules and posteroventral setae more numerous than in *nitidula*. ♀ fore femur with 7-12 anteroventral spinules, without posteroventrals. Mid coxa with a strong seta at middle of fore margin (as in Fig. 327). Mid femur without setae, mid tibia with a short anteroventral at apical third. ♂ hind trochanter bare on inner surface. ♂ hind femur without anterodorsals, with 1 ventral at base and 2-4 posteroventrals in basal half; hind tibia without posteroventrals, with a few erect dorsal setulae just before tip, with a small osmeterium in apical half. ♀ hind femur and tibia without setae.

Wings: Veins bare except for several setulae in basal part of stem-vein on ventral surface. Wing-surface covered with microtrichiae.

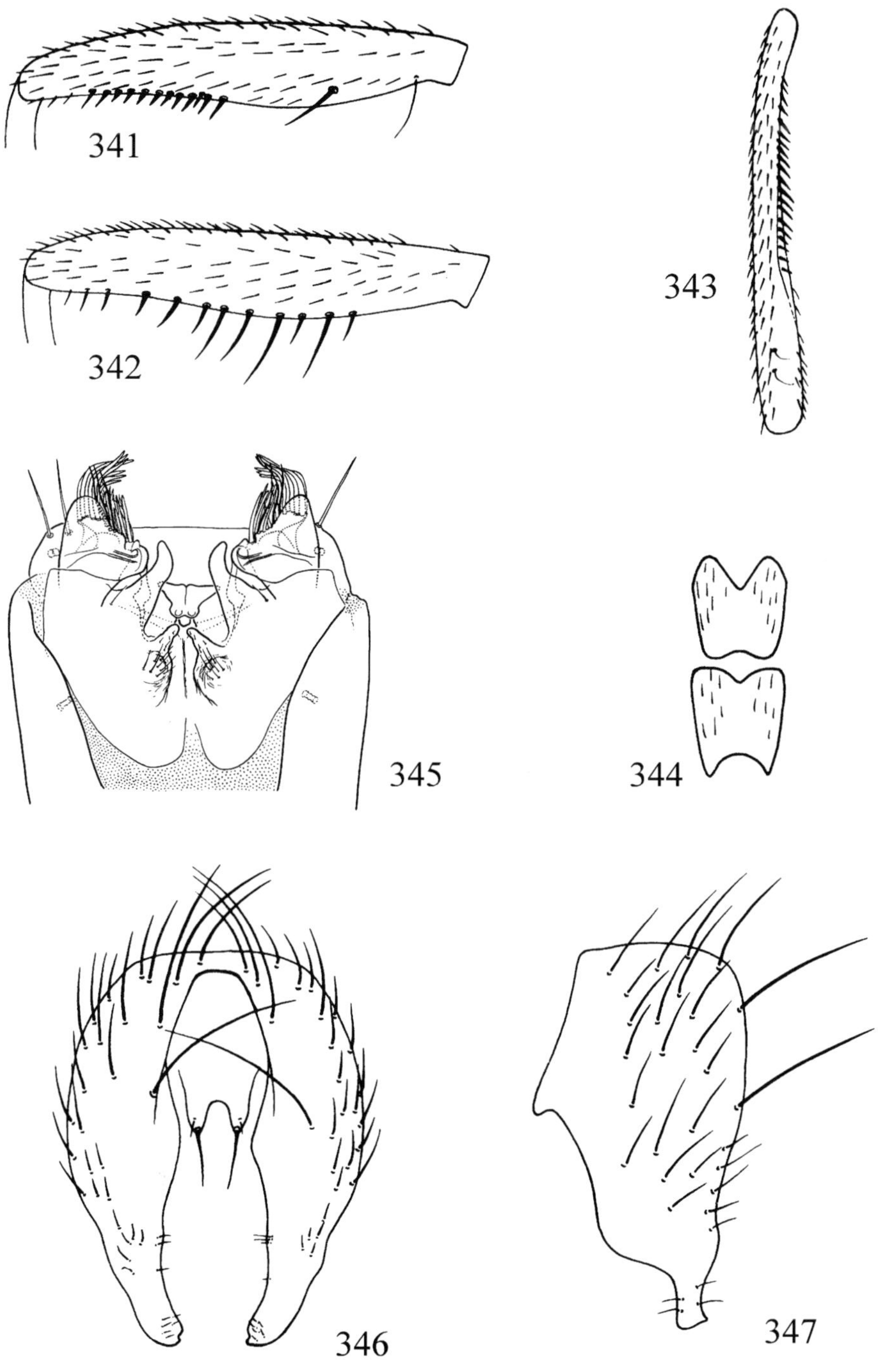

Figs 341-347. *Nemopoda pectinulata* Loew, ♂. 341: fore femur, anterior view; 342: fore femur, posterior view; 343: fore tibia, posterior view; 344: abdominal sternites 2 and 3; 345: sternites 4 and 5; 346: hypopygium, dorsal view; 347: hypopygium, lateral view.

Abdomen: Subshining black, with some violet reflections. Without setae except on margin and sides of ♂ tergite 5; syntergite 1+2 with setulae at sides. ♂ sternites 2-4 complete, sternites 2 and 3 deeply incised on posterior margins (Fig. 344). ♂ sternites 4 and 5 and genitalia in Figs 345-347; genitalia also figured by Gregor & Povolný (1964: figs 66, 68), Zuska (1965: figs 6-7, 11, 15) and Ozerov (1997c: figs 12, 14, but his fig. 12 actually appears to be of *nitidula*).

Wing-length: 2.7-3.6 mm (♂), 2.9-4.1 mm (♀).

Larva. – The larva has been described by Ozerov (1987a: 109, figs 1, 6-7) and Meier (1996: 29-31, figs 64-69).

Distribution. – Widespread, but more northern in its range than *nitidula*. Denmark: not common, but widespread. Norway: Ø and AK to TR (e.g. Duffield, 1938: 165; Ringdahl, 1954a: 52). Sweden: SK to TO (e.g. Ringdahl, 1948: 2; 1951: 142). Finland: most provinces (Frey, 1908: 585).

In northern Europe and upland areas of Central Europe, eastwards to the Far East province of Russia (Ozerov, 1999e: 566) and South Korea (Suh & Kwon, 1994: 61) and Japan (Zuska & Pont, 1984: 156; Iwasa, 1984b: 303; 1995: 792). Oriental: North Pakistan (Iwasa, 1989: 53), Nepal (Zuska, 1977: 177; Iwasa, 1984a: 17), and Taiwan (Iwasa, 1987: 1).

Records are from late May to mid September in the south, late June to August in the north.

Biology. – The habits are apparently similar to those of *nitidula*, but the species appears to be more common in the birch forest zone of Europe and has also been reported to prefer the more mountainous regions in Japan (Iwasa, 1984b: 304). It commonly occurs with *nitidula*, especially in faeces-baited traps, and is known to gather on human excrement (Iwasa, 1984b: 304).

The larvae were found by Lavčiev & Tsankova (1980: 4) in a dunghill. Iwasa et al. (1995: 618) reared one specimen from the nest of *Streptopelia orientalis* (Columbidae).

25. *Nemopoda speiseri* (Duda, 1926)

(Figs 325, 329, 348-356)

Pseudonemopoda speiseri Duda, 1926a: 99, fig. 16. Lectotype ♂, 'Westpreussen, Karthaus' [= POLAND: Kartuzy], in ZMHU [by designation of Ozerov, 1997a: 485].

Diagnosis. – Adult recognised by the virtual absence of postpronotal, orbital and outer vertical setae (Fig. 325).

Larval cephaloskeleton without dental sclerites; anterior spiracles with 14-17 lobes; anal pad with a short median row of spines along anterior margin; lateral lobe smaller than in *pectinulata*.

Adult. – Colour: Frons and occiput dark; gena, face, parafacial and lunular area orange to yellow. Scape, pedicel and base of flagellomere yellow. Thorax, scutellum and abdomen dark, just the proepisternal area (including part of postpronotal lobe and anepisternum) dull yellow to brown. Fore coxa and fore trochanter yellow; fore femur yellow inside, dark outside; fore tibia yellow; fore tarsus yellow at base, apical tarsomeres dark; mid and hind legs with coxae mostly and trochanters wholly yellow, femora yellow at base and tip, tibiae weakly yellow at tip, tarsi mostly brown. Wing clear. Calypter and fringe-hairs dull yellow. Haltere yellow.

Head: Frons in Fig. 325. Inner vertical and postocellar present, outer vertical and orbital absent. Fronto-orbital plate broad and shining. Subvibrissal setae numerous and short.

Thorax: Scutum, with postpronotal lobes and scutellum, matt. Setae reduced: 1 dorsocentral, sometimes with a second one in front, 1 postalar, 1 supraalar, 2 notopleurals; acrostichal setulae minute. Pleura shining, except for a broad band of white pruinosity along upper part of katepisternum (Fig. 329). Setae reduced: anepisternal present, proepisternals reduced to a few fine setulae, katepisternal absent; anterior margin of katepisternum bare. Scutellum short, twice as wide as long, with only a strong pair of apical setae.

Legs: ♂ fore legs in Figs 348-350. ♀ fore femur in apical half with rows of 10-11 anteroventral and 3-6 posteroventral spinulose setulae. Mid femur and mid tibia without setae. Hind femur with 2-3 anterodorsals around middle, without ventral setae. ♂ hind tibia with no trace of osmeterium; at apical sixth with 3-4 close-set decumbent posteroventral setae (Fig. 351), without other setae.

Wings: Veins bare except for 2-3 setulae in basal part of stem-vein on ventral surface. Wing-surface covered with microtrichiae except for basal medial cell and part of posterior cubital cell.

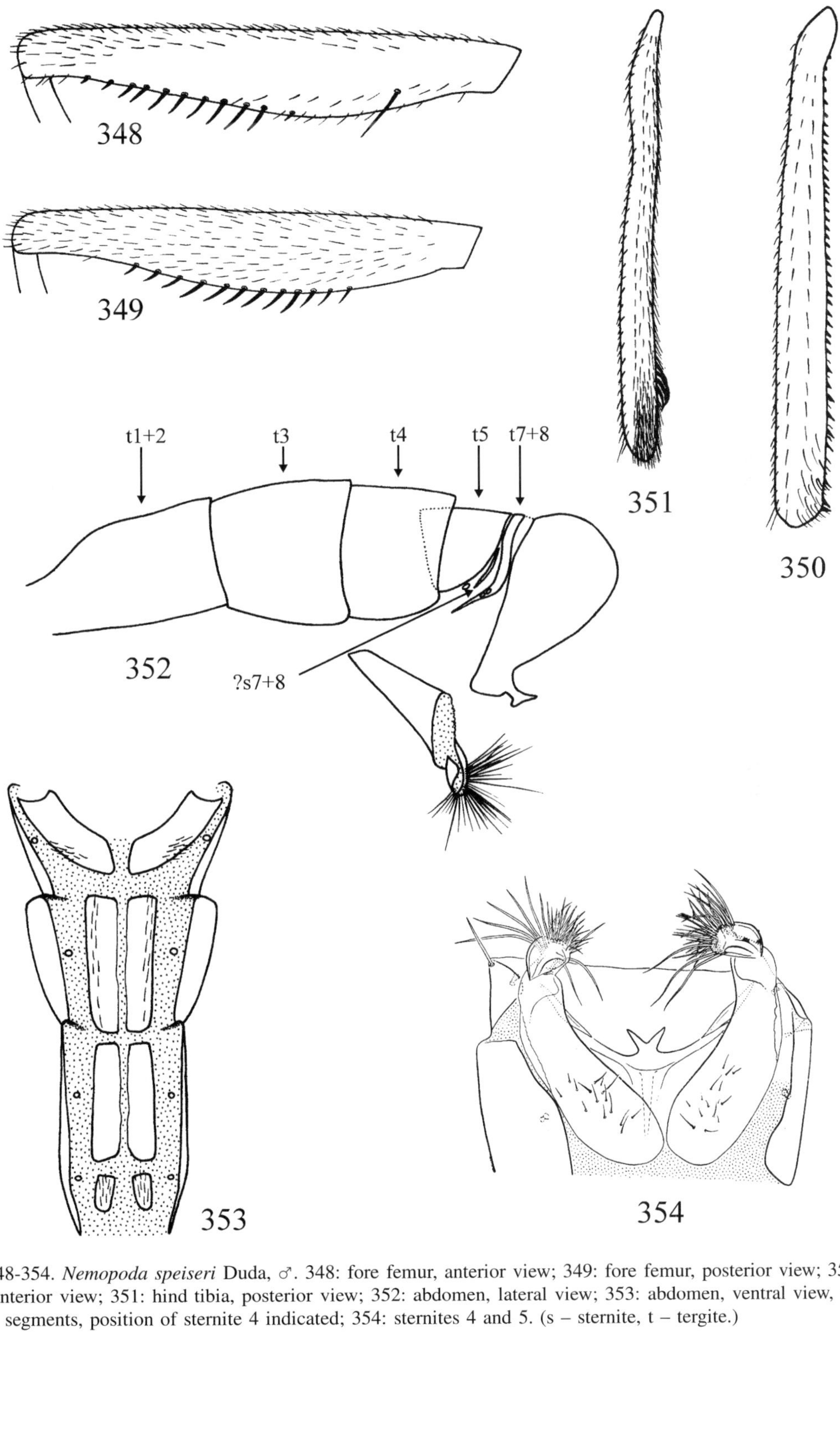

Figs 348-354. *Nemopoda speiseri* Duda, ♂. 348: fore femur, anterior view; 349: fore femur, posterior view; 350: fore tibia, anterior view; 351: hind tibia, posterior view; 352: abdomen, lateral view; 353: abdomen, ventral view, without genital segments, position of sternite 4 indicated; 354: sternites 4 and 5. (s – sternite, t – tergite.)

Abdomen: Subshining black, without metallic reflections, genital segments brownish. In ♂ constricted after segment 1+2, broadest at tip of segment 5. Without setae except on margins and sides of tergite 5; syntergite 1+2 bare along sides. ♂ sternites 2-4 divided longitudinally (Fig. 353). ♂ sternites 4 and 5 and genitalia in Figs 354-356, the surstylus with a very characteristic bifurcate process near tip of inner surface; genitalia also figured by Zuska (1965: figs 10, 13, 16) and Ozerov (1999e: fig. 347.2).

Wing-length: 4.4-4.9 mm (♂), 4.8 mm (♀).

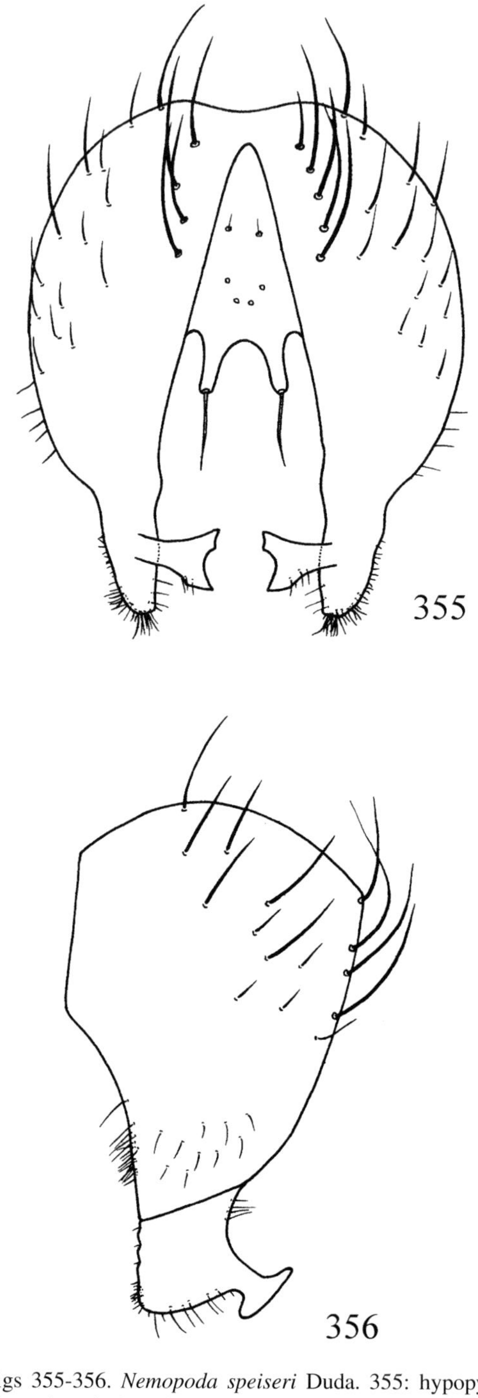

Figs 355-356. *Nemopoda speiseri* Duda. 355: hypopygium, dorsal view; 356: hypopygium, lateral view.

Larva. – The larva has been described by Ozerov (1987a: 110, figs 8-12) and Meier (1996: 31-32, figs 70-76).

Distribution. – Scarce. Denmark: absent. Norway: absent. Sweden: SM (Ringdahl, 1934: 5) and SÖ (UQC). Finland: Ta (MNM; Hedström, 1964: 81) and Kb (MNM).

Outside Fennoscandia, known only from Germany, Poland and the Czech and Slovak Republics (Zuska & Pont, 1984: 157), Latvia (Karps, 1986: 6), Mongolia (Soós, 1972: 360) and the Far East province of Russia (Ozerov, 1983: 635; 1989: 840; 1999e: 566).

Males were collected in June and July, females in August.

Biology. – Ringdahl (1934: 5) collected males on flowers of *Anthriscus*, but adults have also been collected on dog carrion (Barták et al., 1997: 226). Ozerov (1989: 840) found larvae in the carcasses of small mammals.

Genus *Meroplius* Rondani, 1874

Meroplius Rondani, 1874: 175. Type-species: *Nemopoda stercoraria* Robineau-Desvoidy, 1830 [= *Sepsis minuta* Wiedemann], by original designation. Gender: masculine.

Xenosepsis Malloch, 1925: 315. Type-species: *Xenosepsis sydneyensis* Malloch, 1925, by original designation. Gender: feminine.

Pseudomeroplius Duda, 1926a: 25 (as subgenus of *Meroplius* Rondani). Type-species: *Pseudomeroplius acrostichalis* Duda, 1926 [= *Xenosepsis sydneyensis* Malloch, 1925], by monotypy. Gender: masculine.

Adult *Meroplius* can be distinguished from the other genera with inner vertical, outer vertical and orbital setae present (Fig. 17) by the linear gena and the absence of a basal scutellar seta.

Adult. – Head roundish, facial carina small, face receding, gena linear. Postocellar seta present (*minutus*, Fig. 17) or absent (*fukuharai*); orbital, inner vertical and outer vertical present. Scutum and scutellum matt. Postpronotal present. 1 dorsocentral. 0 intraalar, 1 supraalar, 1 strong postalar. Proepisternal hairs present, proepimeral area bare or with a few tiny hairs. Anepisternal present. Katepisternal absent. Anepisternum and meron shining; anepimeron thinly grey pruinose in posterior half, matt; katepisternal pruinosity limited to a patch in upper posterior corner. Prosternum setulose. 1 scutellar seta, apical. Fore femur without posterodorsal setae. Tibiae without anterodorsal or posterodorsal setae. ♂ fore femur and tibia modified (e.g. Figs 359-360). ♂ hind tibia with the osmeterium on anterodorsal surface, large, field-like, occupying most of basal half (*minutus*), or narrow and occupying basal third (*fukuharai*). Wing without a dark spot. Abdominal setae absent except on tergites 4 and 5.

♂ abdomen. Tergites 1+2 to 5 normal; tergite 6 absent; tergite 7+8 present (Figs 361, 367-368). Spiracles 1-5 in the intersegmental membrane; spiracles 7 and 8 in tergite 7+8. Sternites 1-3 complete; sternite 4 modified, and sternite 5 reduced and mainly concealed beneath sternite 4 (Figs 361, 368); sternite 7+8 present as a narrow band (*minutus*) or as a rather more enlarged plate (*fukuharai*). Sternite 1 with tiny hairs at sides. Cercal plate with 2 weak lobes and several long or weak setae (Figs 363, 370). Surstylus simple, directed backwards; epandrium with (*minutus*, Fig. 370) or without (*fukuharai*, Fig. 363) an internal spinose process inside near base of surstylus. Sperm pump large, 'dumb-bell' shaped.

♀ abdomen. Sternites 2-5 normal in shape, plate-like. Spiracles 1-5 in the intersegmental membrane, spiracles 6 and 7 in the tergites. Postabdominal segments 6 and 7 with the tergites and sternites separated on sides of the segments, the sternites (like the tergites) thus very broad and short, the sternites with several very strongly sclerotised areas in *minutus*; each tergite and sternite with an anterior extension (*fukuharai*), or only the tergites with an anterior extension, and that on tergite 6 very short (*minutus*). Segment 8, when not invaginated, long, extended posteriorly and ventrally, with a ventral element (sternite 8) on each side that remains separated at tip and a dorsal element (tergite 8) that forms the usual pair of ring-like bars that do not quite touch apically. Cerci small, round; behind them with a narrow band with 2 setae that extends laterad and fuses with them on their outer edges (?epiproct). Hypoproct present, bare. 2 spermathecae, one larger than the other.

Egg. – Description based on *fukuharai*. Eggs with a single respiratory filament dorsal to the hatching line. Micropyle fused to dorsal side of filament base (as in Fig. 29). Filament with ribbons along rows of small pores. Ribbons arranged in an irregular polygonal pattern (as in Fig. 31). Polygons very long in middle and at end of filament, so that their parallel sides appear to form long, parallel longitudinal bands. In cross section, filament with a single, apparently continuous central canal and regularly arranged unbranched pillars that support the exochorion of the filament (as in Fig. 24). Pillars thickened towards exochorion of the filament and narrow towards the lumen (as in Fig. 24).

Larva. – Pseudocephalon bilobed, cephalic lobes largely covered by combs; facial mask consisting of large combs.

Anal division short (Figs 51, 357), to slightly elongated (Figs 358, 372); lateral tubercles present (Fig. 358); stigmatophores with a dorsal tubercle; ventral tubercles present, simple; subanal tubercles small; anal pad very small, oval, without furrows; anal tuft present; preanal protuberance absent; anal division between anterior margin of anal pad and origin of stigmatophores either slightly wider than long or about as wide as long (Figs 51, 372); either with two bands of spines on lateral aspect of anal division or with numerous spines not arranged in bands; dorsal surface with three double rows of spines (Figs 357-358), integument between rows densely hirsute (Fig. 358) or bare (Fig. 357).

Cephaloskeleton: labial and dental sclerites absent (Fig. 48); bridge of intermediate sclerite approximately at half length of sclerite (as in Fig. 138), anteriorly with a flat plate; epistomal sclerite longer than wide, with posterior bars; parastomal bars long, only slightly shorter than intermediate sclerite (Fig. 48); basal sclerite 2-2.5 times as long as high, dorsal cornua shorter and 3.5-4 times narrower than ventral cornua.

Anterior spiracles tree-like, 2-2.5 times as long as wide at base, with 6-11 short and large lobes; lobes 2-3 times as long as wide (as in Fig. 41).

Posterior spiracles with almost equal, narrow, weakly arched spiracular openings, situated on a trifoliate plate (Fig. 50).

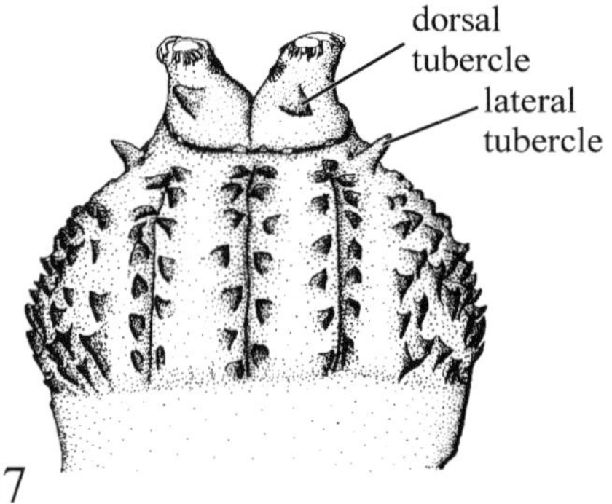

357

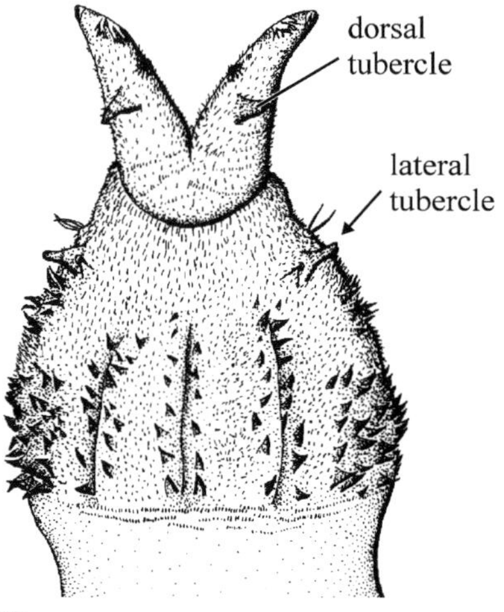

358

Figs 357-358. Larval anal division, dorsal. 357: *Meroplius fukuharai* (Iwasa); 358: *Meroplius minutus* (Wiedemann).

Discussion. – Plesiomorphies (i.e. ground-plan characters of the Sepsinae) include the following: a full complement of head setae (orbital, inner vertical, outer vertical, postocellar, but postocellar absent in *fukuharai*); supraalar seta present; postalar seta present; anepisternal seta present; scutellum reduced in size; wing with basal radial and medial cells distinct; abdominal spiracles 1-5 in the intersegmental membrane; ♂ sternite 7+8 present as a band, narrow or broad; ♀ with postabdominal tergites 6 and 7 each with an anterior prolongation. *Meroplius* is apomorphous relative to the ground-plan of the Sepsinae with regard to the following characters: ♂ fore leg with elaborate and characteristic ornamentation, ♂ hind tibia with an osmeterium, ♂ sperm-pump shaped like a 'dumb-bell.'

No autapomorphy can be given for *Meroplius*. *Meroplius*, like *Nemopoda*, is an easily recognised and apparently monophyletic genus of rather homogeneous species, but it is only weakly founded on adult characters. It is anomalous that the characters of the immature stages support a more derived position, close to *Sepsis*, whereas most of the characters of the adults are plesiomorphous and are ground-plan characters of the Sepsinae. The few apomorphies suggest a closer relationship with *Zuskamira* and *Themira* than with *Sepsis*.

The species *fukuharai* (and a few others) have been placed in a separate genus, *Xenosepsis*, on the basis of the absence of the postocellar setae. However, this character-state and the bare alula (devoid of microtrichiae) are both apomorphous, and recognition of *Xenosepsis* as a separate genus would leave a paraphyletic *Meroplius*. Agreements between *minutus* and *fukuharai* in the form of the ♂ fore leg ornamentation, in the shape and structure of the ♂ epandrium and surstyli, and in the re-acquisition of an epiproct in the ♀ ovipositor support the retention of *Xenosepsis* within *Meroplius*.

One Holarctic species, one Palaearctic, and one East Palaearctic; a number of Afrotropical and Oriental species. The larvae of all three Palaearctic species are known, and have been reared from various kinds of vertebrate carrion (Ozerov, 1991a).

Key to European species of *Meroplius* Rondani

Adults

1 Postocellar seta absent. Alula devoid of microtrichiae except along posterior edge, hyaline. Scutum with a band of brownish-grey dust in front of scutellum and also covering scutellum. ♂: hind tibia with the osmeterium small, confined to basal third of tibia; fore tibia (Fig. 360) more even on ventral surface, without a double protuberance. ♀: fore femur without anteroventral seta; abdominal tergites 4 and 5 without marginal setae .... 26. *fukuharai* (Iwasa)

– Postocellar seta present (Fig. 17). Alula covered with microtrichiae, as on rest of wing. Scutum without a brownish-grey dusted band in front of scutellum, scutum and scutellum uniformly uncolorous matt. ♂: hind tibia (Fig.

366) with the osmeterium large, occupying most of basal half of tibia; fore tibia (Fig. 365) with a double protuberance at middle of ventral surface. ♀: fore femur with an anteroventral seta at apical quarter; abdominal tergites 4 and 5 with lateral marginal setae.......... ....27. *minutus* (Wiedemann)

Larvae

1 With three long double-rows of large spines on dorsal side of anal division (Fig. 357); spines large, with blunt tips; hairs absent; with large warts around bases of ventral tubercles on ventral side of anal division (Fig. 51) ....26. *fukuharai* (Iwasa)

– With three short double-rows of small spines on dorsal side of anal division, rows not reaching lateral tubercles (Fig. 358); spines small, with sharp tips; anal division very hairy, especially around bases of ventral tubercles on ventral side of anal division (Fig. 372)...... ....27. *minutus* (Wiedemann)

26. *Meroplius fukuharai* (Iwasa, 1984)

(Figs 48, 50-51, 357, 359-363)

Xenosepsis fukuharai Iwasa, 1984b: 300, figs 5-8. Holotype ♂, JAPAN: Hokkaido Island, Mori, Oshima, in NSMT.

Diagnosis. – Adult recognised by the absence of postocellar setae, and by the structure of the ♂ fore legs (Figs 359-360).

Anal division on dorsal side with three long double-rows of spines which reach to the level of the lateral tubercles (Fig. 357); spines large, with blunt tips; anal division with large warts around the origin of the ventral tubercles; otherwise without hairs (Fig. 51).

Adult. – Colour: Head mostly dark, only face and usually parafacial reddish-yellow. Antenna mostly brown, translucent reddish around articulation of pedicel and flagellomere and especially on inner surface. Thorax, scutellum and abdomen entirely dark. Fore coxa and fore trochanter yellow; fore femur yellow but darkened above; fore tibia yellow; fore tarsus with tarsomere 1 and base of tarsomere 2 yellow, otherwise dark; mid and hind legs with the coxae and trochanters yellow, femora dark with the bases broadly yellow, knees yellow, tibiae dark but partly yellow apically, especially mid tibia, mid tarsi coloured as fore tarsi, hind tarsi mostly dark. Wing clear. Calypter and fringe-hairs creamy. Haltere yellow.

Head: Orbital, inner vertical and outer vertical setae strong, postocellars absent. Subvibrissal setae numerous but short.

Thorax: Scutum, with postpronotal lobes and scutellum, matt, with a thin band of brownish-grey dust across scutum in front of scutellum and behind dorsocentral seta, and also covering scutellum. Setae: 1 dorsocentral, 1 postpronotal, 2 notopleural, 0 intraalar, 1 supraalar, 1 postalar; acrostichal setulae minute. Pleura shining, except for a patch of white pruinosity in upper posterior corner of katepisternum; anepimeron matt. Setae reduced: anepisternal present, proepisternal represented by a few hairs, no katepisternal. Scutellum short, twice as wide as long, with a pair of apical setae and minute basal setulae.

Legs: ♂ fore legs in Figs 359-360. ♀ fore legs simple, femur without anteroventral seta. Mid and hind femora and tibiae without setae except for a ventral seta on mid tibia.

Wings: Veins bare even on basal part of stem-vein on ventral surface. Wing-surface covered with microtrichiae except near extreme base where, for example, basal medial cell is devoid of microtrichiae; alula without microtrichiae except on posterior edge.

Abdomen: Subshining black, with some violet reflections. Without setae except for lateral marginals on tergites 4 and 5 in ♂; syntergite 1+2 with setulae at sides. ♂ sternites with weak setulae (Fig. 361); genitalia in Figs 362-363, also illustrated by Ozerov (1999e: fig. 347.6).

Wing-length: 2.7-2.8 mm (♂), 2.9-3.0 mm (♀).

Egg. – The egg has been described by Meier (1995b: 427-431).

Larva. – The larva has been described by Ozerov (1991a: 53, figs 1v, 2a, 2g) and Meier (1996: 123-126, figs 454-462).

Distribution. – Not in Fennoscandia.

Widespread but of sporadic occurrence: Great Britain (Ozerov, 1999e: 568), France (Gallis, 1994: 146), Germany (Meier, 1999: 190), Czech and Slovak Republics (Zuska & Barták, 1997: 75), Russia, from the Moscow district and North Ossetia to the Far East province (Ozerov, 1989: 841; 1991a: 50; 1999e: 568), South Korea (Iwasa et al.,

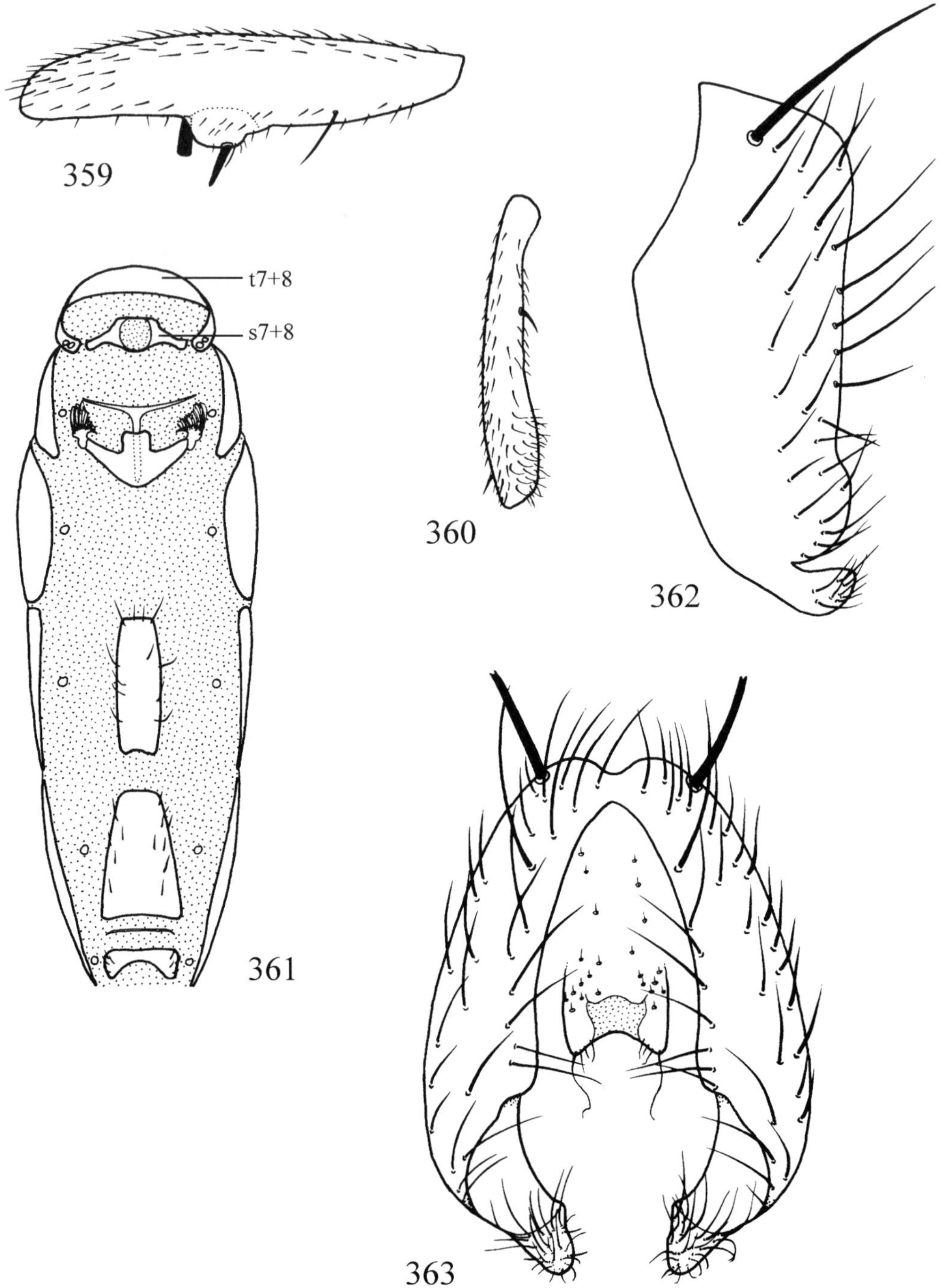

Figs 359-363. *Meroplius fukuharai* (Iwasa), ♂. 359: fore femur, anterior view; 360: fore tibia, anterior view; 361: abdomen, ventral view, hypopygium removed; 362: hypopygium, lateral view; 363: hypopygium, dorsal view. (s – sternite, t – tergite.)

1994: 205; Suh & Kwon, 1994: 62), North-east China (Manchuria; Iwasa, 1995: 791) and Japan (Hokkaido, Honshu, Kyushu) (Iwasa, 1984b: 300; 1995: 791).

Records are from May to September.

Biology. – When describing *fukuharai*, Iwasa (1984b: 302) reported: 'The adult flies are mainly found on the manure heap of pig dung.' Ozerov (1989: 841) found adults on vegetation at a refuse pit and collected larvae and adults from vertebrate carcasses (see also Ozerov, 1991a: 54), an observation that was confirmed by Gallis (1994: 146) who found the adults on a dead calf. In Germany the fly was collected and reared on chicken dung, but the development times were somewhat longer (17 days at 24°C) than the 12-14 days reported by Ozerov (1991a: 54). Under laboratory conditions, adults only reluctantly lay eggs on cow dung, the larvae rarely completing development. *M. fukuharai* is one of the sepsids with precopulatory guarding.

27. *Meroplius minutus* (Wiedemann, 1830)

(Figs 17, 358, 364-372)

Sepsis minuta Wiedemann, 1830: 468. Holotype ♂, USA: New York, in NMW [see below].

Sepsis lutaria Fallén, 1820b: 22. Unavailable name; proposed in synonymy with *cylindrica* and not made available subsequently [see below].

Nemopoda stercoraria Robineau-Desvoidy, 1830: 745. Syntypes ♂♀, locality not stated [FRANCE: probably Saint-Sauveur], not in MNHNP and presumed destroyed.

Nemopoda nigrilatera Macquart, 1835: 481. Holotype/syntypes ♂, FRANCE, not in MHNL and presumed destroyed.

Sepsis rufipes Meigen, 1838: 349; Meigen plate 281 fig. 8 [Morge, 1976c]. Lectotype ♂, 'in hiesiger Gegend' [GERMANY, Stolberg], in MNHNP [see designation below].

Nemopoda varipes Walker, 1871: 345. [Junior primary homonym of *Nemopoda varipes* Meigen, 1838.] Holotype ♀, EGYPT: Cairo, not in BMNH and presumed destroyed.

Nemopoda polita Duda, 1926a: 96, 98. Unavailable name; citation of a Meigen MS name, in synonymy with *Meroplius stercorarius* (Robineau-Desvoidy).

Diagnosis. – Adult recognised by the presence of postocellar setae, and by the structure of the ♂ fore legs (Figs 364-365).

Anal division on dorsal side with three short double-rows of spines not reaching the level of the lateral tubercles (Fig. 358); spines small, with distinct tips; anal division without warts but very hairy around the origin of the ventral tubercles (Fig. 372).

Adult. – Colour: Head mostly dark, only lunula and vibrissal area reddish. Antenna mostly brown, translucent reddish around articulation of pedicel and flagellomere. Thorax, scutellum and abdomen entirely dark. Fore coxa and fore trochanter yellow; fore femur yellow but dark above in ♂, entirely dark in ♀; fore tibia yellow in ♂, reddish-brown in ♀; fore tarsus mostly dark, basal tarsomeres often paler; mid and hind legs with the coxae yellow (♂) or mostly dark (♀), trochanters yellow, femora dark with the bases broadly yellow, tibiae dark but partly yellow apically in ♂, tarsi mostly dark but basal tarsomeres often pale. Wing clear. Calypter and fringe-hairs yellow. Haltere yellow.

Head: Frons in Fig. 17. A full complement of strong setae. Subvibrissal setae numerous but short.

Thorax: Scutum, with postpronotal lobes and scutellum, matt. Setae: 1 dorsocentral, 1 postpronotal, 2 notopleural, 0 intraalar, 1 supraalar, 1 postalar; acrostichal setulae minute. Pleura shining, except for a patch of white pruinosity in upper posterior corner of katepisternum; anepimeron matt. Setae reduced: anepisternal present, proepisternal represented by a few hairs, no katepisternal. Scutellum short, twice as wide as long, with a pair of apical setae and minute basal setulae.

Legs: ♂ fore legs in Figs 364-365. ♀ fore legs simple, femur with a short anteroventral seta at apical quarter. Mid and hind femora and tibiae without setae except for a ventral seta on mid tibia. ♂ hind tibia with osmeterium in Fig. 366.

Wings: Veins bare except for several setulae on basal part of stem-vein on ventral surface. Wing-surface covered with microtrichiae except near extreme base where, for example, basal medial cell is devoid of microtrichiae.

Abdomen: Subshining black, with some violet reflections. Without setae except for lateral marginals on tergites 4 and 5, stronger in ♀ than in

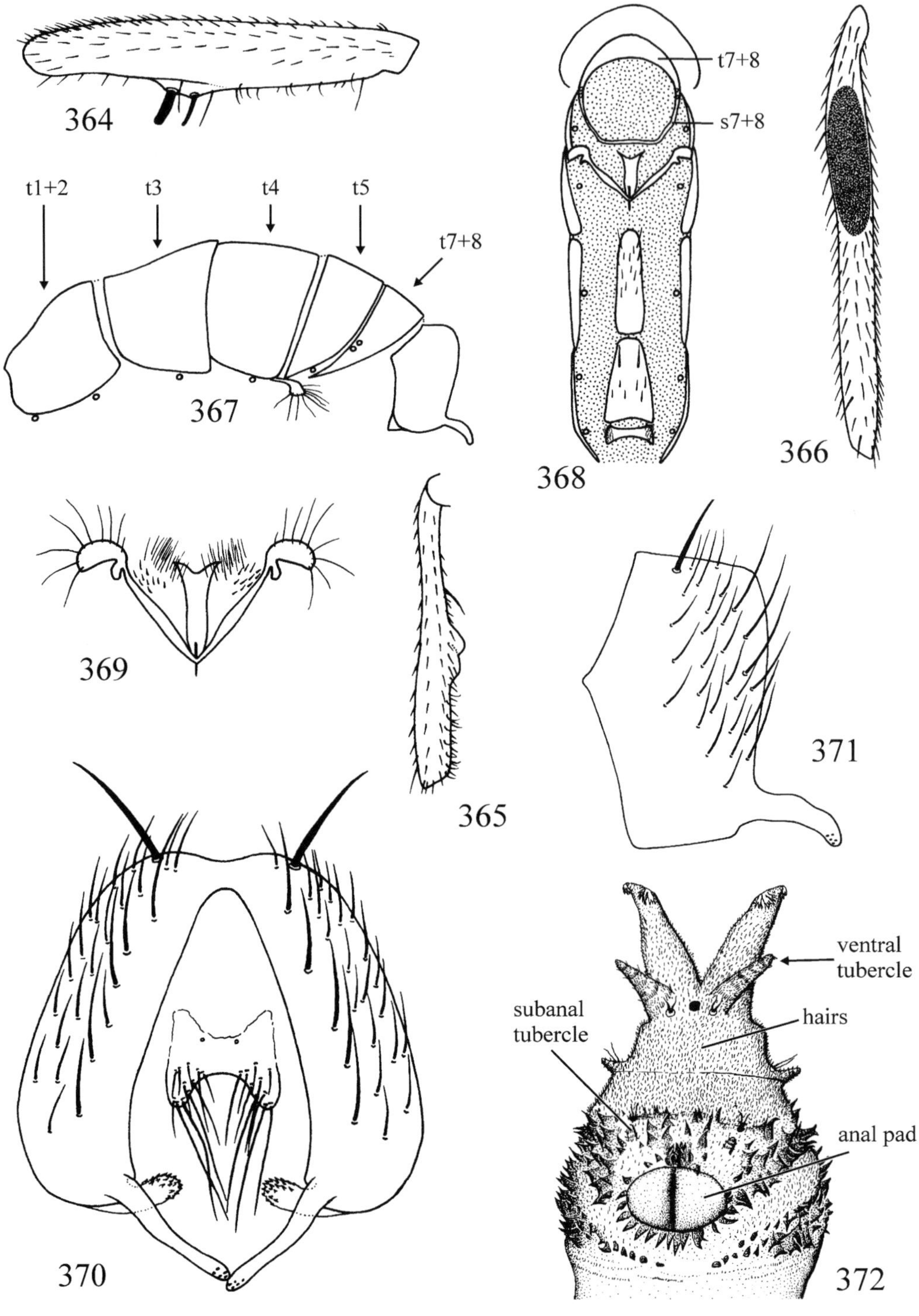

Figs 364-372. *Meroplius minutus* (Wiedemann), ♂. 364: fore femur, anterior view; 365: fore tibia, anterior view; 366: hind tibia, dorsal view; 367: abdomen, lateral view; 368: abdomen, ventral view, hypopygium removed; 369: sternites 4 and 5; 370: hypopygium, dorsal view; 371: hypopygium, lateral view. 372: larval anal division, ventral. (s – sternite, t – tergite.)

♂; syntergite 1+2 with setulae at sides. ♂ sternites with weak setulae (Fig. 368); sternites 4 and 5 and genitalia in Figs 369-371; genitalia also illustrated by Ozerov (1999e: fig. 347.3).

Wing-length: 2.7-3.1 mm (♂), 3.0-3.2 mm (♀).

Larva. – The larva has been described by Hennig (1949: 18, figs 22, 29, 34, 44, 52, 56, 64, 65), Ozerov (1991a: 52, figs 1d, 2v, 2e), and was included in a key to sepsid larvae by Mangan (1977: 341). It has recently been fully described by Meier (1996: 22-25, figs 37-45),

Distribution. – Uncommon, with few captures in modern times. Denmark: a few old records. Norway: a few old records (Siebke, 1877: 153). Sweden: SK to DR (e.g. Zetterstedt, 1847: 2298; Ringdahl, 1948: 2). Finland: northwards to Tb and Kb (e.g. Frey, 1908: 585). Longitude 63°N seems to mark its northern limit, but it is mentioned from Arkhangelsk at almost 65°N by Ozerov (1999e: 568).

Widespread but uncommon throughout Europe, North Africa (Egypt), Caucasus Mts, eastwards to North China and the Far East province of the Russia (Zuska & Pont, 1984: 156; Ozerov, 1989: 841; 1999e: 568), South Korea (Iwasa et al., 1994: 204; Suh & Kwon, 1994: 62), Japan (Iwasa, 1984b: 299; 1995: 791). It is declining through most of its range, and was declared extinct in Belgium and the Netherlands by Van der Goot (1987b: 82), though Debry (1978: 135) had recently recorded adults on pig slurry in Belgium. Nearctic: British Columbia to Quebec, south to Washington, Utah and Florida (Steyskal, 1965: 682). Oriental: Nepal (Iwasa & Thapa, 1994: 13), and Taiwan (Iwasa, 1987: 2).

Records are from May to September, with a more restricted flight period towards the north of its range.

Biology. – *M. minutus* is a generalist with a preference for exceptionally filthy habitats. The adults are apparently particularly attracted to human excrement (Melander & Spuler, 1917: 35; Howard in Hennig, 1949: 15; Lobanov, 1962a: 37; Iwasa, 1984b: 299) and favour the more liquid variety ('ich fand sie massenhaft an Abortjauchepfützen': Duda, 1926a: 19; Van der Goot, 1985b: 179-180; 1987b: 82). Adults have also been found around faecal masses in cattle pens, on rotting vegetation (Mangan, 1977: 339), compost (RM, pers. obs.), on small mammal and fish carrion (Reed, 1958: 238; Mangan, 1977: 339), pig dung (Iwasa, 1984b: 299), rotting fungi (Ozerov, 1991a: 54), and near rabbit hutches and heaps of decaying cabbages (Minder, 1963: 388). Ozerov (1991a: 54) observed that large, old quantities of human faeces are the preferred substrate and that fresh faeces and smaller patches are avoided.

Adults are known to visit flowers. ACP observed a mass occurrence of several thousand individuals in southern England on 26 August 1985, on a wild parsnip (*Pastinaca sativa*). Adults carry the mite *Macrocheles subbadius* Berlese.

Larvae have been reported from human faeces (Howard in Hennig, 1949: 15; Iwasa, 1984b: 299), a rabbit cage (Hennig, 1949: 18-19; Lobanov, 1962a: 37), pig dung (Lobanov, 1962a: 37; Iwasa, 1984b: 299), and rodent carcasses (Ozerov, 1989: 841). The development time from egg to adult is 12-14 days. Under laboratory conditions the species can be reared on cow dung (RM, pers. obs.). The species overwinters as a puparium (Lobanov, 1962a: 37).

Van der Goot (1987b: 82) suggested that the decline of this species might be due to improved methods of sewage management and the liberal use of poisonous toilet-cleaning chemicals.

Type-material. – *Sepsis minuta*: The holotype ♂ is in NMW, where it was seen by Duda (1926a: 98) and also by ACP. It is labelled as from New York. It is in poor condition, with the abdomen and most legs missing. However, details of the fore leg and head leave no doubt as to its identity. A ♀ in NMW, also labelled as *minuta* by Wiedemann (Duda, l.c.), cannot be a syntype as Wiedemann only described the ♂ sex. It is customary to give Wiedemann (1830) priority over Robineau-Desvoidy (1830) (see Aubertin, 1932), which is why the well-known name *stercorarius* has recently been replaced by *minutus*.

Sepsis lutaria: Fallén published this name in synonymy with *cylindrica* Fabricius (now *Nemopoda nitidula* Fallén). However, specimens seen by ACP under this name in the Fallén collection in NRS, drawer 22 (4♂ 2♀, including 1♂ 1♀ labelled *lutaria* by Fallén) belong to *Meroplius minutus* except for 1♂ which is *Themira nigricornis*. ACP has also seen a ♀ labelled *lutaria* from Fallén's collection in ZML, in the Diptera Scandinaviae collection.

Sepsis rufipes was described from both sexes, 'mehrere Exemplare in hiesiger Gegend'. ACP has seen 4♂ 1♀ syntypes in MNHNP, Meigen collection no. 2242 (see also Becker, 1902: 230, and

Duda, 1926a: 98, who gave notes only on the ♂ and ♀ with Meigen's labels). The species was described as having a dark wing spot, yet also with the femur [presumably fore femur] in ♂ 'unten mit einem stacheligen Zahne in der Mitte' and in ♀ 'etwas dornig'. As already noted by Becker, the original description and the type-series are an amalgam of two different genera. 1♂, labelled '*Sepsis rufipes* ♂' by Meigen, is *Meroplius minutus*; the description, especially the fore leg but not the wing spot, fits *minutus* very well; ACP has labelled this ♂ and herewith designates it as lectotype. The other 3♂ and 1♀ have been labelled as paralectotypes: 2♂ without labels, and 1♀, labelled 'rufipes ♀' by Meigen with no head, are *Sepsis violacea*; 1♂, labelled 'rufipes' by Meigen is *Sepsis fulgens*.

Genus *Sepsis* Fallén, 1810

Sepsis Fallén, 1810: 17. Type-species: *Musca cynipsea* Linnaeus, 1758, by designation of Curtis, 1829: plate 245. Gender: feminine.

Threx Gistel, 1848: x. Unjustified replacement name for *Sepsis* Fallén. Type-species: *Musca cynipsea* Linnaeus, 1758, automatic. Gender: masculine.

Acrometopia Lioy, 1864: 1088. Type-species: *Sepsis cornuta* Meigen, 1826 [= *Musca cynipsea* Linnaeus, 1758], by monotypy. Gender: feminine.

Beggiatia Lioy, 1864: 1088. Type-species: *Sepsis barbipes* Meigen, 1826 [= *Musca cynipsea* Linnaeus, 1758], by monotypy. Gender: feminine.

Sepsidimorpha Frey, 1908: 578, 584. Type-species: *Sepsis loewi* Hendel, 1902 [= *Sepsis duplicata* Haliday, 1838], by monotypy. Gender: feminine.

Australosepsis Malloch, 1925: 314. Type-species: *Australosepsis fulvescens* Malloch, 1925 [= *Sepsis niveipennis* Becker, 1903], by original designation. Gender: feminine.

Saltelliseps Duda, 1926a: 25. Type-species: *Sepsis niveipennis* Becker, 1903, by designation of Hennig, 1949: 63. Gender: feminine.

Lasiosepsis Duda, 1926a: 28. Type-species: *Sepsis hirtipes* Becker, 1903 [= *Sepsis melanota* Bigot, 1866], by monotypy. Gender: feminine.

Note. – Fallén's genus *Sepsis* is based on the species *cynipsea* Linnaeus. However, the series of *cynipsea* in the Fallén collection (drawer 22) in NRS consists of the species *fulgens* Meigen (2♂ 2♀) and *flavimana* Meigen (1♂ 1♀). The nominal species *cynipsea* is retained as the type-species of this genus, as has always been done, and application should be made to the *International Commission on Zoological Nomenclature* to confirm this. See also the comments below on the type-series of *cynipsea*.

Adult *Sepsis* can be recognised in both sexes by the very weak postalar seta and, in Central and North Europe, by the conspicuous dark wing-spot. Males have ornate fore legs, and abdominal sternite 4 absent.

Adult. – Head oval to elongate-oval (Fig. 375), facial carina small but distinct, face slightly receding or vertical or sloping forwards, gena narrow (1/3 to 2/3 width of antennal flagellomere). Outer vertical present but short. Postocellar present. Orbital minute or absent. Scutum and scutellum matt to pruinose, rarely subshining. Postpronotal present. 1-2 dorsocentral setae. 0 intraalar. 1 strong supraalar. Postalar very short to minute, only conspicuous in some large *punctum* and *lateralis*. Proepisternals and proepimerals represented only by tiny setulae. Anepisternal present, strong. Katepisternal absent. Anepisternum shagreened, the other pleural plates varying in extent of gloss or whitish-grey pruinosity. Prosternum narrow, usually bare but sometimes with some short and inconspicuous setulae. 1 scutellar (apical) (Fig. 376). Fore femur without posterodorsal setae. Mid tibia with 0 anterodorsal, 0-4 posterodorsal, 0-1 anteroventral. Hind tibia with 0-1 anterodorsal (present in *cynipsea* group and some others), 0 posterodorsal (1 in large specimens). ♂ fore femur and tibia modified, only weakly in *duplicata*. ♂ hind tibia without an osmeterium. Wing with a dark spot at tip of vein R_{2+3} (Fig. 377), except in *lateralis*. Abdominal marginal setae on tergites 3 or 4-5 in ♂, except in *duplicata*; absent in ♀, except in *punctum* group, *barbata*, *lateralis*.

♂ abdomen. Constricted after syntergite 1+2, base of tergite 3 much narrower than tip of syntergite 1+2; other segments normal in shape. Tergite 6 absent; tergite 7+8 with 2 spiracles in lower corner. Spiracles 1 and 2 in the intersegmental membrane; spiracles 3, 4 and 5 at extreme ventral edge of tergites, 3 sometimes in the membrane (Figs 373-374). Sternite 1 band-shaped and setulose at sides; sternites 2 and 3 elongate, sternite 2 short in *duplicata*; sternite 4 absent, at most represented by 2 pores; sternite 5 small and V-shaped, the setae rather stout except in *duplicata*

(Figs 373-374); sternite 7+8 present as a very narrow weak band connecting tergite 7+8 across the ventral part of the abdomen. Cercal plate simple at tip, or weakly or distinctly bilobed. Surstylus usually simple, with an anterior basal lobe in *cynipsea* group (*cynipsea*, *neocynipsea*, *thoracica*, *barbata*). Sperm pump large, 'dumb-bell' in shape.

♀ abdomen (based on *punctum*, *duplicata*, Figs 5-7). Sternites 2-5 relatively narrow, longer than broad, and weakly bristled. Spiracles 1-5 in the intersegmental membrane, close to the tergites (*punctum*), unclear in *duplicata* where spiracles 4 and 5 may be in the tergites; spiracles 6 and 7 in the tergites. Postabdominal segments 6 and 7 with the tergites and sternites well developed, tergite 7 with a long anterior extension, tergite 6 with a short narrow one, sternite 6 divided into 2 plates in *duplicata*; segment 8, when not invaginated, long, extended posteriorly and ventrally, with a ventral element (sternite 8) on each side that remains separated at tip and a dorsal element (tergite 8) that forms the usual pair of ring-like bars that do not quite touch apically. Cerci small, rounded-triangular. Epiproct absent. Hypoproct present, bare. 2 spermathecae, one slightly larger than the other.

Egg. – Description based on *duplicata*, *flavimana*, *fulgens*, *orthocnemis*, *punctum*, *thoracica*, *violacea*; a single filament also present in *biflexuosa*, *cynipsea*, *lateralis*, *neocynipsea*, *niveipennis*.

The generic diagnosis is identical to that for *Meroplius*. Hinton (1981: plate 124, E-F) illustrates some eggs of '*Sepsis* spp.' showing respiratory filaments without the ribbons. We consider it likely that these eggs were misidentified. They seem to be of uncertain origin, as indicated by the 'spp.' on the figure legend.

In all the eggs of *Sepsis cynipsea* that have been studied, the pillars were arranged in longitudinal rows (Fig. 25); they were not staggered as in the other species of *Sepsis* (Fig. 24).

Larva. – Pseudocephalon bilobed (Fig. 401) or monolobed (Fig. 400), cephalic lobes with few combs; facial mask consisting of a moderately large to small number of ridges (Figs 400-401).

Anal division short (Figs 53, 397, 403) or elongated (Figs 395-396); in most species, hairs on ventral surface transformed into small warts (Figs 53, 395-396, 403); lateral tubercles present; stigmatophores with a dorsal tubercle; ventral tubercles present; with a small bare subanal tubercle; anal pad small to moderately large, often convex near anal opening, more or less strongly wing-shaped (Figs 53, 395-397, 403); anal tuft present; preanal protuberance absent; anal division between anterior margin of anal pad and origin of stigmatophores 1-2.5 times as long as wide (Figs 53, 395-396, 403); with two bands of spines on lateral aspect of anal division and three double rows of spines on dorsal surface (as in Fig. 44); integument between rows hirsute or bare.

Cephaloskeleton: labial and dental sclerites absent (as in Fig. 48); bridge of intermediate sclerite approximately at half length of sclerite (as in Fig. 138), anteriorly with a flat plate; epistomal sclerite longer than wide, with or without posterior bars; parastomal bars long, only slightly shorter than intermediate sclerite; basal sclerite 2-3.2 times as long as high, dorsal cornua shorter and 2.5-4.1 times narrower than ventral cornua.

Anterior spiracles tree-like, not narrowing towards apex, 2-4 times as long as wide at base, with 3-8 short and wide lobes; lobes 2-2.5 (seldom 3-3.5) times as long as wide (Fig. 41).

Posterior spiracles with three spiracular openings, one very long and extending on to median side of stigmatophore (Figs 49, 398-399, 402); openings either unequal in length: two of them very long, the third one short (Fig. 402: *flavimana*-group: *duplicata*, *biflexuosa*, *flavimana*), or spiracular openings weakly differentiated and one located on a small tubercle (Figs 49, 398-399: most species, e.g. *punctum*, *fulgens*, *cynipsea*).

Discussion. – Plesiomorphies (i.e. ground-plan characters of the Sepsinae) include the following: outer vertical seta present; postocellar seta present; supraalar seta present; anepisternal seta present; scutellum reduced in size; wing with basal radial and medial cells distinct (except in *niveipennis*); ♂ hind tibia without osmeterium; ♂ sternite 7+8 present as a narrow band; ♂ surstylus fused with epandrium; ♀ abdominal spiracles 1-5 in the membrane; ♀ with postabdominal tergite 7 with an anterior prolongation; *Sepsis* is apomorphous relative to the ground-plan of the Sepsinae with regard to the following characters: orbital seta very short, ♂ sperm-pump shaped like a 'dumb-bell'.

Autapomorphies for *Sepsis* are:

1. Postalar seta weak or absent
2. ♂ sternite 4 absent

3. ♂ abdominal spiracles 3-5 in the tergites, not in the intersegmental membrane

Sepsis is a well-defined monophyletic genus, and is also by far the largest genus in the family. It is well-founded on characters in the adult and immature stages. However, as in the other sepsid genera discussed here, it is not clear what is the sister-group of *Sepsis*, although the genus itself is clearly recognised by several autapomorphies. Among the external characters, the postalar seta is weak or absent, and is only developed in some large *punctum* and *lateralis*. It is also very weak in *Australosepsis* and *Lasiosepsis*, which we have synonymised with *Sepsis*, but is strongly developed in *Allosepsis* Ozerov.

A few species, including the European *duplicata*, have sometimes been placed in a separate genus, *Sepsidimorpha* Frey, on the basis of the absence of strong setae on the ♂ abdomen and the reduced secondary sexual ornamentation of the ♂ fore legs. These are indeed apomorphies in terms of the ground-plan of *Sepsis*. In the ♀ postabdomen, the small and divided sternite 6 of *duplicata* is also an apomorphy, whereas the short anterior extension on tergite 6 is a plesiomorphy. However, as is so often the case, recognising *Sepsidimorpha* as a separate genus would leave a paraphyletic *Sepsis* as a rest-group. So far as *duplicata* is concerned, the absence of the postalar seta, the absence of ♂ abdominal sternite 4, and spiracles 3-5 in the tergites in the ♂ abdomen are all synapomorphies of '*Sepsidimorpha*' + *Sepsis*, and in the absence of any convincing evidence of a sister-group relationship between '*Sepsidimorpha*' and *Sepsis* we see no basis for maintaining them as separate genera.

Numerous specimens of *Sepsis* have been seen in which the ovipositor displays a relatively long, eversible, membraneous, finger-like extension, which may be withdrawn within the abdomen or may be partly or completely everted at the apex of segment 8, ventrad of the hypoproct. This structure is briefly discussed and illustrated by Kiontke (1989: 97, fig. 18), and is similar to the 'secondary ovipositor' of the Milichiidae (I. Brake, pers. comm.). Further study is needed to elucidate the structure and function of this organ, and to establish its distribution within the Sepsidae.

Distribution in all zoogeographic regions, with some 35 Palaearctic species. There is a key to all the Russian species (21 species) in Ozerov (1999e).

Key to European species of *Sepsis* Fallén

Adults

1 Alula absolutely hyaline, without any microtrichiae. Meron entirely shining, without pruinosity. Katepisternum with a large glossy area on anterior lower part. Hind tibia with 1 anterodorsal 2

– Alula entirely covered with microtrichiae. Meron at least partly pruinose. Katepisternum usually entirely white pruinose (except *fissa*, *lateralis*, *thoracica*)..3

2(1) Wing-tip milky-white beyond the dark spot, around apical parts of veins R_{4+5} and M_{1+2}. Basal medial and basal radial cells distinct (as in Fig. 377). ♂: fore legs in Figs 404-406, the femur with a patch of dense erect black hairs on basal third of anterior to anteroventral surfaces; genitalia in Figs 407-408. ♀: tergites 4 and 5 with lateral marginal setae..28. *barbata* Becker

– Wing-tip clear beyond the dark spot. Basal medial and basal radial cells fused (as in Fig. 13). ♂: fore legs in Figs 455-456, the femur without any dense black hairs; genitalia in Figs 457-458. ♀: tergites 4 and 5 without setae..39. *niveipennis* Becker

3(1) Wing clear, without a dark spot at the tip of vein R_{2+3}. Katepisternum extensively glossy on anterior and lower parts. Meron entirely pruinose, only narrowly glossy on hind-margin. Hind tibia usually with 1 anteroventral. ♂: fore legs in Figs 436-438, the large tubercle placed close to middle of femur; mid femur with several submedian anterior setae; hind femur with 1-2 anterodorsals; hind tibia with 1 anterodorsal and 1 posterodorsal in addition to the anteroventral; genitalia in Figs 439-440. ♀: tergites 3 and 4 with lateral marginal setae ....35. *lateralis* Wiedemann

– Wing with a dark spot, at the tip of vein R_{2+3} 4

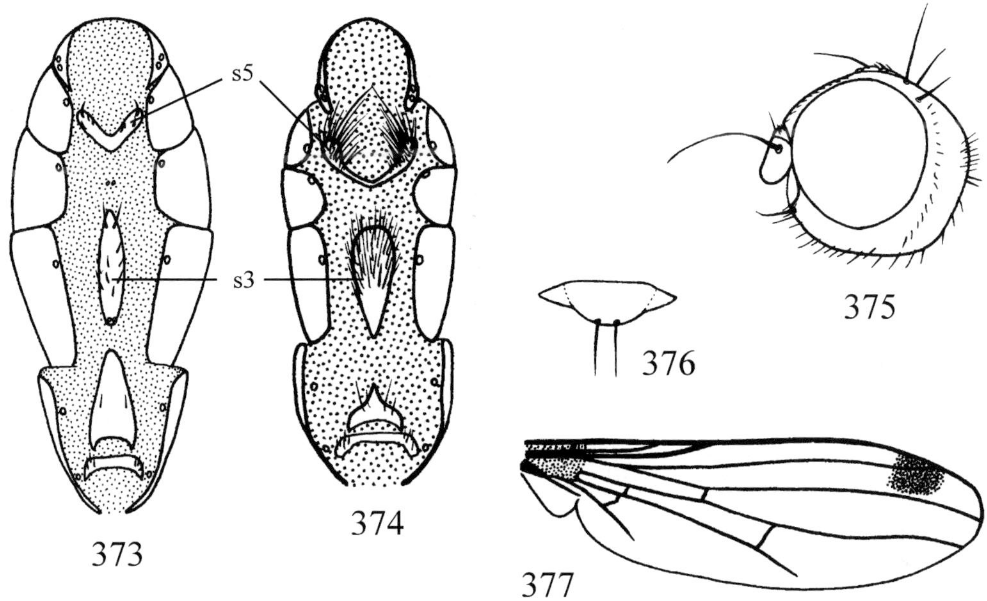

Figs 373-377. ♂ abdomen, ventral view, hypopygium removed. 373: *Sepsis cynipsea* (Linnaeus); 374: *Sepsis duplicata* Haliday. *Sepsis fulgens* Meigen. 375: head, lateral view; 376: scutellum, dorsal view; 377: wing. (s – sternite.)

4(3) Wing, in addition to the dark spot at tip of vein R_{2+3}, with a milky-white patch at the tip that includes apical parts of veins R_{4+5} and M_{1+2}. Gena relatively broad, equal to width of antennal flagellomere, and very few subvibrissal setae present. Katepisternum white pruinose, but narrowly devoid of pruinosity on extreme anterior and lower areas (in front of mid coxa and behind fore coxa). Meron pruinose anteriorly, broadly glossy behind (as in *cynipsea*). Hind tibia with 1 anteroventral, absent only in some ♀. ♂: fore legs in Figs 423-425; mid femur with 1-2 submedian anterior setae; hind tibia with 1 anterodorsal; genitalia in Figs 426-427. ♀: tergites without setae .... 32. *fissa* Becker

– Wing with only a dark spot at tip of vein R_{2+3}, without a milky-white patch at tip. Gena narrower, no more than 2/3 width of antennal flagellomere and often considerably less, and with numerous short subvibrissal setae 5

5(4) Meron with the anterior half glossy black, strongly contrasting with the white pruinose katepisternum and posterior meron (Fig. 380). ♂: mid femur with long anteroventral and posteroventral setae on basal 2/3, that are up to 3 times as long as femoral depth (Fig. 378); hind femur with short dense ventral to posteroventral setae on basal half, as long as femoral depth (Fig. 379); fore legs virtually unmodified (Figs 419-420), the femur with only a few (3-4) stronger but inconspicuous spinules; sternites 3 and 5 with dense fine hairs (Fig. 374); mid and hind tibiae without setae; abdominal setae entirely absent (as in Fig. 383); genitalia in Figs 421-422. ♀: mid tibia often without an anteroventral seta....... ....31. *duplicata* Haliday

– Meron densely pruinose on anterior half, like the katepisternum, the posterior half usually pruinose (Fig. 387), sometimes contrasting glossy black (Figs 381-382). ♂: mid and hind femora simple, without long setae on any of the ventral surfaces; fore legs conspicuously modified, the femora with groups of stout spines or tubercles; sternites 3 and 5 with the normal short sparse setulae (Fig. 373); mid tibia always with an anteroventral seta, usually also with a posterodorsal, and hind tibia

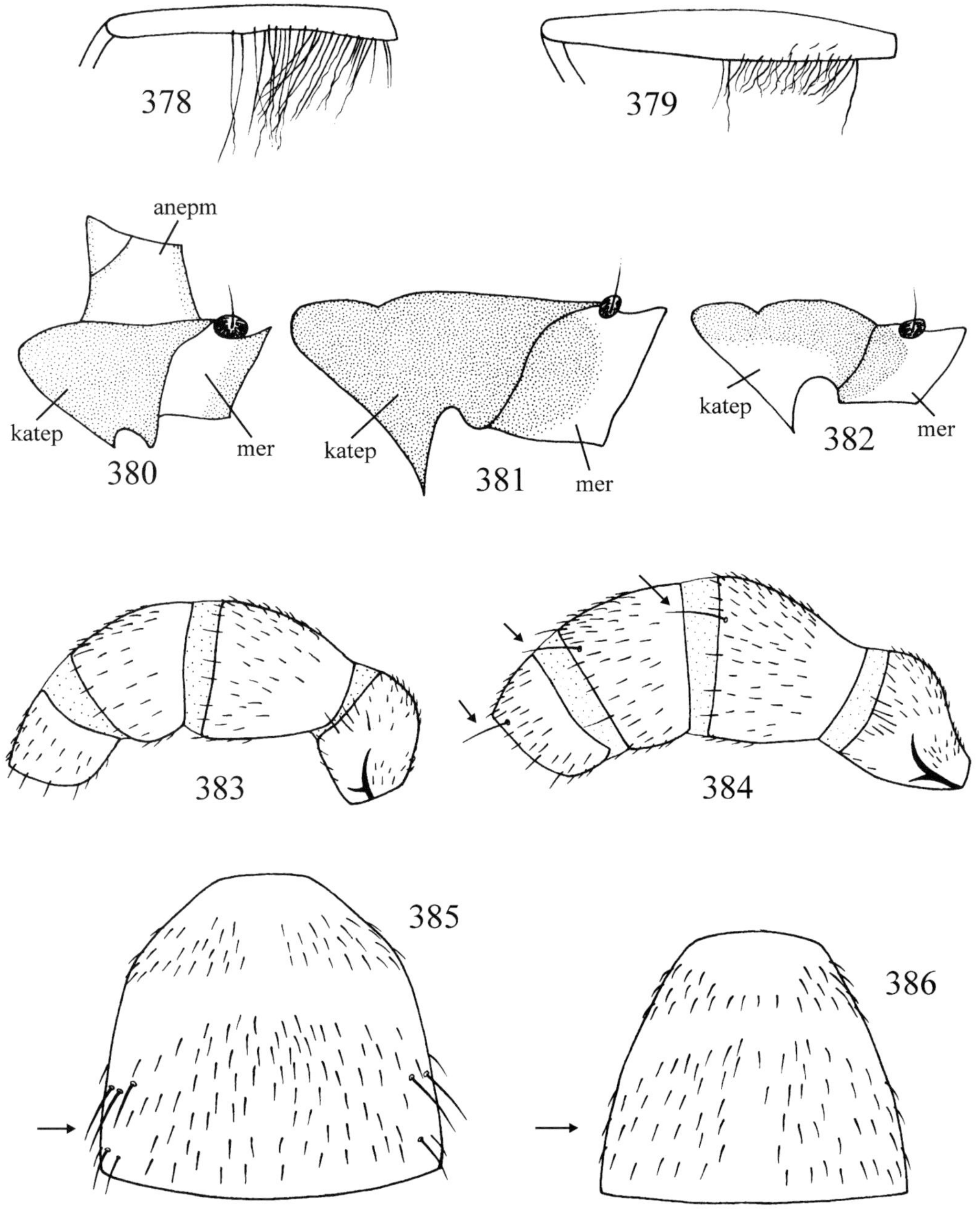

Figs 378-386. *Sepsis duplicata* Haliday, ♂. 378: mid femur, posterior view; 379: hind femur, posterior view. Katepisternum and meron. 380: *Sepsis duplicata* Haliday; 381: *Sepsis cynipsea* (Linnaeus); 382: *Sepsis thoracica* (Robineau-Desvoidy). ♀ abdomen, lateral view. 383: *Sepsis cynipsea* (Linnaeus); 384: *Sepsis violacea* Meigen. Abdominal syntergite 1+2, dorsal view: 385: *Sepsis violacea* Meigen; 386: *Sepsis fulgens* Meigen. (anepm – anepimeron, katep – katepisternum, mer – meron.)

sometimes with an anterodorsal; abdominal setae usually present on tergites 3-5 (as in Fig. 384), even if weak on one or two of these tergites. ♀: mid tibia with an anteroventral seta 6

6(5) Meron pruinose only on anterior part, from just in front of the spiracle, with quite a distinct division between the anterior pruinose and posterior glossy areas (Figs 381-382). Hind tibia with a submedian anterodorsal seta. Basal medial cell generally partly devoid of microtrichiae 7

– Meron pruinose like the katepisternum, at most extreme posterior edge narrowly shining and merging into the pruinose area (Fig. 387). Hind tibia without an anterodorsal seta in ♀ (except in *punctum*) 9

7(6) Katepisternum extensively glossy black on lower anterior part, the pruinosity confined to upper and posterior part (Fig. 382). Mid tibia with 2-4 posterodorsal setae. ♂: hind tibia with 1 posterodorsal seta; legs very variable in colour; fore legs in Figs 471-472; genitalia in Figs 473-474 .... 43. *thoracica* (Robineau-Desvoidy)

– Katepisternum entirely white pruinose (Fig. 381). Mid tibia rarely with more than 2 posterodorsal setae. ♂: hind tibia without a posterodorsal seta 8

8(7) Mid femur without a submedian anterior seta. ♂: fore tibia with a deep ventral notch at basal third (Fig. 416); fore femur usually with 2 stout setae preceding the submedian group (Fig. 415); genitalia in Figs 417-418, the surstyli rather narrower .... 30. *cynipsea* (Linnaeus)

– Mid femur with a distinct submedian anterior seta, which is shorter in smaller specimens. ♂: fore tibia with a conspicuous ventral swelling at basal quarter, that is covered with short spinules (Fig. 446); fore femur with only 1 stout seta preceding the submedian group (Fig. 445); genitalia in Figs 447-448, surstyli rather broader .... 37. *neocynipsea* Melander & Spuler

9(6) Basal medial cell partly or wholly devoid of microtrichiae. ♂: fore femur on ventral surface with 1 or 2 tubercles bearing spinules, situated between the submedian group of spinules and the femoral tip (e.g. Fig. 432). ♀: tergites 4 and 5 each with a pair of lateral marginal setae, rarely only twice as long as one of the tergal groundsetulae and usually much longer than this (Fig. 384); if these setae are weakly developed or absent, then sides of syntergite 1+2 without stronger setulae posteriorly (Fig. 386) and femora and tibiae black except for fore tibia which is brown (some *fulgens*), or anepimeron pruinose broadly on hind margin and along part of lower margin (Fig. 391) (*luteipes*) 10

– Basal medial cell usually entirely covered with microtrichiae. ♂: fore femur on ventral surface without any tubercles between the submedian group of spinules and the femoral tip (e.g. Fig. 428). ♀: tergites 4 and 5 without any marginal setae (as in Fig. 383), these present only in very exceptional cases; sides of syntergite 1+2 with some stronger setulae posteriorly (as in Fig. 385) and legs with at least some parts of femora and tibiae yellow 14

10(9) Syntergite 1+2 posteriorly at sides with setulae that are short and weak, as on the rest of tergites 1+2 and 3 (Fig. 386) (best seen when the abdomen is viewed dorsally against a white background). ♂: fore femur with the ventral tubercle longer, straight, and with 2 spinules at tip (Fig. 432); tergite 3 usually without marginal setae; mid and hind femora and tibiae virtually entirely dark brown; mid femur without a submedian anterior seta, and mid tibia with 0-1 posterodorsal seta (very rarely 2); fore tibia in Fig. 433; genitalia in Figs 434-435, surstylus rather truncate at tip. ♀: fore femur mainly dark, mid and hind femora almost wholly dark; fore tibia partly dark .... 34. *fulgens* Meigen

– Syntergite 1+2 posteriorly at sides with several setulae that are stronger than the adjacent ones and those in the same position on tergite 3 (Fig. 385) (best seen when the abdomen is viewed dorsally against a white background). ♂: fore femur with the ventral tubercle shorter, finger-shaped or straight, usually with 1 spinule at tip (e.g. Fig. 467); tergite 3 usually with a pair of lateral marginal setae; at least parts of mid

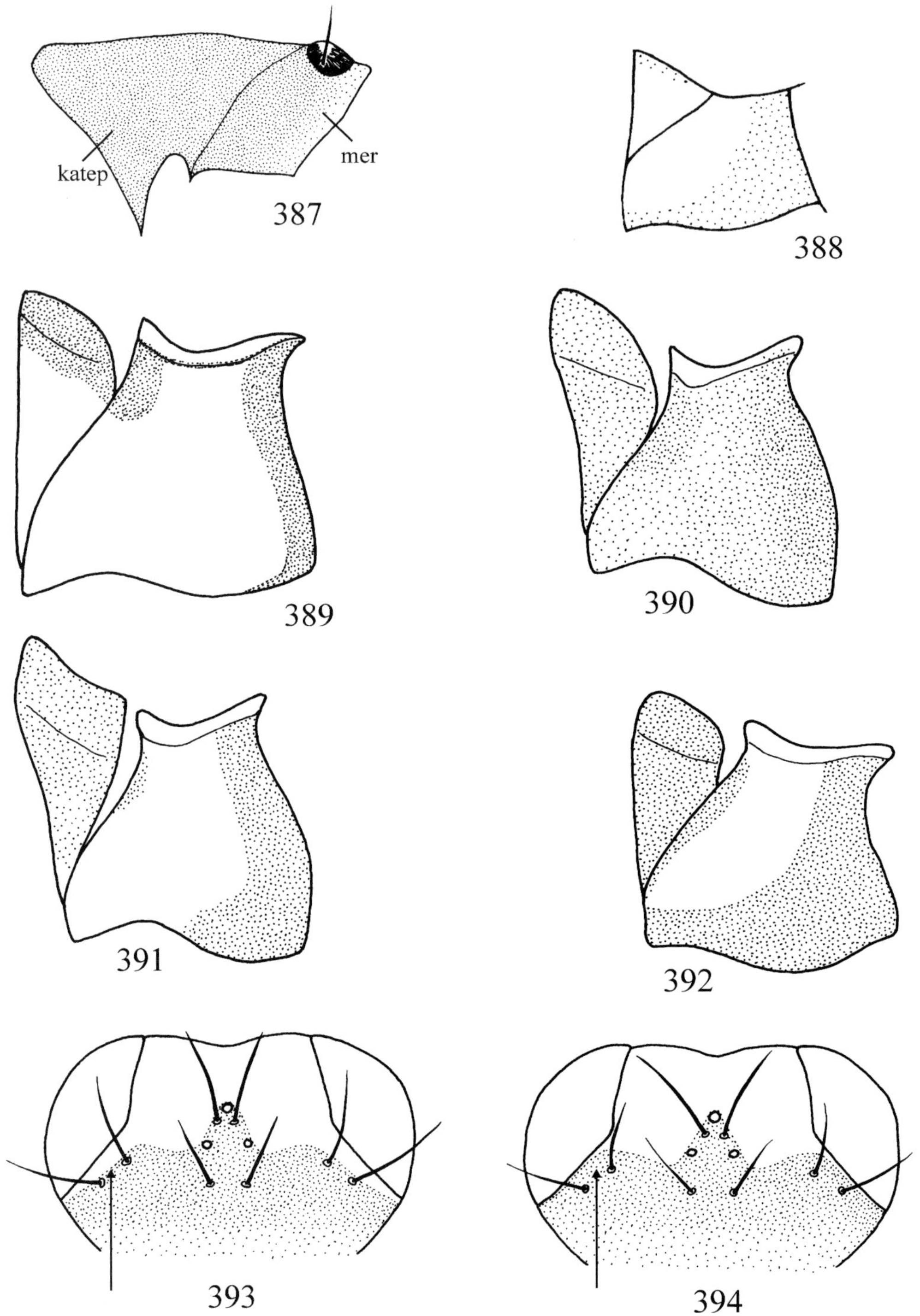

Figs 387-394. 387: *Sepsis fulgens* Meigen, katepisternum and meron. Anepimeron. 388: *Sepsis orthocnemis* Frey; 389: *Sepsis punctum* (Fabricius); 390: *Sepsis pseudomonostigma* Ursu; 391: *Sepsis luteipes* Melander & Spuler; 392: *Sepsis violacea* Meigen. Vertex of head. 393: *Sepsis orthocnemis* Frey; 394: *Sepsis punctum* Fabricius. (katep – katepisternum, mer – meron.)

and hind legs yellow, usually the femora; mid femur with 1-4 submedian anterior setae, and mid tibia with (1-) 2-5 posterodorsal setae. ♀: fore femur usually mainly yellow, mid and hind femora usually yellow on basal third or more, sometimes wholly yellow; fore tibia yellow 11

11(10) Anepimeron with the pruinosity more or less weak, irregular in intensity, but covering at least sparsely the entire anepimeron (Fig. 390). 2 dorsocentral setae. ♂: fore femur with the submedian group of spinules preceded by 1 stout seta (Fig. 463); surstyli subtriangular in form and, in dorsal view, partly crossed (Figs 465-466)........... ....41. *pseudomonostigma* Ursu

– Anepimeron with the pruinosity clearly visible but partial, restricted to one part, leaving most of anepimeron (Fig. 389) or just the central part (Figs 391-392) of anepimeron shining. ♂: surstyli not shaped as above and never crossed............. 12

12(11) Anepimeron pruinose along posterior margin and, more thinly so, on upper part of the anterodorsal triangular plate, otherwise glossy black (Fig. 389). Usually only 1 dorsocentral seta. ♂: fore femur with the submedian group of spinules preceded by 2 or rarely 3 stout setae (Fig. 467); femora and tibiae usually yellow in the larger forms; genitalia in Figs 469-470, the surstyli short, stouter, less pointed. ♀: tergite 3 without lateral marginal setae; all femora and the two posterior pairs of tibiae usually mainly or entirely yellow ....42. *punctum* (Fabricius)

– Anepimeron more extensively pruinose along lower and posterior margins, and the anterodorsal triangular plate entirely pruinose (Figs 391-392). 2 dorsocentral setae. ♂: fore femur with the submedian group of spinules preceded by only 1 stout seta (Figs 441, 475); femora and tibiae always at least partly darkened. ♀: tergite 3 with lateral marginal setae; legs darker; mid and hind femora usually darker on apical half or more.............................. 13

13(12) Anepimeron with the pruinosity extending far along lower margin, leaving only a small median area shining (Fig. 392). Anterior dorsocentral seta equal to or only slightly shorter than posterior one. ♂: fore legs in Figs. 475-476; genitalia in Figs. 477-478, surstyli pointed, in dorsal view their inner margins with many long setae. ....44. *violacea* Meigen

– Anepimeron with the pruinosity not extending halfway along lower margin, broadly shining medially (Fig. 391). Anterior dorsocentral seta always very short, at most half as long as posterior one. ♂: fore legs in Figs. 441-442; genitalia in Figs. 443-444, surstyli rounded in dorsal view, their inner margins without long setae but with short setulae....................... ....36. *luteipes* Melander & Spuler

14(9) Anepimeron rather broadly pruinose along lower and posterior parts (Fig. 388). ♂: tergite 3 usually with lateral marginal setae; fore femur in apical third with 2 ventrally-directed and usually diverging spinules (Fig. 459); fore tibia in Fig. 460; mid tibia with 0-1 posterodorsal seta; hind tibia without setae; femora dark brown, yellow at tips; fore tibia mainly yellow, mid and hind tibiae brown but yellow at tips; genitalia in Figs 461-462.......................... ....40. *orthocnemis* Frey

– Anepimeron with at most a weakly pruinose line at pleural sutures anteriorly and posteriorly. ♂: tergite 3 usually without lateral marginal setae; fore femur in apical third without 2 ventrally-directed spinules, at most with 2 apically directed hairs or setulae (Figs 409-410, 428, 449-450) 15

15(14) Proepisternal depression mostly pruinose. Inner and outer vertical setae well inside the pruinose area on occiput (as in Fig. 394). Meron with the narrow glossy hind-marginal streak very reduced, only present on upper part and there hardly if at all visible. Anepimeron with the anterodorsal triangular plate dusted, like the scutum. Hind tibia without an anteroventral seta. ♂: fore tibia with a very low ventral swelling in basal quarter that is covered with short setulae, the narrowest part of the tibia situated within basal half of tibia (Figs 451-452); fore femur in Figs 449-450; hind tibia without anterodorsal seta; femora and tibia mostly dark, yellow at tips, fore tibia yellow; genitalia in Figs 453-454 . . . . 38. *nigripes* Meigen

– Proepisternal depression almost completely shining, glossy. Inner and outer vertical setae on the edge of the pruinose area on occiput (as in Fig. 393) 16

16(15) Anepimeron with the anterodorsal triangular plate dusted, like the scutum. Meron with the narrow glossy hind-marginal streak virtually complete. Gena very narrow, the depth below lowest eye-margin at most 1/3 width of antennal flagellomere. Hind tibia without an anteroventral seta. ♂: fore tibia with a conspicuous ventral swelling at basal quarter that is covered with short spinules, the narrowest part of the tibia situated within basal half of tibia (Fig. 429); fore femur in Fig. 428; hind tibia without anterodorsal seta; femora and fore tibia variable in colour; mid and hind tibiae dark, partly yellow in at least apical part; genitalia in Figs 430-431. ♀: abdomen, when viewed in profile, with the ground-setulae normal, short and sparse . . . . 33. *flavimana* Meigen

– Anepimeron with the anterodorsal triangular plate mostly or completely shining. Meron with the narrow glossy hind-marginal streak reduced, only present on upper part. Gena broader, the depth below lowest eye-margin at least half width of antennal flagellomere. Hind tibia usually with 1 anteroventral seta. ♂: fore tibia curved, and covered with spinules on basal half of ventral surface, the narrowest part of tibia situated within apical half of tibia (Figs 411-412); fore femur in Figs 409-410; hind tibia with 1 anterodorsal seta; femora and fore tibia variable in colour; mid and hind tibia usually wholly dark; genitalia in Figs 413-414. ♀: abdomen, when viewed in profile, with the ground-setulae tending to be longer and denser . . . . 29. *biflexuosa* Strobl

Larvae

1 Anal division elongated and usually constricted between ventral tubercles and anal pad (Figs 395-396) . . . .
. . . . 33. *flavimana* Meigen
. . . . 31. *duplicata* Haliday
. . . . 29. *biflexuosa* Strobl

– Anal division not elongated and never constricted between ventral tubercles and anal pad (Figs 53, 397, 403) 2

2(1) All four peristigmatic tufts with more than 2 hairs (Fig. 398); ventral side of anal division between base of ventral tubercles and anal pad hairy (Fig. 397) . . . . 44. *violacea* Meigen

– One peristigmatic tuft with only 2 hairs (Figs 49, 399, 402); ventral side of anal division between base of ventral tubercles and anal pad with small warts (Figs 53, 395-396, 403) 3

3(2) Cephalic lobes fused (Fig. 400) . . . . 39. *niveipennis* Becker

– Cephalic lobes separate (Fig. 401) . . . . remaining *Sepsis* spp.

Short key to adults of Fennoscandian species of *Sepsis* Fallén

1 Meron with the anterior half glossy black, strongly contrasting with the white pruinose katepisternum and posterior meron (Fig. 380). ♂: mid femur with long anteroventral and posteroventral setae on basal 2/3, that are up to 3 times as long as femoral depth (Fig. 378); hind femur with short dense ventral to posteroventral setae on basal half, as long as femoral depth (Fig. 379); fore legs virtually unmodified (Figs 419-420), the femur with only a few (3-4) stronger but inconspicuous spinules; sternites 3 and 5 with dense fine hairs (Fig. 374); genitalia in Figs 421-422. ♀: mid tibia often without an anteroventral seta . . . . 31. *duplicata* Haliday

– Meron densely pruinose on anterior half, like the katepisternum, the posterior half usually pruinose (Fig. 387), sometimes contrasting glossy black (Figs 381-382). ♂: mid and hind femora simple, without long setae on any of the ventral surfaces; fore legs conspicuously modified, the femora with groups of stout spines or tubercles;

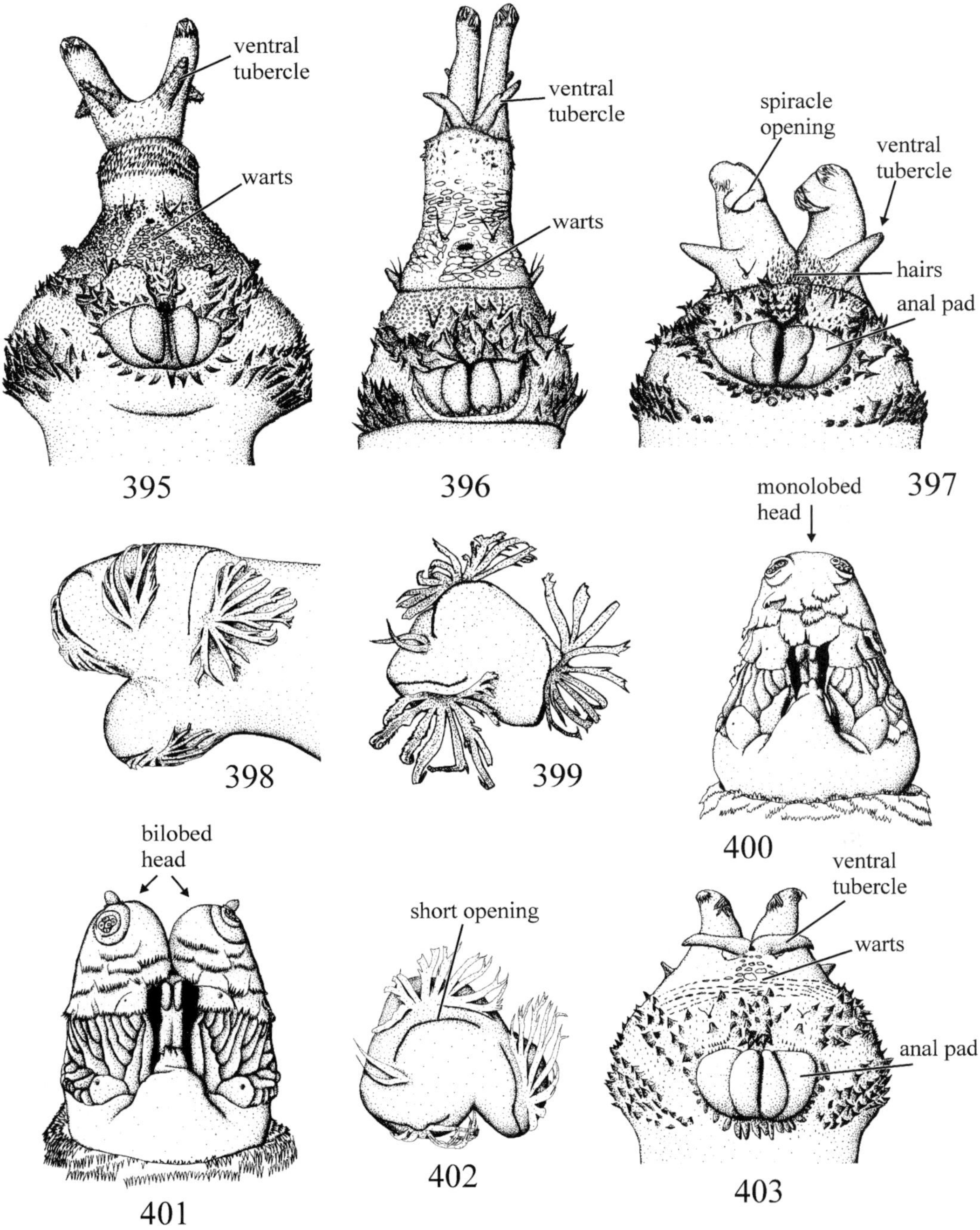

Figs 395-403. Larval morphology. Anal division ventral. 395: *Sepsis biflexuosa* Strobl; 396: *Sepsis flavimana* Meigen; 397: *Sepsis violacea* Meigen. Posterior spiracles. 398: *Sepsis violacea* Meigen; 399: *Sepsis cynipsea* (Linnaeus); 400: *Sepsis niveipennis* Becker, pseudocephalon, ventral; 401: *Sepsis lateralis* Wiedemann, pseudocephalon, ventral; 402 *Sepsis biflexuosa* Strobl, posterior spiracle; 403 *Sepsis niveipennis* Becker, anal division, ventral.

sternites 3 and 5 with the normal short sparse setulae (Fig. 373); ♀: mid tibia with an anteroventral seta 2

2(1) Meron pruinose only on anterior part, from just in front of the spiracle, with quite a distinct division between the anterior pruinose and posterior glossy areas (Figs 381-382). Hind tibia with a submedian anterodorsal seta 3

– Meron pruinose like the katepisternum, at most extreme posterior edge narrowly shining and merging into the pruinose area (Fig. 387). Hind tibia without an anterodorsal seta in ♀ (except in *punctum*) 4

3(2) Katepisternum extensively glossy black on lower anterior part, the pruinosity confined to upper and posterior part (Fig. 382). Mid tibia with 2-4 posterodorsal setae. ♂: hind tibia with 1 posterodorsal seta; fore legs in Figs 471-472; genitalia in Figs 473-474 .. .. 43. *thoracica* (Robineau-Desvoidy)

– Katepisternum entirely white pruinose (Fig. 381). Mid tibia rarely with more than 2 posterodorsal setae. ♂: hind tibia without a posterodorsal seta; fore legs in Figs 415-416; genitalia in Figs 417-418 .... 30. *cynipsea* (Linnaeus)

4(2) Basal medial cell partly or wholly devoid of microtrichiae. ♂: fore femur on ventral surface with 1 or 2 tubercles bearing spinules, situated between the submedian group of spinules and the femoral tip (e.g. Fig. 432). ♀: tergites 4 and 5 each with a pair of lateral marginal setae, rarely only twice as long as one of the tergal groundsetulae and usually much longer than this (Fig. 384); if these setae are weakly developed or absent, then sides of syntergite 1+2 without stronger setulae posteriorly (Fig. 386) and femora and tibiae black except for fore tibia which is brown (some *fulgens*), or anepimeron pruinose broadly on hind margin and along part of lower margin (Fig. 391) (*luteipes*) 5

– Basal medial cell usually entirely covered with microtrichiae. ♂: fore femur on ventral surface without any tubercles between the submedian group of spinules and the femoral tip (e.g. Fig. 428). ♀: tergites 4 and 5 without any marginal setae (as in Fig. 383), these present only in very exceptional cases; sides of syntergite 1+2 with some stronger setulae posteriorly (as in Fig. 385) and legs with at least some parts of femora and tibiae yellow 8

5(4) Syntergite 1+2 posteriorly at sides with setulae that are short and weak, as on the rest of tergites 1+2 and 3 (Fig. 386) (best seen when the abdomen is viewed dorsally against a white background). ♂: fore femur with the ventral tubercle longer, straight, and with 2 spinules at tip (Fig. 432); mid and hind femora and tibiae virtually entirely dark brown; fore tibia in Fig. 433; genitalia in Figs 434-435, surstylus rather truncate at tip. ♀: fore femur mainly dark, mid and hind femora almost wholly dark; fore tibia partly dark .... 34. *fulgens* Meigen

– Syntergite 1+2 posteriorly at sides with several setulae that are stronger than the adjacent ones and those in the same position on tergite 3 (Fig. 385) (best seen when the abdomen is viewed dorsally against a white background). ♂: fore femur with the ventral tubercle shorter, finger-shaped or straight, usually with 1 spinule at tip (e.g. Fig. 467); at least parts of mid and hind legs yellow, usually the femora. ♀: legs not usually as dark as this 6

6(5) Anepimeron pruinose along posterior margin and, more thinly so, on upper part of the anterodorsal triangular plate, otherwise glossy black (Fig. 389). Usually only 1 dorsocentral seta. ♂: fore femur with the submedian group of spinules preceded by 2 or rarely 3 stout setae (Fig. 467); femora and tibiae usually yellow in the larger forms; genitalia in Figs 469-470, the surstyli short, stouter, less pointed. ♀: tergite 3 without lateral marginal setae; all femora and the two posterior pairs of tibiae usually mainly or entirely yellow .... 42. *punctum* (Fabricius)

– Anepimeron more extensively pruinose along lower and posterior margins, and the anterodorsal triangular plate entirely pruinose (Figs 391-392). 2 dorsocentral setae. ♂: fore femur with the submedian group of spinules preceded by only 1 stout seta

(Figs 441, 475); femora and tibiae always at least partly darkened. ♀: tergite 3 with lateral marginal setae; legs darker; mid and hind femora usually darker on apical half or more 7

7(6) Anepimeron with the pruinosity extending far along lower margin, leaving only a small median area shining (Fig. 392). Anterior dorsocentral seta equal to or only slightly shorter than posterior one. ♂: fore legs in Figs 475-476; genitalia in Figs 477-478, surstyli pointed, in dorsal view their inner margins with many long setae . . . . 44. *violacea* Meigen

– Anepimeron with the pruinosity not extending halfway along lower margin, broadly shining medially (Fig. 391). Anterior dorsocentral seta always very short, at most half as long as posterior one. ♂: fore legs in Figs 441-442; genitalia in Figs 443-444, surstyli rounded in dorsal view, their inner margins without long setae but with short setulae . . . . 36. *luteipes* Melander & Spuler

8(4) Anepimeron rather broadly pruinose along lower and posterior parts (Fig. 388). ♂: fore femur in apical third with 2 ventrally-directed and usually diverging spinules (Fig. 459); fore tibia in Fig. 460; genitalia in Figs 461-462 . . . . 40. *orthocnemis* Frey

– Anepimeron with at most a weakly pruinose line at pleural sutures anteriorly and posteriorly. ♂: fore femur in apical third without 2 ventrally-directed spinules, at most with 2 apically directed hairs or setulae (Figs 409-410, 428, 449-450) 9

9(8) Proepisternal depression mostly pruinose. Inner and outer vertical setae well inside the pruinose area on occiput (as in Fig. 394). Meron with the narrow glossy hind-marginal streak very reduced, only present on upper part and there hardly if at all visible. Anepimeron with the anterodorsal triangular plate dusted, like the scutum. Hind tibia without an anteroventral seta. ♂: fore tibia with a very low ventral swelling in basal quarter that is covered with short setulae, the narrowest part of the tibia situated within basal half of tibia (Figs 451-452); fore femur in Figs 449-450; genitalia in Figs 453-454 . . . . 38. *nigripes* Meigen

– Proepisternal depression almost completely shining, glossy. Inner and outer vertical setae on the edge of the pruinose area on occiput (as in Fig. 393) 10

10(9) Anepimeron with the anterodorsal triangular plate dusted, like the scutum. Meron with the narrow glossy hind-marginal streak virtually complete. Gena very narrow, the depth below lowest eye-margin at most 1/3 width of antennal flagellomere. Hind tibia without an anteroventral seta. ♂: fore tibia with a conspicuous ventral swelling at basal quarter that is covered with short spinules, the narrowest part of the tibia situated within basal half of tibia (Fig. 429); fore femur in Fig. 428; genitalia in Figs 430-431. ♀: abdomen, when viewed in profile, with the ground-setulae normal, short and sparse . . . . 33. *flavimana* Meigen

– Anepimeron with the anterodorsal triangular plate mostly or completely shining. Meron with the narrow glossy hind-marginal streak reduced, only present on upper part. Gena broader, the depth below lowest eye-margin at least half width of antennal flagellomere. Hind tibia usually with 1 anteroventral seta. ♂: fore tibia curved, and covered with spinules on basal half of ventral surface, the narrowest part of tibia situated within apical half of tibia (Figs 411-412); fore femur in Figs 409-410; genitalia in Figs 413-414. ♀: abdomen, when viewed in profile, with the ground-setulae tending to be longer and denser . . . . 29. *biflexuosa* Strobl

28. *Sepsis barbata* Becker, 1907
(Figs 404-408)

Sepsis barbata Becker, 1907: 292; and 1908: 145. Lectotype ♀, CHINA: Xinjiang, south-east Tien Shan Mts, River Bugas' near Hami, in ASSP [by designation of Ozerov, 1986b: 12].

Sepsis chopardi Séguy, 1932: 184, figs 107-108. Syntypes ♂, FRANCE: Gard, L'Aigoual, and GREECE: Salonica, in MNHNP [see below].

Sepsis pamirensis Enderlein, 1934: 137. Holotype ♀, TADZHIKISTAN: 'Djol-Boeruljuk', in ZMHU [Ozerov, 1986b: 12; 1997a: 483].

Diagnosis. – Adults of *barbata* are easily recognised by the complete absence of microtrichiae on alula and the wholly shining meron; ♂ fore femur with a patch of dense erect black hairs on basal third of anterior to anteroventral surfaces (Fig. 404).

Adult. – Colour: Head mostly black; parafacial, face and entire area below eyes yellow, and anterior part of fronto-orbital plate sometimes brownish. Antenna brown, the articulation between pedicel and flagellomere, and part of inner surface of the latter, yellow; yellow colour often more extensive. Thorax, scutellum and abdomen entirely black. Coxae wholly yellow to partly dark; trochanters yellow; femora yellow, with dark dorsal streaks, to dark, with narrowly or broadly yellow bases and yellow knees; fore tibia yellow; mid and hind tibiae brown to black; tarsi dark, tarsomeres 1-3 yellow. Wing clear except for a round to squarish spot at tip of vein R_{2+3}; calypter, margin and fringe-hairs creamy; haltere yellow.

Head: Frontal plate mostly shining, occiput weakly grey pruinose. Outer vertical 1/2 and postocellar 2/3 length of inner vertical. Gena 1/2 to 2/3 width of antennal flagellomere. Subvibrissal setae numerous and short.

Thorax: Scutum and postpronotal lobes only weakly matt, subshining and shagreened rather than pruinose; scutellum brownish-grey matt. Setae: notopleurals subequal; 2 dorsocentrals, anterior one short, especially in ♀; acrostichal and dorsocentral ground-setulae minute. Pleural pruinosity: anepisternum subshining, entirely shagreened; katepisternum white pruinose, but with a large glossy area on anterior lower part (as in *thoracica*, Fig. 382); anepimeron glossy black, with a thinly and weakly pruinose hind-margin; meron glossy black, anterior half weakly shagreened but nowhere appearing pruinose. Setae reduced: no katepisternal setulae. Scutellum short, twice as wide as long, usually with a pair of basal setulae.

Legs: Femora subshining. ♂ fore legs in Figs 404-406; ♀ fore legs simple. Mid femur without setae except for a submedian anterior in most ♂. Mid tibia with 0 anterodorsal, 1-2 posterodorsal and 1 anteroventral setae. Hind femur without setae. Hind tibia without setae except for 1 submedian anterodorsal.

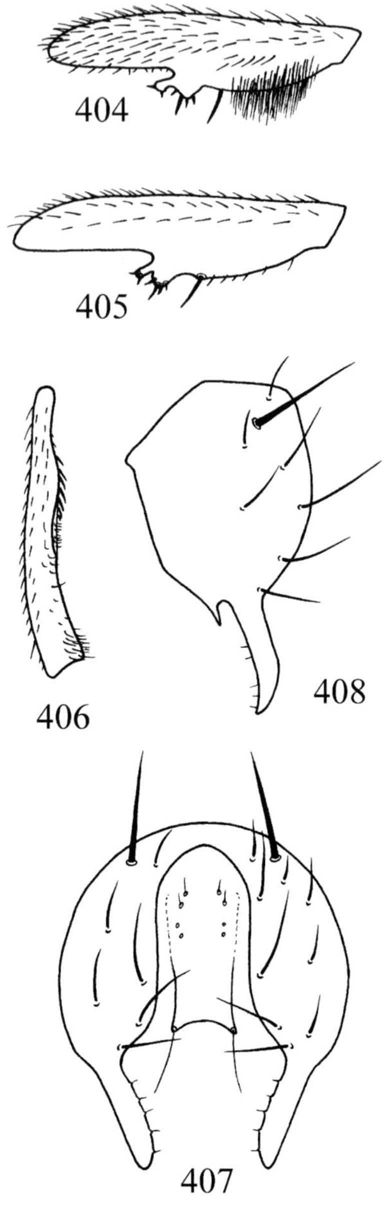

Figs 404-408. *Sepsis barbata* Becker, ♂. 404: fore femur, anterior view; 405: fore femur, posterior view; 406: fore tibia, posterior view; 407: hypopygium, dorsal view; 408: hypopygium, lateral view.

Wing: Narrow. Veins bare except for a few setulae in basal part of stem-vein on ventral surface. Covered with microtrichiae except for basal medial and posterior cubital cells, base of discal medial, basal radial, r_{2+3} and r_1 cells, part of costal cell, most of subcostal cell, and entire alula.

Abdomen: Subshining but shagreened, with violet reflections. ♂ tergites only weakly expanded behind. Without setae, except for lateral marginals, or even some median marginals too, on tergites 3-5 in ♂ and tergites 4-5 in ♀; syntergite 1+2 with erect setulae at sides. ♂ sternites with short setulae, strong on sternite 5; genitalia in Figs 407-408.

Wing-length: 2.2-2.4 mm (♂), 2.3-2.8 mm (♀).

Egg and larva. – Not known.

Distribution. – Not in Fennoscandia.

Southern Europe: Canary Is (Báez, 1982: 70), France, Austria (Vienna-Mauerbach, Franz, 1989: 152), Slovakia (Zuska & Barták, 1997: 74), Italy (Munari & Rivosecchi, 1984: 48), possibly Sardinia and Sicily (Munari, 1995: 5), Romania (Ursu, 1984: 228), Greece, Turkey; eastwards through the Middle East, across Uzbekistan (Sychevskaya, 1972c: 542), Tadzhikistan and northern China (Zuska & Pont, 1984: 159). Oriental: Taiwan, and Pakistan (Iwasa, 1989: 56). The species may be extending its range northward (Barták et al., 1997: 226).

Records are from May, June, and July.

Biology. – Adults have been collected on cow and buffalo dung, in which the larvae presumably live (Papp, 1976: 128). Adults have also been taken 'on waste accumulations from a nearby village decaying in small water pools' (Barták et al., 1997: 226). Adults carry the mite *Bonomoia sphaerocerae* Vitzthum.

Type-material. – *Sepsis chopardi*: ACP has seen 3♂ syntypes in MNHNP: Gard, Aigoual, 1200-1400 m, viii.1926 (L. Chopard), 1♂; Camp de Zeitenlik, prês de Salonique, 1917 (Rivet), 2♂. All three are *barbata*.

29. *Sepsis biflexuosa* Strobl, 1893
(Figs 41, 395, 402, 409-414)

Sepsis biflexuosa Strobl, 1893: 225. Syntypes ♂♂, AUSTRIA: Styria and HUNGARY, in BAA, Strobl collection [see below].

Sepsis biflexuosa var. *flavipes* Goetghebuer & Bastin, 1925: 132. Lectotype ♂, BELGIUM, Rouge-Cloître (Brux.), in IRSNB [see designation below].

Sepsis desultor Séguy, 1932: 185, figs 109-110. Holotype ♂, FRANCE: Var, Callian, in MNHNP [seen by ACP].

Diagnosis. – Adult *biflexuosa* is most reliably recognised in the ♂ sex, by the form of the fore tibia (Figs 411-412), the armature of the fore femur (Figs 409-410), and the shape of the surstylus (Fig. 414).

Larval pseudocephalon with fused cephalic lobes (as in Fig. 400); anal division elongated and constricted behind anal pad (Fig. 395); posterior spiracle with one opening much shorter than the remaining two particularly long ones that extend on to the stigmatophore (Fig. 402); one peristigmatic tuft with only two hairs. The larvae cannot be reliably separated from those of *duplicata* and *flavimana*.

Adult. – Colour: Head mostly black; parafacial, facial carina (or sometimes entire face) and gena brownish to yellow. Antenna dark brown, the articulation between pedicel and flagellomere, and sometimes part of inner surface of the latter, orange-yellow. Thorax, scutellum and abdomen entirely black. Fore coxa yellow except at base, mid and hind coxae mostly brown; trochanters yellow to dark brown; fore femur dark, sometimes partly lightened and yellow at base and tip, or, in some ♀, wholly yellow; mid and hind femora wholly dark, to broadly yellow at base and on knees, to almost wholly yellow in some ♀; fore tibia wholly yellow to mostly dark; mid and hind tibiae wholly dark, but yellow towards tip in paler ♀; tarsi dark, at most tarsomere 1 yellow. Wing clear except for a dark elongate oval to oblong spot at tip of vein R_{2+3}; calypter, margin and fringe-hairs brown; haltere creamy to yellow.

Head: Frontal plate mostly shining, occiput weakly grey pruinose. Outer vertical 1/2 and postocellar 3/4 length of inner vertical. Gena at least half width of antennal flagellomere. Subvibrissal setae numerous and short.

Thorax: Scutum, postpronotal lobes and scutellum brownish-grey matt and nowhere shining. Setae: notopleurals subequal; 2 dorsocentrals; acrostichal and dorsocentral ground-setulae minute. Pleural pruinosity: anepisternum subshining, entirely shagreened; katepisternum entirely white pruinose; anepimeron glossy, mostly or completely shining on upper anterior corner but pruinose on posterior margin; meron pruinose, very narrowly glossy on upper part of hind-margin. Setae reduced: no katepisternal setulae. Scutellum short, twice as wide as long, with a pair of basal setulae.

Legs: Femora subshining. ♂ fore legs in Figs 409-412; ♀ fore legs simple. Mid femur without setae or, in more robust ♂, with 1 or more submedian anterior setae. Mid tibia with 0 anterodorsal, 0-2 posterodorsal and 1 anteroventral setae. Hind femur without setae. Hind tibia with 1 anterodorsal in ♂ and, in robust ♂ and all ♀, with 1 anteroventral.

Wing: Narrow. Veins bare except for a few setulae in basal part of stem-vein on ventral surface. Entirely covered with microtrichiae.

Abdomen: Subshining but shagreened, with violet reflections. ♂ tergites only weakly expanded behind. Without setae, except for lateral marginals on tergites 3-5 in ♂; syntergite 1+2 with erect setulae at sides. ♂ sternites with short setulae, strong on sternite 5; genitalia in Figs 413-414.

Wing-length: 2.0-2.5 mm (♂), 2.3-2.7 mm (♀).

Remarks. – *S. biflexuosa* was fully described and illustrated by Ozerov (2000e: 162-164, fig. 1); the ♂ genitalia were also illustrated by Ozerov (1999e: fig. 350.9).

The ♀ can only be separated with difficulty from that of *flavimana* and *nigripes*, and the characters given in the key are only a guide. Single females in this group of species cannot be named with certainty.

S. biflexuosa is perhaps even more variable than *flavimana* in size and colour.

Larva. – The larva has been described by Hennig (1949: 20, figs 27, 36, 40, 59, 60, 61), Ursu (1973: plate 1 fig. 9), Mangan (1977: 341, figs 7-8). A full description is given by Meier (1996: 57-60, figs 176-184). Schumann (1962: 427) reported that the larva was identical with that of *duplicata*.

Distribution. – Uncommon in Fennoscandia. Denmark: EJ, NEJ, NEZ (e.g. Hammer, 1941: passim). Norway: TR. Sweden: SK, GO, GS (Ringdahl, 1934: 5; 1954b: 231; 1959b: 45). Finland: Ab, Ks (Pont, 1990: 154).

Sparsely throughout Europe, Canary Is, Morocco, Turkey, Mongolia (Zuska & Pont, 1984: 159); in Russia from the Moscow district, through the Ukraine, North Ossetia, Turkmenistan, Transbaikalia to the Far East province (Ozerov, 1999e: 570). Nearctic: British Columbia to Nova Scotia,

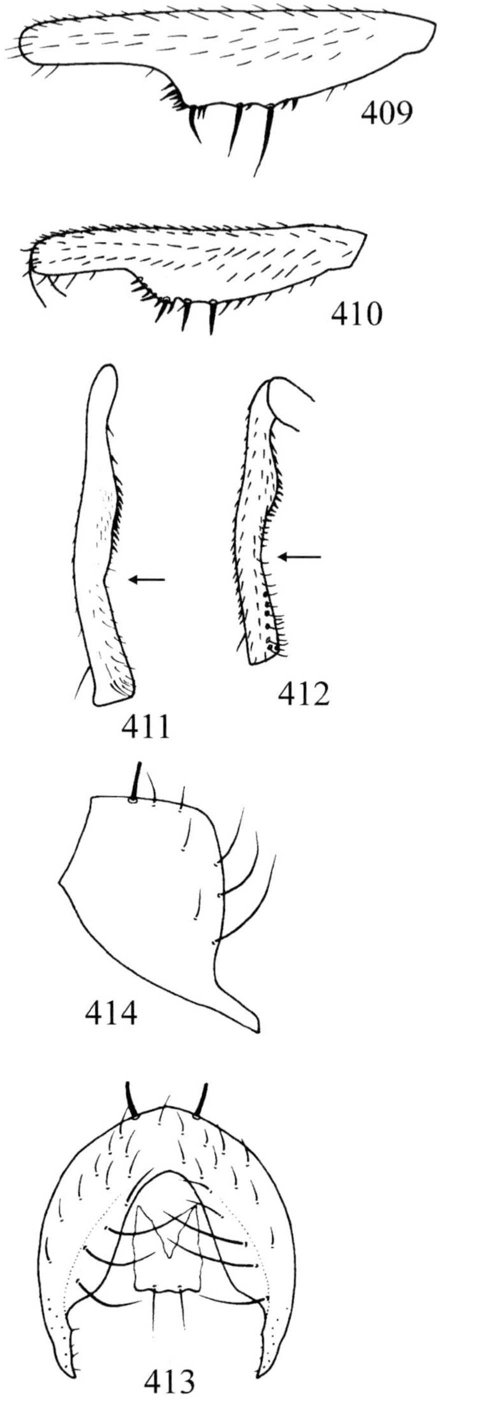

Figs 409-414. *Sepsis biflexuosa* Strobl, ♂. 409, 410: fore femur, posterior view; 411, 412: fore tibia, posterior view; 413: hypopygium, dorsal view; 414: hypopygium, lateral view.

south to California, Texas and Virginia, also Mexico (Steyskal, 1965: 683; Ozerov, 2000e: 164). Hawaii (Hardy, 1975b: 12).

Records are from July and August in Fennoscandia, elsewhere from May to September.

Biology. – Adults are most common on buffalo and cow dung (Naswaty, 1960: 97-98; Valiela, 1969: 213; Papp, 1971a: table 1; 1976: 128; Mangan, 1977: 339) However, they also frequent small mammal faeces (Mangan, 1977: 339), sewage (Duda, 1926a: 19; Mangan, 1977: 339), horse dung (Coffey, 1966: 214), pig dung (Coffey, 1966: 214), stable manure (references in Schulz, 1989), and are found in cattlesheds (Minder, 1963: 387). Van der Goot (1986a: 4) found that *biflexuosa* is one of the *Sepsis* species that visit flowers. Adults have been found carrying 2 unnamed species of Parasitidae mites.

The species has been repeatedly reared from cow and buffalo dung (Hammer, 1941; Hennig, 1949: 17; Laurence, 1954: 252; Coffey, 1966: 214; Papp, 1976: table 1, table 3; Wharton & Moon, 1979: 88; Schweiger, 1988). Larvae have also been found in cow dung (Laurence, 1954: 252; Schumann, 1962: 427; Papp, 1971a: 94; 1971b: 430; 1975: tables 1, 2; 1976: 122, 126; Mangan, 1977: 339; Papp & Garzó, 1985: 156) as well as in sewage leaks and overflows (Mangan, 1977: 339). The development time from egg to adult is 25-32 days in the field (Hammer, 1941: 273; Laurence, 1954: 252).

S. biflexuosa belongs to a group of sepsid species that specialize on older cow pads and has a long development time. It is very common in the eastern United States but rather rare throughout Europe where it is largely replaced by *flavimana*, which appears to have a similar natural history. Under laboratory conditions it can be reared in cow dung, but the cultures generally do poorly (RM, pers. obs.). Schulz (1999: 115) has described precopulatory guarding for some populations of this species.

Type-material. – *Sepsis biflexuosa*: Although the holotype should be in Strobl's collection in BAA, it was not listed by Morge (1974a, 1984) in his accounts of Strobl's types. However, the species was listed by Strobl himself as being present in his collection (Morge, 1976b: 354; Morge, 1978: 82).

Sepsis flavipes Goetghebuer & Bastin. 2♂ syntypes were found by ACP in IRSNB. 1♂ in excellent condition, from Rouge-Cloître, 23.vii.1920, has been labelled and is herewith designated by ACP as lectotype. The second ♂, without a head, from Anderlves, viii.1921, has been labelled as paralectotype.

30. *Sepsis cynipsea* (Linnaeus, 1758)

(Figs 25, 373, 381, 383, 399, 415-418)

Musca cynipsea Linnaeus, 1758: 599. Holotype [?]♂, 'habitat in Europa' [EUROPE], in LSL [see below].

Musca elongata Müller, 1764: 85. Syntypes, sex not stated, 'in hortis' [from title: DENMARK, Sjælland, Frederiksdal], not in ZMUC and presumed destroyed [see below].

Sepsis hilaris Meigen, 1826: 288; Meigen plate 281 fig. 5 [Morge, 1976c]. Holotype ♀, locality not stated [probably GERMANY: Stolberg], in MNHNP [see below].

Sepsis barbipes Meigen, 1826: 289; Meigen plate 157 fig. 5 [Morge, 1976a]. Holotype ♀, locality not stated [probably GERMANY: Stolberg], in MNHNP [see below].

Micropeza vivida Robineau-Desvoidy, 1830: 742. Syntypes, sex not stated, locality not stated, 'elle est excessivement commune' [FRANCE, probably Saint-Sauveur], not in MNHNP and presumed destroyed.

Sepsis cynipsea var. *incisa* Strobl, 1894: 106. Syntypes 10♂, AUSTRIA: Kaiserau, Scheiblegger-hochalpe, and Bösenstein, in BAA, Strobl collection [see below].

Diagnosis. – Adult *cynipsea* can be recognised in the ♂ sex by the characteristically incised fore tibia (Fig. 416). The ♀ has katepisternum wholly pruinose and meron glossy on posterior half (Fig. 381), and can usually be distinguished from *neocynipsea* ♀ by the absence of a submedian anterior seta on mid femur.

Larval pseudocephalon with two separate cephalic lobes (as in Fig. 401); anal division not elongated and constricted behind anal pad (as in Figs 53, 397, 403); posterior spiracle with all three spiracular openings of approximately the same length (as in Figs 49, 398-399); one peristigmatic tuft with only two hairs. The larvae of this species cannot be reliably separated from those of *fulgens*, *lateralis*, *luteipes*, *neocynipsea*, *orthocnemis*, *punctum*, *thoracica*, and *violacea* (but see Meier, 1996: table 2).

Adult. – Colour: Head mostly black; parafacial, facial carina and gena reddish to yellow, but sometimes more extensively pale with entire genal area yellow and frontal plate brownish. Antenna dark brown, the articulation between pedicel and flagellomere, and sometimes part of inner surface of the latter, yellow. Thorax, scutellum and abdomen entirely black. Fore coxa almost wholly yellow, mid and hind coxae mostly brown; trochanters yellowish-brown to brown; femora dark, but yellow at extreme base and on knees, broadly so in large ♀; fore tibia mostly yellow; mid and hind tibiae dark brown but, especially on mid tibia, tending to be yellow towards tip; tarsi with tarsomere 1, and tarsomere 2 in ♀, mostly yellow, the rest dark brown. Wing clear except for a dark elongate oval to oblong spot at tip of vein R_{2+3}; calypter and margin brown, fringe-hairs pale; haltere creamy to yellow.

Head: Frontal plate mostly shining, occiput weakly brownish-grey pruinose. Outer vertical 1/2 and postocellar 2/3 length of inner vertical. Gena 1/3 to 2/3 width of antennal flagellomere. Subvibrissal setae numerous and short.

Thorax: Scutum, postpronotal lobes and scutellum brownish-grey matt and nowhere shining. Setae: notopleurals subequal; 2 dorsocentrals; acrostichal and dorsocentral ground-setulae minute. Pleural pruinosity: anepisternum subshining, entirely shagreened; katepisternum entirely white pruinose; anepimeron glossy, thinly pruinose on upper anterior corner and on posterior margin; meron pruinose anteriorly, broadly glossy posteriorly (Fig. 381). Setae reduced: no katepisternal setulae. Scutellum short, twice as wide as long, with or without a pair of basal setulae.

Legs: Femora subshining. ♂ fore legs in Figs 415-416; ♀ fore legs simple. Mid femur without setae. Mid tibia with 0 anterodorsal, 2 short posterodorsal and 1 anteroventral setae. Hind femur without setae. Hind tibia with 1 submedian anterodorsal.

Wing: Narrow. Veins bare except for a few setulae in basal part of stem-vein on ventral surface. Covered with microtrichiae except in basal medial cell.

Abdomen: Subshining but shagreened, with violet reflections. ♂ tergites only weakly expanded behind. Without setae, except for lateral marginals on tergites 3-5 in ♂; syntergite 1+2 with erect setulae at sides. ♂ sternites with short setulae, strong on sternite 5 (Fig. 373); genitalia in Figs 417-418.

Wing-length: 2.4-3.2 mm (♂), 2.3-3.4 mm (♀).

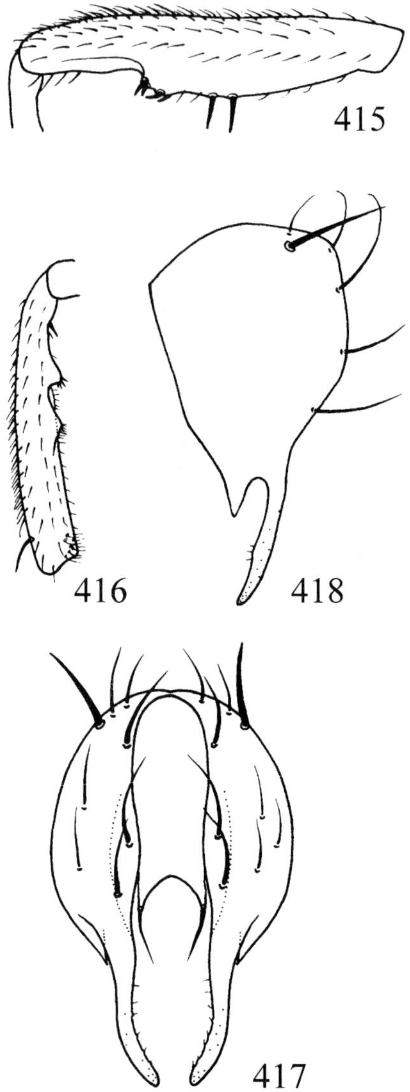

Figs 415-418. *Sepsis cynipsea* (Linnaeus), ♂. 415: fore femur, posterior view; 416: fore tibia, posterior view; 417: hypopygium, dorsal view; 418: hypopygium, lateral view.

Egg. – The egg is fully described by Meier (1995b: 427-431: fig 7). There is also an illustration in Schumann (1961: fig. 12).

Larva. – The larva was described and illustrated by Hennig (1949: 19, figs 19, 25, 32, 39, 50, 51, 53), Schumann (1962: 428-429, figs 52-55), and, most fully, by Meier (1996: 60-63, figs 185-193).

Distribution. – Widespread and common throughout Fennoscandia. Denmark: all provinces (e.g. Hammer, 1941: passim; Lyneborg, 1965: 231). Norway: Ø and A to F, but with few published post-Siebke records (e.g. Strand, 1913: 325; Hennig, 1949: 79). Sweden: SK to TO (e.g. Zetterstedt, 1847: 2284; Ringdahl, 1931: 25; 1951: 142; 1954b: 231). Finland: Al and Ab to north Lapland (e.g. Frey, 1908: 583; Krogerus, 1960: 77). Also in the adjacent areas of Russia as far north as Arkhangelsk (Shtakel'berg, 1958: 134).

Throughout Europe, North Africa, but not the Middle East except Turkey and Syria, eastwards to Mongolia (Zuska & Pont, 1984: 159), through Siberia to the Far East province of Russia (Ozerov, 1999e: 568) and Hokkaido (Iwasa, 1980: 404; 1995: 796).

Records are from May to early September in the north, April to October in the south, with occasional records from the winter months of overwintering adults. In Austria, Czerny (1903: 239) found adult *cynipsea* active during mild weather on 6-8 January 1903, but his specimens were most probably *fulgens*.

Biology. – *Sepsis cynipsea* is one of the most abundant and best studied European sepsids. Adults are found up to 1400 m in the Alps (Timon-David, 1937: 19) and over 2550 m in the Tien Shan Mts (Sychevskaya & Vtorov, 1969: 820). They have been collected on snowfields in Scotland (Ashmole et al., 1983: 611). They are common in the coniferous and birch forest zones (Ringdahl, 1931: 25; Krogerus, 1960: 77), and even coastal dune systems (Krogerus, 1932: 116; Ardö, 1957: 121). They are most abundant in pastures and meadow habitats and apparently prefer dry and sunny habitats (Bährmann, 1993a: 104-105). The females of high-altitude populations have been shown to be on average smaller, to be shorter-lived, and to lay fewer clutches of eggs than females from lower altitude populations (Blanckenhorn, 1997: 342). Adults of *cynipsea* are found most commonly on cow dung and especially fresh dung (Hammer, 1941; Naswaty, 1960: 99; Schumann, 1962: 428; Minder, 1963: 387; Parker, 1972a: 173; Iwasa, 1980: 405; Schulz, 1989; Ward, 1983: 75). They are rare on other kinds of dung, but there are references to horse dung (Pont, 1979: 19; Tamarina et al., 1982: 86), sheep dung (Randall et al., 1981: 187), and stable manure (references in Schulz, 1989).

Adults are also commonly found on various flowers: *Potentilla fruticosa* (Wahlgren, 1915: 53), *Inula helenium* (Payne, 1975: 221), *Smyrnium olusatrum* (Payne, 1979: 233), *Heracleum* (Van der Goot, 1986a: 4); *Saxifraga hirculus* (Warncke et al., 1993: 145); *Cornus mas* and *Prunus spinosa* (Bährmann, 1993a: 105); older records are in Silén (1909: 37, 43, 45) and Poppius (1903: passim), and yet others in Alfken (1924: 453-475). A mass occurrence was recorded on *Polygonum cuspidatum* (Van der Goot, 1986b: 33).

S. cynipsea has been recorded as the prey of *Empis livida* L. (Tuomikoski, 1952: 171). Adults carry the mite *Bonomoia sphaerocerae* Vitzthum and have been found with 2 unnamed species of Parasitidae. They are also carriers for the phoretic stage of the coprophagous nematode *Diplogaster coprophila* Sudhaus & Rehfeld, 1990 (Kiontke, 1996: 356).

S. cynipsea is among the first insects to arrive on a freshly deposited cow pad (Hammer, 1941; Minder, 1963: 387; Parker, 1972a: 173; Schulz, 1989). Minder (1963: 387) reports arrival within 7-10 minutes of deposition and the flies remain 1-3 hours, or 5-6 hours on dull days. The first flies to arrive are usually males. They are mostly found in the vegetation around the dung, and the few on the pad defend territories. Males are usually much more abundant than females. As soon as females arrive, males mount the females and engage in precopulatory guarding. Females deposit their eggs while carrying a male, and only a few unmounted ovipositing females will be found on a fresh cow pad. After oviposition, the precopula pair leaves the cow pad and, in 60% of all cases, the male will copulate with the female in the surrounding vegetation (Parker, 1972a: 203). At this stage females no longer carry mature ova in their ovaries (Parker, 1972a: 194; see Eberhard, 1999: 129-130, for similar findings in American sepsids). The take-over of a female by another male is rare. However, Ward (1983: 75) reports that large males have a three-fold advantage over small males: (1) they are more likely to engage in a precopula, (2) they copulate with larger females and (3) they are more successful in fights with competitors. He also found that smaller males search for longer at any one cow pad before they abandon it as a potential mating site. McLachlan & Allen (1987: 13) argue that small males have an advantage over large males when mating takes place on the wing. *S. cynipsea* is cited as an example of this, although this species never mates on the wing. Further information on the quantitative genetics and sexual selection in *cynipsea* can be found in Blanckenhorn et al. (1998, 1999), and in Reusch & Blanckenhorn (1998).

Rearing records of *cynipsea* from cow dung are abundant in the literature (Hammer, 1941; Laurence, 1954: 250; Schumann, 1962: 428; Minder, 1963: 387; Papp, 1976: table 1; Iwasa, 1980: 405; Kirk, 1992: 510, 514; Schweiger, 1988; Schulz, 1989) and larvae in cow dung have additionally been reported by Brindle (1965: 139), Papp (1971a: table 2), Papp & Garzó (1985: 156) and Lavčiev & Tsankova (1980: 4: in dung heaps). The only rearing record from a substrate other than cow dung is from sheep dung (Randall et al., 1981: 187).

Development times have been reported by several workers. Schumann (1962: 428-429) reports the egg stage to last 24-30 hours, 1st and 2nd larval instars each 1 day, 3rd larval instar 3 days, and the pupa as little as 3 days. The reported total development times vary considerably, presumably due to temperature differences: 20 days (Pont, 1979: 4, in midsummer), 12-25 days (Laurence, 1954: 252, with a median of 14 days), 14 days (Schulz, 1989, in the laboratory), 8 days (at 24°C, RM, pers. obs.), 8 or 10-14 days (Schumann, 1962: 428, first results from a laboratory culture), 17-22 days in Denmark (Hammer, 1941: 273), and 18-20 days in the Moscow area (Minder, 1963: 387). Laboratory cultures of the same population take approximately 8 days at 30°C and 26 days at 15°C (Blanckenhorn, 1997: 348). Females hatch approximately one day earlier than males, although they pupariate later than males (Schulz, 1989; Blanckenhorn, 1997: 344). In the laboratory a batch of about 30 eggs is laid in 15-20 minutes (Parker, 1972a: 174; Schulz, 1989). The first batch is deposited one week after eclosion and every 2.2 days a new batch can be produced under laboratory conditions (Schulz, 1989). Both puparia (Hammer, 1941) and adults have been reported to hibernate (Minder, 1963: 387; ACP, pers. obs.). However, Blanckenhorn (1998: 6-7) has shown in experiments that overwintering of puparia is highly unlikely for at least Swiss populations. Here, hibernation is induced mostly by temperature and not so much by day length, and Blanckenhorn's conclusion that this species overwinters in the adult stage is in agreement with Minder (1963: l.c.) who observed adults in October and Van der Goot (1986b: 33-34) who reported a swarm of sepsids including *cynipsea* in September. Such swarms are otherwise only well-known for *fulgens* (see Pont, 1987a: 309) and have been interpreted as the aggregation of flies preparing for hibernation.

The following hymenopterous parasitoids have been reared from puparia: *Cothonaspis gracilis* Hartig and *Cothonaspis (Erisphagia)* sp. (Eucoilidae).

Type-material. – *Musca cynipsea*: ACP has studied the two specimens under this name in Linnaeus' collection (LSL), under *Musca* no. 113 in Diptera box 15. Both are on identical, medium-sized pins. One is a ♂, which at some time became detached from its pin and has been glued back on. The left wing and left mid leg are missing. It has Linnaeus' own label 'cynipsea 87.', which is the number of this species in his *Systema Naturae* (1758). This specimen was not seen by Haliday (1851). There is every indication that this is Linnaeus' original specimen, as his brief description fits very well: he mentions a ventral tooth on fore femur, and so was clearly looking at a ♂, and the constricted syntergite 1+2 matches his descriptions (1758: 599; 1761: 459). The second specimen is a female, has no Linnaean label, and does not agree with Linnaeus' description as the fore femur is simple and not dentate. The ♂ is therefore considered to be the holotype.

Unfortunately, however, the holotype does not belong to *cynipsea*, as understood for well over half a century and as defined in the present paper, but to *fulgens*: the fore femur has the characteristic armature of *fulgens* (Fig. 432), and the fore tibia lacks the ventral incision that characterises *cynipsea* (Fig. 416).

The past decades have seen a burgeoning of the literature on the two species *cynipsea* and *fulgens*, as aspects of their anatomy, behaviour and reproductive strategies have been studied, not to mention the very considerable faunistic, taxonomic and biological literature, and we can see no useful purpose in now seeking to change the names of these two well-known species. We propose to continue using the name *cynipsea* in its traditional sense, for the most abundant and widespread European coprophilous species with incised male fore tibia, and a ruling by the *International Commission for Zoological Nomenclature* should be sought to conserve this usage.

Musca elongata: The synonymy of *elongata* and *cynipsea* was proposed by Thompson (1981: 472), who also noted that no Müller collection is known to exist (op.cit.: 476). It is certainly expedient to accept this synonymy, since Müller's name is older than any *Sepsis* name other than *cynipsea*, but it is highly unlikely to be correct: Müller described his species as 'alis apice puncto nigris: abdomine aeneo, basi flauo: pedibus longis luteis', and the mention of an abdomen with yellow base and yellow legs indicates that he was describing *punctum* (Fabricius) or even *thoracica* (Robineau-Desvoidy).

Sepsis hilaris: Meigen described this species without any statement of locality, from the ♀ sex only. ACP has seen 1♀ in MNHNP, Meigen collection no. 2239 (see also Becker, 1902: 228). It is labelled 'hilaris ♀' by Meigen. It belongs to *cynipsea* as currently defined. Duda (1926a: 110, 132) studied material in NMW, and stated that 3♂ 3♀ are *cynipsea* and 1♂ 2♀ are *tonsa* [i.e. *fulgens*].

Sepsis barbipes: Meigen described this from a single ♀, without any locality. It must have been an aberrant or deformed specimen: 'Die hintere Füsse haben am dritten Gliede eine Haarflokke'. ACP has seen the holotype ♀ in MNHNP, Meigen collection no. 2235. It is labelled 'Barbipes ♂ [sic]' by Meigen. There is no trace of any fasciculus of hairs on hind tarsomere 3, nor could Becker (1902: 229) or Zuska (1971: 58) find one. It belongs to *cynipsea* as currently defined (see also Zuska, 1971: 58).

Sepsis cynipsea var. *incisa*: Although the syntypes should be in Strobl's collection, the name was not listed by Morge (1974a, 1984) in his accounts of Strobl's types. The species was listed by Strobl himself as being present in his collection (Morge, 1978: 113; but not in Morge, 1976b).

31. ***Sepsis duplicata*** Haliday, 1838
(Figs 374, 378-380, 419-422)

Sepsis duplicata Haliday, 1838: 186. Holotype ♂, ENGLAND: Kent, in NMI [see below].

Sepsis pilipes Van der Wulp, 1871a: 189 (and 1871b: plate 8 fig. 1). Syntypes ♂♀, NETHERLANDS: 'Oosterdoksdijk te Amsterdam' and Utrecht, not in ZMAN [see below].

Sepsis pilipes Loew, 1873: 304. [Junior primary homonym of *Sepsis pilipes* Van der Wulp, 1871.] Lectotype ♂, without data, presumably 'Ungarn' [HUNGARY], in ZMHU [by designation of Ozerov, 1997a: 483].

Sepsis loewi Hendel, 1902: 265. Replacement name for *pilipes* Loew.

Diagnosis. – Adult *duplicata* is a small species, immediately recognised by having the anterior half of meron glossy black (Fig. 380), contrasting with the white pruinose katepisternum and posterior half of meron, and, in the ♂ sex, by the long ventral hairs on mid and hind femora (Figs 378-379).

Larval morphology as diagnosed for *biflexuosa*. Larvae cannot be reliably separated from those of *biflexuosa* and *flavimana*.

Adult. – Colour: Head mostly black; parafacial, facial keel and gena brownish-yellow, and sometimes also lower part of frontal plate. Antenna dark brown, usually the articulation between pedicel and flagellomere, and sometimes part of inner surface of the latter, yellow. Thorax, scutellum and abdomen entirely black. Coxae and trochanters dark, at most fore coxa partly yellow; femora, tibiae and tarsi dark, at most fore knees weakly to conspicuously yellow. Wing clear except for a large, elongate spot on apical part of vein R_{2+3}; calypter, margin and fringe-hairs dark brown; haltere creamy to yellow.

Head: Frontal plate mostly shining, occiput weakly grey pruinose. Outer vertical 1/2 and postocellar 3/4 length of inner vertical. Gena about half width of antennal flagellomere. Subvibrissal setae short and sparse.

Thorax: Scutum, postpronotal lobes and scutellum brownish-grey matt and nowhere shining. Setae: notopleurals subequal; 2 dorsocentrals; acrostichal and dorsocentral ground-setulae minute. Pleural pruinosity: anepisternum subshining, entirely shagreened; katepisternum entirely white pruinose; anepimeron glossy, very thinly pruinose on upper anterior corner and on posterior margin; meron glossy anteriorly and pruinose posteriorly (Fig. 380). Setae reduced: no katepisternal setulae. Scutellum short, twice as wide as long; with a pair of basal setulae.

Legs: Femora matt. ♂ fore legs hardly ornamented, in Figs 419-420; ♀ fore legs simple. Mid femur without anterior setae; in ♂ with long anteroventral and posteroventral setae on basal 2/3, the longest up to 3 times as long as femoral depth (Fig. 378). Mid tibia without setae, except for a weak anteroventral in ♀. Hind femur without setae, but in ♂ with short dense rather curled setae

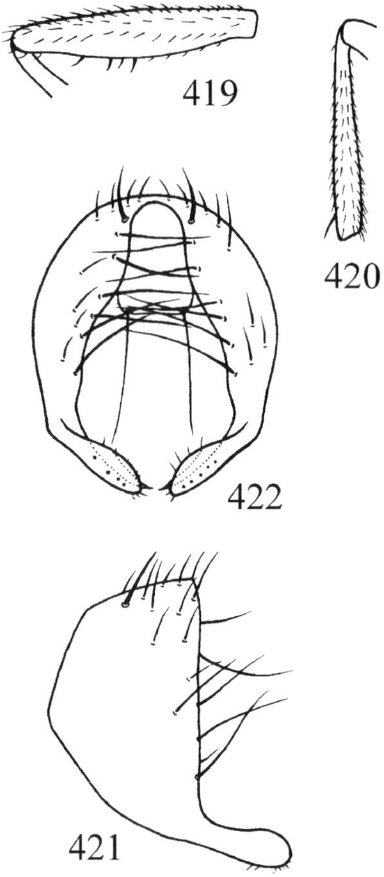

Figs 419-422. *Sepsis duplicata* Haliday, ♂. 419: fore femur, posterior view; 420: fore tibia, posterior view; 421: hypopygium, lateral view; 422: hypopygium, dorsal view.

on basal half of ventral to posteroventral surfaces, the longest ones as long as femoral depth (Fig. 379). Hind tibia without setae.

Wing: Narrow. Veins bare except for a few setulae in basal part of stem-vein on ventral surface. Entirely covered with microtrichiae.

Abdomen: Subshining but shagreened (♂) to wholly matt (♀), with violet reflections; syntergite 1+2 partly, and tergites 3-5 entirely, covered with weak punctures, this being particularly obvious when macerated. ♂ tergites only weakly expanded behind. Without setae, except for a short lateral on tergite 5 in ♂; syntergite 1+2 with erect setulae at sides. ♂ sternite 2 rather short, sternites 3 and 5 with long fine hairs (Fig. 374); genitalia in Figs 421-422.

Wing-length: 1.6-2.0 mm (♂), 1.4-2.1 mm (♀).

Remarks. – This is one of the smallest species of the genus, and has sometimes been placed in a separate genus (*Sepsidimorpha* Frey) on the basis of its reduced ♂ secondary sexual characters (see discussion above, p. 138).

Egg. – The egg has been fully described by Meier (1995b: 427-431).

Larva. – The larva has been described by Schumann (1962: 426-427, figs 45-51) and in detail by Meier (1996: 67-69, figs 211-219).

Distribution. – Widespread in Fennoscandia, but possibly overlooked. Denmark: Jutland peninsula, B. Norway: absent. Sweden: SK to DR, also TO. Finland: northwards to 63°N.

Throughout northern and central Europe (Zuska & Pont, 1984: 160), eastwards to the Far East province of Russia (Ozerov, 1999e: 568) and Hokkaido (Iwasa, 1980: 408; 1995: 795).

Records are from April (in the south) and May to August.

Biology. – Adults are commonly found on cow dung and in cattle pastures (e.g. Frey, 1908: 585; Duda, 1926a: 105; Ringdahl, 1934: 5; Hammer, 1941: passim; Naswaty, 1960: 97; Van der Goot, 1986a: 4; Schulz, 1989). They have also been found in coastal sand dunes (Ardö, 1957: 121), and even in gardens and gravel pits (Püchel, 1992: 421), but in general are only rarely found away from cow dung. Randall et al. (1981: 188) considered altitude to be no barrier to the species, so long as cow dung was available. In the North Moravian peat-bogs, Roháček & Zuska (1983: 12) found adults on the droppings of red deer in which they thought the species probably bred in such habitats. The ecology and biology have been discussed in detail by Hammer (1941: passim), and the ecology in the grasslands of Thuringia by Bährmann (1993a: 110). This is one of the few species of Sepsidae that exhibits courtship behaviour (Hammer, 1941). It is also a potential carrier for the phoretic stage of the coprophagous nematode *Diplogaster coprophila* Sudhaus & Rehfeld, 1990, although no nematodes were found on flies caught in the field (Kiontke, 1996: 356).

Rearing records from cow dung are found in Hammer (1941), Laurence (1954: 250), Minder (1963: 387), Schweiger (1988) and Schulz (1989), and in the laboratory the species also breeds in this substrate. In addition, larvae have been reported in cow dung by, for example, Gaunitz (1951: 76), Papp (1971a: table 2; 1971b: 427-431), and Papp & Garzó (1985: 156). There is general agreement that this species is a specialist on older cow pads (e.g. Hammer, 1941; Minder, 1963: 387; Schweiger, 1988). Minder (1963: 387) reports that it oviposits on cow pads as young as 1-2 hours old, but most authors observed that the first adults of this species were present after four hours (Schulz, 1989) and that most adults visit cow pads that are 1-2 days old (Hammer, 1941). Like all other sepsid species that specialize on older dung, the development time is very long. It is 25 days in midsummer according to Pont (1979: 4), 20-31 days according to Hammer (1941), and 30 days (26-39 days) according to Laurence (1954: 252). Under laboratory conditions at 24°C, the first adults emerged 15 days after oviposition (RM, pers. obs.). Schumann (1962: 426) recorded that the egg, 1st and 2nd instars last each one day, while the 3rd instar stage took 4-5 days. Hammer (1941: 273) gave 20-31 days for the cycle in Denmark. Larvae pupariate in or below the cow pad (Hammer, 1941). Minder (1963: 387) reported that copulation takes place in the early morning hours, and the eggs are deposited in beetle tunnels 0.5-1.5 cm from the surface of the cow pad (Hammer, 1941; Minder, 1963: 387). Females carry about 50-60 maturing eggs in their ovaries (Hammer, 1941). The species hibernates as a puparium (Hammer, 1941).

Type-material. – *Sepsis duplicata*: Haliday described this simply from 'Kent', without giving any indication of the sex or number of specimens. ACP has seen his holotype, which is labelled 'England' and has his label 'duplicata'. It is gummed to card and is in very good condition. The wing has an adventitious cross-vein between veins R_{2+3} and R_{4+5}, in addition to the usual cross-veins r-m and dm-cu, situated slightly apicad of cross-vein dm-cu. This was noted by Haliday in his description (hence the name '*duplicata*'), and so there can be no doubt that this is the holotype. Other features of the holotype are typical for this small and characteristic species, previously known as *pilipes* Van der Wulp.

Sepsis pilipes Van der Wulp: According to H. de Jong (pers. comm.), no syntypes can be found in ZMAN.

32. ***Sepsis fissa*** Becker, 1903

(Figs 423-427)

Sepsis fissa Becker, 1903b: 143. Lectotype ♂, EGYPT: Cairo, in ZMHU [by designation of Ozerov, 1997a: 480].

Diagnosis. – The adult is easily recognised by the milky-white wing-tip, at the tip of veins R_{4+5} and M_{1+2} and beyond the dark spot.

Adult. – Colour: Head mostly yellow, frontal plate brown, median occipital sclerite (♂) or occiput (♀) black. Antenna reddish-brown, the articulation between pedicel and flagellomere, and part of inner surface of the latter, yellow. Thorax dark, with postpronotal lobes and proepisternal area yellow, anepisternum and anepimeron brown; to only central part of scutum, between postalar line, black with rest of scutum and pleura yellow, postnotum brown. Scutellum black. Abdomen wholly dark; to mostly yellow, with only most of tergite 3 black. Legs wholly yellow, with tarsi black, mid and hind tibiae brownish dorsally; to more extensively darkened with mid and hind femora brown streaked, mid and hind tibiae and all tarsi dark. Wing clear except for a small round to squarish dark spot at tip of vein R_{2+3}; wing-tip including apical portions of veins R_{4+5} and M_{1+2} milky-white, contrasting with rest of wing; calypter, margin and fringe-hairs brown; haltere yellow.

Head: Frontal plate mostly shining, occiput weakly grey pruinose. Outer vertical 1/2 and postocellar 2/3 length of inner vertical. Gena relatively broad, equal to width of antennal flagellomere. Subvibrissal setae few in number.

Thorax: Scutum, postpronotal lobes and scutellum brownish-grey pruinose and nowhere shining. Setae: notopleurals subequal; 2 dorsocentrals; acrostichal and dorsocentral ground-setulae conspicuous in ♂, minute in ♀. Pleural pruinosity (not easy to see in specimens with yellow pleura): anepisternum subshining, entirely shagreened; katepisternum white pruinose, narrowly devoid of pruinosity and subshining on extreme anterior and lower areas (in front of mid coxa and behind fore coxa); anepimeron glossy, narrowly pruinose on hind-margin; meron pruinose anteriorly, broadly glossy posteriorly (as in *cynipsea*, Fig. 381). Setae reduced: with a few tiny scattered

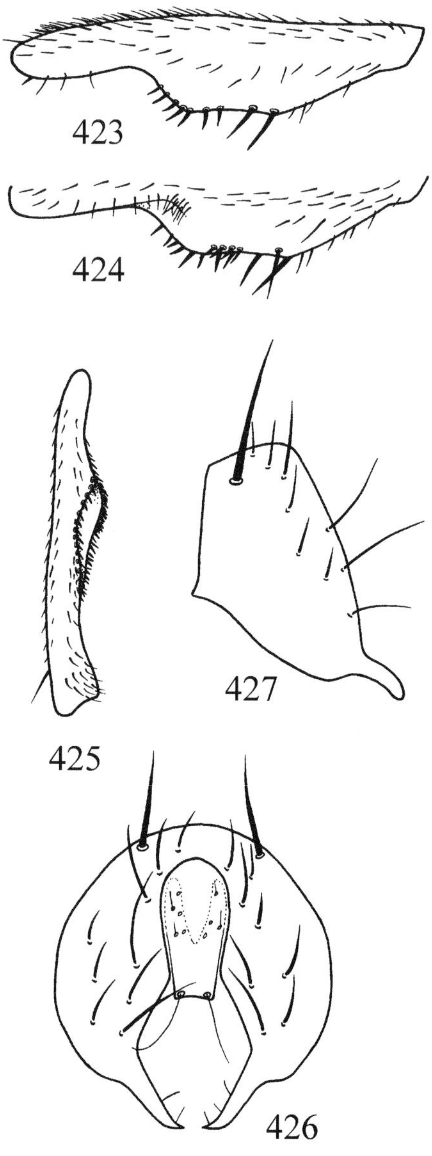

Figs 423-427. *Sepsis fissa* Becker, ♂. 423: fore femur, posterior view; 424: fore femur, anterior view; 425: fore tibia, posterior view; 426: hypopygium, dorsal view; 427: hypopygium, lateral view.

katepisternal setulae. Scutellum twice as wide as long, with a pair of basal setulae.

Legs: Femora subshining. ♂ fore legs in Figs 423-425; ♀ fore legs simple. Mid femur without setae except for 1-2 submedian anterior setae in ♂. Mid tibia with 0 anterodorsal, 2-4 posterodorsal and 1 anteroventral seta. Hind femur without setae. Hind tibia with 1 anteroventral, except in some ♀, and in ♂ also with 1 anterodorsal.

Wing: Narrow. Veins bare except for a few setulae in basal part of stem-vein on ventral surface. Covered with microtrichiae except in basal medial cell.

Abdomen: Subshining but shagreened, with some violet reflections. ♂ tergites only weakly expanded behind. Without setae, except for lateral marginals, and even some median marginals, on tergites 3-5 in ♂; syntergite 1+2 with long erect setulae at sides. ♂ sternites with short setulae, strong on sternite 5; genitalia in Figs 426-427.

Wing-length: 2.0-2.5 mm (♂), 1.7-2.4 mm (♀).

Remarks. – This species is also variable in size and colour, with the ♀ tending to be smaller and much darker than the ♂.

Egg and larva. – Not known.

Distribution. – Not in Fennoscandia.

A mainly East Mediterranean species, from Egypt, the Middle East and Turkey (Zuska & Pont, 1984: 160), but more recent work indicates an expanding range: Malta (Schembri et al., 1991: 265), Sicily (Munari & Vienna, 1979b: 23), Slovakia (Zuska & Barták, 1997: 74; Barták et al., 1997: 226), Romania (Ursu, 1984: 229), and Pakistan (Iwasa, 1989: 55).

Records are from August and September, and, at least in the Mediterranean area, this is a late-autumn species. There are also two records of the species in July (Barták et al., 1997: 226).

Biology. – Adults have been collected 'on waste accumulations from a nearby village decaying in small water pools' (Barták et al., 1997: 226).

33. *Sepsis flavimana* Meigen, 1826
(Figs 396, 428-431)

Sepsis flavimana Meigen, 1826: 288; Meigen plate 157 fig. 6 [Morge, 1976a]. Lectotype ♀, locality not stated [probably GERMANY: Stolberg], in MNHNP [see designation below].

Sepsis ruficornis Meigen, 1826: 288; Meigen plate 281 fig. 7 [Morge, 1976c]. Lectotype ♂, locality not stated [probably GERMANY: Stolberg], in MNHNP [see designation below].

Micropeza pygmaea Robineau-Desvoidy, 1830: 743. Syntypes, sex not stated, FRANCE: Saint-Sauveur, not in MNHNP and presumed destroyed. **Syn. n.**

Sepsis maculipes Walker, 1833: 248. Syntypes ♂♀, ENGLAND: near London, and near Ambleside, Westmoreland, not in BMNH and presumed destroyed [see below].
Sepsis simplex Goetghebuer & Bastin, 1925: 129, 132. Syntypes ♂, BELGIUM: Heusden (Fl.) and Schooten (Camp.), not found in IRSNB.
Sepsis borealis Frey, 1925: 71. 'New name' for *nigripes* sensu Frey (1908: 584), not Meigen [misidentification]. Lectotype ♂, FINLAND: Aaland Islands, Eckerö, in UZMH [see designation below].
Sepsis melanopoda Duda, 1926a: 58, 121. Lectotype ♂, GERMANY: Harz Mts, Brocken, in ZMHU [by designation of Ozerov, 1997a: 482].
Sepsis melanopoda var. *kerteszi* Duda, 1926a: 58, 121. Lectotype ♂, CZECH REPUBLIC: Tatra Mts, in ZMHU [by designation of Ozerov, 1997a: 481].
Sepsis meijerei Duda, 1926a: 60, 128. Holotype ♂, NETHERLANDS: Beetsterzwaag, in ZMHU [Ozerov, 1997a: 482].

Diagnosis. – Adult *flavimana* is most reliably recognised in the ♂ sex, by the form of the fore tibia (Fig. 429), the armature of fore femur (Fig. 428), and the characteristic, almost beak-like surstylus (Figs 430-431).

Larval morphology as diagnosed for *biflexuosa*. Larvae cannot be reliably separated from those of *biflexuosa* and *duplicata*.

Adult. – Colour: Head mostly black; parafacial, facial carina and gena brownish to yellow. Antenna dark brown, the articulation between pedicel and flagellomere, and also part of inner surface of the latter, yellow. Thorax, scutellum and abdomen entirely black. Fore coxa yellow except at base, mid and hind coxae brown to extensively yellow; trochanters yellow; fore femur yellow with a dark dorsal streak, to wholly dark; mid and hind femora mainly yellow, to mainly dark with only yellow bases and knees; fore tibia wholly yellow to partly infuscated; mid and hind tibiae dark, at most yellowish in apical half or just before tip; tarsi dark, or with tarsomeres 1-2 yellow. Wing clear except for a dark elongate oval to oblong spot at tip of vein R_{2+3}; calypter, margin and fringe-hairs brown; haltere creamy to yellow.

Head: Frontal plate mostly shining, occiput weakly grey pruinose. Outer vertical 1/2 and post-ocellar at least 3/4 length of inner vertical. Gena narrow, only 1/3 width of antennal flagellomere. Subvibrissal setae numerous and short.

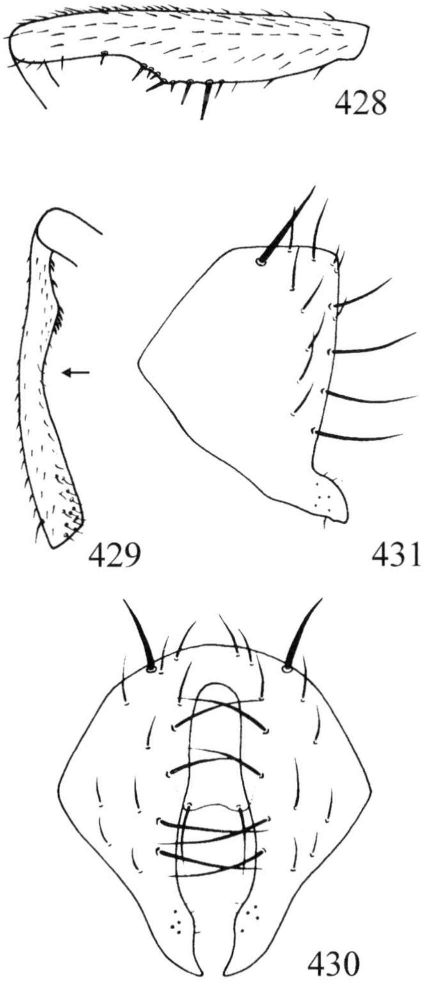

Figs 428-431. *Sepsis flavimana* Meigen, ♂. 428: fore femur, posterior view; 429: fore tibia, posterior view; 430: hypopygium, dorsal view; 431: hypopygium, lateral view.

Thorax: Scutum, postpronotal lobes and scutellum brownish-grey matt and nowhere shining. Setae: notopleurals subequal; 2 dorsocentrals; acrostichal and dorsocentral ground-setulae minute. Pleural pruinosity: anepisternum subshining, entirely shagreened; katepisternum entirely white pruinose; anepimeron glossy, pruinose on upper anterior corner and thinly so on posterior margin; meron pruinose, narrowly glossy on entire hindmargin. Setae reduced: no katepisternal setulae. Scutellum short, twice as wide as long, with a pair of basal setulae.

Legs: Femora subshining. ♂ fore legs in Figs 428-429; ♀ fore legs simple. Mid femur without setae or with a submedian anterior seta. Mid tibia with 0 anterodorsal, 0-2 posterodorsal and 1 anteroventral setae. Hind femur and tibia without setae, but larger ♂ sometimes with an anterodorsal on hind tibia.

Wing: Narrow. Veins bare except for a few setulae in basal part of stem-vein on ventral surface. Entirely covered with microtrichiae.

Abdomen: Subshining but shagreened, with violet reflections. ♂ tergites only weakly expanded behind. Without setae except for lateral marginals on tergites 3-5 in ♂; syntergite 1+2 with erect setulae at sides. ♂ sternites with short setulae, strong on sternite 5; genitalia in Figs 430-431.

Wing-length: 2.2-3.0 mm (♂), 2.9-3.2 mm (♀).

Remarks. – *S. flavimana* was fully described and illustrated by Ozerov (2000e: 164-166, fig. 2); the ♂ genitalia were also illustrated by Ozerov (1999e: fig. 350.10).

The ♀ can only be separated with difficulty from that of *biflexuosa* and *nigripes*, and the characters given in the key are only a guide. Single females in this group of species cannot be named with certainty.

This species is extremely variable in size and colour, especially in leg-colour. The legs may vary from wholly yellow to wholly black except for fore coxa, and it is not always easily distinguished from *biflexuosa*, even in the ♂. This variation has given rise to many misidentifications, and also to the description of the extreme forms as distinct species. The variability of *flavimana* has been discussed in some detail by Munari (1987c: 78, figs 2-4; 1991b: 48-49). The same author (Munari, 1993: 12-13, fig. 1) has recently suggested that *melanopoda* may be a good species, and there is no disputing the differences in leg colour and ♂ surstylus shape that he discusses, but we regard these as part of the variable phenotype of *flavimana*.

Some very small males have the fore femoral armature reduced and the subbasal ventral swelling on fore tibia reduced or absent, thus resembling *nigripes*. They can easily be recognised as *flavimana* by the genitalia.

The ♂ surstylus is sometimes more elongate, like that of *biflexuosa*. In both these species the surstylus is pointed at tip.

A number of *flavimana* have been seen from the northern part of Fennoscandia that have the posterior part of meron rather extensively glossy, and such specimens can be confused with *neocynipsea*, particularly because of the similarities of ♂ fore leg armature (compare Figs 429 and 446) and the presence of a submedian anterior seta on mid femur (see also Van der Goot, 1985b: 177). However, the genitalia are very different, and *flavimana* lacks the anterodorsal seta on hind tibia and has the basal medial cell entirely covered with microtrichiae.

Sepsis simplex Goetghebuer & Bastin, as identified by Frey (1925: 72, 75, fig. 9), is no more than a rather small form of *flavimana* with rather pale legs (usually) and rather reduced fore leg armature in the ♂. As Frey himself remarks, the ♂ surstylus is hardly different from that of *flavimana*. *Sepsis borealis* Frey is also a small form of *flavimana*, with very dark legs; the ♂ fore leg armature and surstyli are exactly as in *flavimana*, as is apparent from Frey's illustration (Frey, 1925: fig 10). *Sepsis atripes* Robineau-Desvoidy of Frey (1908: 584, 1♂ from Aaland Islands) appears to be *flavimana* with entirely black legs. Duda (1926a: 123) identified it as his *melanopoda*, which we consider to be an extreme dark-legged form of *flavimana*. Hennig (1949: 77) suggested that Frey's 1908 *flavimana* was actually *biflexuosa*, and Frey's figure of the ♂ fore leg does does indeed suggest this. However, the UZMH series of *flavimana* and *ruficornis* identified by Frey prove to be the same species and to be *flavimana* as defined here. Frey's '*flavimana*' is a somewhat paler-legged form than his '*ruficornis*'.

Larva. – The 3rd instar larva of *flavimana* has been described by Meier (1996: 69-71, figs 220-228).

Distribution. – A widespread and common species in Fennoscandia. Denmark: most provinces (e.g. Hammer, 1941: passim). Norway: Ø and A to F. Sweden: SK to T0 (e.g. Ringdahl, 1951: 142; Ringdahl, 1954b: 231). Finland: Al and Ab to Li (e.g. Hedström, 1964: 81; Hedström & Nuorteva, 1971: 123). Also present in the adjacent parts of Russia.

Throughout Europe, Iran, China (Zuska & Pont, 1984: 160), Russia eastwards to the Far East province (Ozerov, 1989: 842; 1999e: 570), and Japan (Iwasa, 1980: 407; 1995: 796). Oriental: Pakistan (Ozerov, 1999e: 570). Nearctic: Alaska to Quebec, south to California, New Mexico and Virginia (Ozerov, 2000e: 166).

Records are from late May to August in the north, and April to early September in the south.

Biology. – Adults occur in a variety of habitats, and are found well into the birch forest zone in the north. They are also recorded from coastal meadows (Dahl, 1968: 22, 23), and peat-bogs in Moravia (Roháček & Zuska, 1983: 12). Their ecology in grasslands in Thuringia is discussed by Bährmann (1993a: 110). *S. flavimana* is a cow dung specialist (Minder, 1963: 387; Tamarina et al., 1982: 87; Van der Goot, 1986a: 4), although adults have also been reported on stable manure (Hennig, 1949: 17), yak, buffalo and horse dung (Papp, 1976: 128, 130), and red deer dung (Roháček & Zuska, 1983: 12).

Van der Goot recorded this species from flowers of *Heracleum* (1986a: 4) and *Polygonum* (1986b: 33-34), and Randall et al. (1981: 186) from *Rheum*. Van der Goot (1986b: 33-34) observed the participation of *flavimana* in a sepsid swarm (see comments under *fulgens*). *S. flavimana* is a carrier for the phoretic stage of the coprophagous nematode *Diplogaster coprophila* Sudhaus & Rehfeld, 1990 (Kiontke, 1996: 356).

Rearing records or reports of larvae in cow dung can be found in Hammer (1941), Gaunitz (1951: 76), Laurence (1954: 250), Schweiger (1988), Ozerov (1989: 842), and Schulz (1989), but larvae have also been found once in human faeces (Sychevskaya, 1972a: 150). There is general agreement that this species is a specialist on older cow pads (Minder, 1963: 387; Schulz, 1989). The most detailed study has been made by Schulz (1989) who collected the first males six hours after deposition of the pad, though Minder (1963: 387) found them already after 1-2 hours. In many European countries the species becomes the dominant sepsid after two days. Under laboratory conditions females can lay about 40 eggs every 1.5 days. In the laboratory, development is relatively long, with all stages developing more slowly than *punctum* and *cynipsea*. Laurence (1954: 252) reported a development time of 22-33 days (average of about 18 days). According to Minder (1963: 387), it is much shorter (15-17 days), and in the laboratory at 24°C the first adults hatch after 14 days. These discrepancies may be explained by Schweiger's (1988) observation that *flavimana* is particularly sensitive to low temperatures. Eggs are deposited in beetle tunnels. Copulation take place in the vegetation surrounding the dung. No pre- or postcopulatory guarding has been observed.

Type-material. – *Sepsis flavimana*: Meigen described this without giving any locality: 'nur einige Male das Weibchen gefangen'. ACP has studied 1♀ in MNHNP, Meigen collection no. 2232 (see also Becker, 1902: 228), which is *flavimana* as currently understood. ACP has labelled it and designates it herewith as lectotype. Duda (1926a: 125) saw 2♂ 1♀ in NMW which are also *flavimana*.

Sepsis ruficornis: Meigen described *ruficornis* simply from 'beide Geschlechter', without giving any locality. ACP has seen 1♂ 1♀ syntypes in MNHNP, Meigen collection no. 2236, labelled 'ruficornis ♂' and 'ruficornis ♀ ' by Meigen (see also Becker, 1902: 228; Collin, 1910: 175). ACP has labelled the ♂ and herewith designates it as lectotype. It belongs to *flavimana* as currently understood; the surstyli are exposed and match those of *flavimana*. The ♀ has been labelled as paralectotype, and appears also to be *flavimana*. In NMW Duda (1926a: 121, 125) found 4♂ 1♀ under this name: 1♂ is *biflexuosa*, 2♂ are *flavimana*, 1♂ is *pectoralis* (i.e. *thoracica*), and 1♀ is *violacea*.

Sepsis maculipes: Walker described a typical form of *maculipes* together with a var. β ♂ and var. γ ♀. Nothing has been found in any of the sources for Walker material known to ACP. Walker's description is rather generalised, but his description of the leg colour actually fits *flavimana* rather well and better than other species. 2♀ in the Curtis collection in NMV (not syntypes) (seen by ACP) are *punctum*.

Sepsis borealis: Although proposed as a 'new name' for *nigripes* sensu Frey (1908: 584) [misidentification], strictly speaking this was a new species. The syntype series is therefore the material in UZMH listed by Frey in his two papers (Frey, 1908: 584, 1♂ 1♀ from Aaland; Frey, 1925: 75, 'eine nördliche Art, in Lappland häufig'). ACP has labelled and designates herewith the Aaland ♂ as lectotype; the head is missing, and the fore legs are mounted separately on a slip of paper; it is labelled as from Eckerö island, collected by Frey. It is conspecific with *flavimana*, an identification also recorded by A. L. Ozerov on a label dated 1985. Other Finnish specimens in UZMH that have been recognised by ACP as syntypes have been labelled as paralectotypes: 1♀, Al, Sund, R.Frey (Frey's 1908 ♀); 2♂, LkW, Muonio; 1♂ 1♀, LkW, Kittilä.

34. *Sepsis fulgens* Meigen, 1826
(Figs 14, 375-377, 386-387, 432-435)

Musca vibrans Harris, 1780b: 122, plate 35 fig. 43. [Junior primary homonym of *Musca vibrans* Linnaeus, 1758.] Syntypes, sex not stated, locality not stated [from title: ENGLAND], not found and presumed destroyed [see below]. **Syn. n.**

Musca vibrans minimus Harris, 1780c: index, plate 35 fig. 43. [Junior primary homonym of *Musca vomitoria minima* Harris, 1780.] Syntypes, sex not stated, locality not stated [from title: ENGLAND], not found and presumed destroyed [see below]. **Syn. n.**

Sepsis fulgens Meigen, 1826: 287; Meigen plate 281 fig. 4 [Morge, 1976c]. Lectotype ♂, locality not stated [probably GERMANY: Hamburg], in NMW [see designation below].

Micropeza fulgida Robineau-Desvoidy, 1830: 742. Syntypes ♂♀, locality not stated, 'cette espèce est excessivement commune le long des haies fraîches et parmi les gazons des lieux un peu humides' [FRANCE: probably Saint-Sauveur], not in MNHNP and presumed destroyed. **Syn. n.**

Micropeza fulvida Robineau-Desvoidy, 1830: 743. Erroneous spelling of *fulgida*.

Micropeza nitida Robineau-Desvoidy, 1830: 743. Syntypes ♂♀, locality not stated, 'c'est une espèce bien distincte et très-commune, principalement sur les fleurs des légumineuses herbacées' [FRANCE: probably Saint-Sauveur], not in MNHNP and presumed destroyed. **Syn. n.**

Micropeza atripes Robineau-Desvoidy, 1830: 743. Syntypes ♂♀, FRANCE: Saint-Sauveur, not in MNHNP and presumed destroyed. **Syn. n.**

Sepsis concinna Walker, 1833: 249. Syntypes ♀, ENGLAND: near London, not in BMNH and presumed destroyed [see below].

Sepsis communis Frey, 1925: 71. 'New name' for *cynipsea* sensu Frey (1908: 583) and *nigripes* sensu Collin (1910: 175) [misidentifications]. Lectotype ♂, FINLAND: Ab, Uskela, in UZMH [see designation below].

Sepsis tonsa Duda, 1926a: 56, 108, figs 22a-b. Syntypes ♂♀, 'Mähren, Österreich, Tirol, Polen, Ungarn, Albanien, Euböa, Kreta, Korfu, Kleinasien, Tunis, Algier', in ZMHU, MNM, etc.

Note. – Meigen described this species as '*fulgens*. Hgg.', and the name has subsequently been commonly cited as '*fulgens* Hoffmannsegg in Meigen', but there is no justification for this. Meigen used a manuscript name of Hoffmannsegg's, but the description is quite clearly his own and so, according to the *International Code of Zoological Nomenclature*, 4th edition (Art. 50.1), Meigen is sole author of the name as published.

Diagnosis. – Adult *fulgens* can be recognised in both sexes by having syntergite 1+2 without outstanding setulae posteriorly at sides (Fig. 386), and, in the ♂, by having a ventral tubercle at apical quarter of fore femur with 2 slightly diverging spinules at its tip (Fig. 432).

Larval morphology as diagnosed for *cynipsea*; i.e. the larvae of this species cannot be reliably separated from those of *cynipsea*, *lateralis*, *luteipes*, *neocynipsea*, *orthocnemis*, *punctum*, *thoracica*, and *violacea* (but see Meier, 1996: table 2).

Adult. – Colour: Head mostly black; parafacial, facial carina and gena brownish. Antenna dark brown, the articulation between pedicel and flagellomere, and sometimes part of inner surface of the latter, orange-yellow. Thorax, scutellum and abdomen entirely black. Fore coxa largely to mostly yellow, mid and hind coxae dark brown; trochanters brown, sometimes yellow; femora dark brown, at most extreme bases and fore knees yellow; fore tibia partly yellow to wholly brown, mid and hind tibiae wholly brown; tarsi brown, only tarsomeres 1-2 yellowish. Wing clear except for a dark round to squarish spot at tip of vein R_{2+3}; calypter, margin and fringe-hairs dark brown; haltere creamy to yellow.

Head: Frontal plate mostly shining, occiput weakly grey pruinose. Outer vertical 1/2 and postocellar 2/3 length of outer vertical. Gena half width of antennal flagellomere. Subvibrissal setae numerous and short.

Thorax: Scutum, postpronotal lobes and scutellum brownish-grey matt and nowhere shining. Setae: anterior notopleural a little shorter than posterior; 2 dorsocentrals; acrostichal and dorsocentral ground-setulae minute. Pleural pruinosity: anepisternum subshining, entirely shagreened; katepisternum entirely white pruinose; anepimeron glossy, thinly pruinose on upper anterior corner and, more conspicuously, on posterior and lower margins; meron pruinose, narrowly glossy on hind-margin. Setae reduced: no katepisternal setulae. Scutellum short, twice as wide as long, with or without a pair of basal setulae.

Legs: Femora subshining. ♂ fore legs in Figs 432-433; ♀ fore legs simple. Mid femur without setae. Mid tibia with 0 anterodorsal, 0-1 posterodorsal and 1 anteroventral setae. Hind femur and tibia without setae.

Wing: Narrow. Veins bare except for a few setulae in basal part of stem-vein on ventral surface. Covered with microtrichiae except on part of basal medial cell.

Abdomen: Subshining but shagreened, with violet reflections. ♂ tergites only weakly expanded behind. Without setae, except for lateral marginals on tergite 4 and 5 in both sexes (as in Fig. 384); syntergite 1+2 without erect setulae at sides (Fig. 386). ♂ sternites with short setulae, strong on sternite 5; genitalia in Figs 434-435; also illustrated by Ozerov (1999e: fig. 351.1).

Wing-length: 1.8-3.3 mm (♂), 2.7-3.4 mm (♀).

Remarks. – *Sepsis fulgens* is rather constant in colour and structure. Sometimes the fore femoral tubercle of the male is partly or completely reduced (see Munari, 1987c: 79, figs 5-9). Some females lack the strong abdominal setae, but can be recognised by the absence of erect setulae at the sides of syntergite 1+2 (Fig. 386).

Egg. – The egg morphology has been dealt with in Meier (1995b: 427-431).

Larva. – The larva has been described by Meier (1996: 73-74, figs 229-237).

Distribution. – Common and widespread throughout Fennoscandia except in the far north, although there are relatively few published records. Denmark: all provinces (e.g. Hammer, 1941: passim; Lyneborg, 1965: 231). Norway: Ø and A to TR (e.g. Hennig, 1949: 82). Sweden: SK to LY (e.g. Ringdahl, 1951: 142, as *nigripes*). Finland: Al and Ab to 64°N, also LkW (e.g. Frey, 1908: 583, as *cynipsea*; Hedström, 1964: 81). In the adjacent areas of Russia, also Arkhangelsk (Shtakel'berg, 1958: 134), and Murmansk and Kandalaksha (Zvereva, 1982: 486).

Throughout Europe, North Africa and the Middle East to Afghanistan (Zuska & Pont, 1984: 160) and the Far East province of Russia (Ozerov, 1989: 843; 1999e: 570).

Records are from May to early October in the north, and almost every month of the year in the south as the adults overwinter.

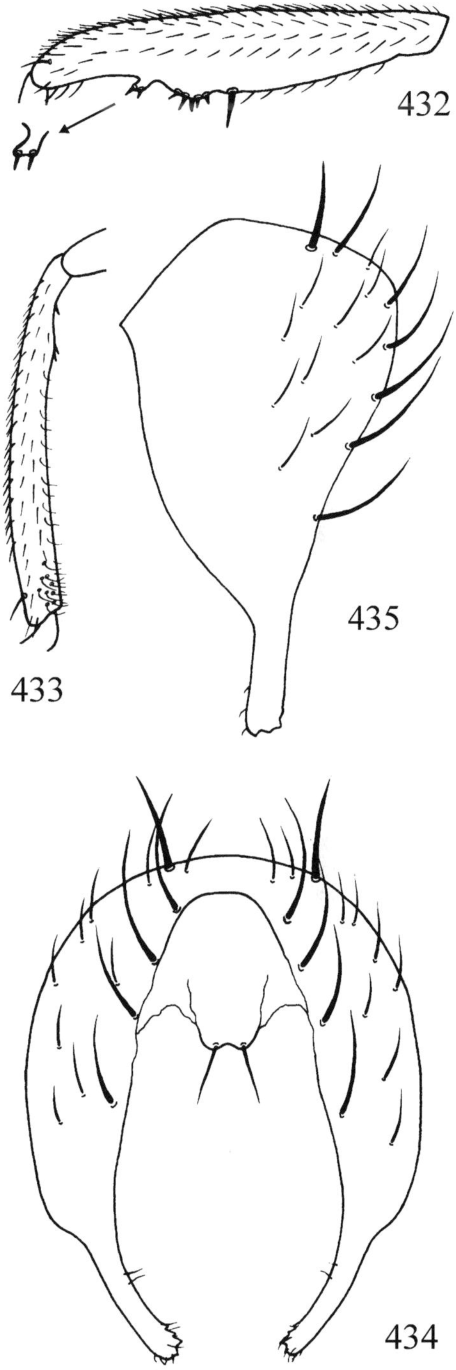

Figs 432-435. *Sepsis fulgens* Meigen, ♂. 432: fore femur, posterior view (detail: ventral protuberance); 433: fore tibia, posterior view; 434: hypopygium, dorsal view; 435: hypopygium, lateral view.

Biology. – This is a very eurytopic species and often the most abundant sepsid throughout large parts of Europe (e.g. Minder, 1963: 386; Püchel, 1992: 420; Bährmann, 1993a: 104-105). It is most abundant in meadows and damp woodlands (Naswaty, 1960: 101), and its ecology in grasslands in Thuringia is discussed by Bährmann (1993a: 104-105, 110). Adults have been found in coastal dune systems (Krogerus, 1932: 116; Ardö, 1957: 121), and peat bogs in Italy (Munari & Vienna, 1977: 46). It is commonly associated with farm animals and their sheds (Veselkin, 1966: 789). It has been collected on dung of various kinds, such as cow dung (see below), horse dung (Papp, 1974a: 172, 174; Papp, 1976: table 5; Pont, 1979: 19; 1987a: 19), compost (Van der Goot, 1986a: 4), pig swill (Pont, 1979: 19), 'steaming semiliquid manure' (Minder, 1963: 386), silage (Hennig, 1949: 17), sewage (Duda, 1926a: 19), and a sheep carcass (Gregor, 1966: 81).

S. fulgens is dependent upon a source of carbohydrate (Draeger & Scheurer, 1999: 68) and there are a large number of flower visits reported in the literature, such as Tiensuu (1948: 213, 216), Randall et al. (1981: 186) on *Rheum*, Van der Goot (1986a: 4-5) on *Heracleum* and *Veronica*, Van der Goot (1986b: 33-34) on *Polygonum*, Bährmann (1993a: 105) on *Cornus mas* and *Prunus spinosa*, and Roper (1993: 193) on *Hamamelis*.

S. fulgens displays a spectacular swarming behaviour which has spawned many notes in European entomology journals (summarized in Pont, 1987a). Swarms comprise up to at least 50,000-60,000 individuals and can be stationary for 40-90 days. The sex ratio is usually male biased. No copulation has been observed and the flies are sometimes seen feeding on flowers. Because the ovaries of the females are undeveloped, the spermathecae are empty, swarming is usually observed in late summer/autumn, and the swarm commonly leaves an odour at the swarming site, Pont (1987a: 305-309) interpreted the swarming as a hibernation phenomenon. In the literature there is some confusion over the composition of the swarms. Pont (1987a: 309) believed that only *fulgens* regularly aggregates, but Van der Goot (1986b: 33-34) observed an aberrant swarm consisting mainly of *cynipsea* and *orthocnemis*. Schweiger (1988) found that in some years *fulgens* can be the dominant sepsid on cow pads in Germany while in other years it is very rare. It would be interesting to know whether swarming only occurs in the years of dominance.

It has been reared from or immature stages have been found in horse, sheep and pig dung (Papp, 1974a: 172; 1974b: 140; 1975: table 1; 1985: 372; Papp & Garzó, 1985: 156), dunghills (Lavčiev & Tsankova, 1980: 4), hen manure (Zvereva, 1982: 486), semi-liquid manure but not faeces (Minder, 1963: 386), cow pads (Hammer, 1941; Laurence, 1954: 250; Papp, 1976: table 1; Van der Goot, 1986a: 4-5; Schulz, 1989) where it is not usually very abundant (but see Schweiger, 1988), and decaying stems of spring barley (Krusteva & Beschovski, 1998: 46). Adults start to arrive at cow dung five hours after deposition (Schulz, 1989). Development time in the Moscow area is approximately 15-17 days (Minder, 1963: 386), and in cow dung about 25 days (23-30 days) (Laurence, 1954: 252). At 24°C, 13 days are required in the laboratory when the species is bred in cow dung (RM, pers. obs.). This species can hibernate as a puparium (Lobanov, 1962a: 37) or an adult (Minder, 1963: 386; Roper, 1993: 193; Bährmann, 1996: 263), and in the Moscow area *fulgens* is thought to produce about 2-3 generations per year (Minder, 1963: 386).

Adults have been found carrying the mites *Bonomoia sphaerocerae* Vitzthum, *Macrocheles insignitus* Berlese, *Veigaia* sp., and 2 unidentified species of Parasitidae. *S. fulgens* is also a potential carrier for the phoretic stage of the coprophagous nematode *Diplogaster coprophila* Sudhaus & Rehfeld, although no nematodes were found on flies caught in the field (Kiontke, 1996: 356). The following hymenopterous parasitoids have been reared from puparia (Ionescu & Ursu, 1971: 526; Fabritius, 1981: 143): *Kleidotoma geniculata* Hartig (Eucoilidae) and *Coptera depressus* Kieffer (Diapriidae).

Type-material. – *Sepsis fulgens*: Meigen described *fulgens* without any indication of sex, number or locality. He wrote simply 'eben so gemein wie jene [i.e. *cynipsea*]'. ACP has seen 1♀ in MNHNP, Meigen collection no. 2238 (see also Becker, 1902: 228). It is labelled by Meigen 'fulgens Hg., g. v. .... Hgg.', which could mean 'Graf or 'Gabe' or 'Geschenk' or 'gegeben von Hoffmannsegg'. It is in poor condition, but the absence of abdominal setae and the pattern of anepimeral dust clearly indicates that it is *orthocnemis*.

There is no material of *fulgens* from Hoffmannsegg's collection in ZMHU (H. Schumann, pers. comm.).

In NMW, Becker (1902: 228) found 5 specimens. Duda (1926a: 110, 130, 132) saw only 4: 1 '♂' is *orthocnemis* ♀, 2♀ are *incisa* (i.e. *cynipsea*), 1♀ is *tonsa* (i.e. *fulgens*). ACP has seen a number of specimens in NMW labelled as *fulgens* from Winthem's collection, including 1♂ labelled by Meigen 'fulgens / ♂' which he has labelled and designates herewith as lectotype.

Musca vibrans and *minimus*: No material from Harris' collection is known to exist (Pont & Michelsen, 1982: 25-27). The name *minimus* appears only in the index of Harris' work, i.e. the same species is discussed as *vibrans* ('the lesser *vibrans*') in the text but is listed as *vibrans minimus* in the index. The names *vibrans* and *minimus* are both preoccupied: *vibrans* by *Musca vibrans* Linnaeus, 1758, and *minimus* by *Musca vomitoria minima* Harris, 1780a (see Thompson & Pont, 1994: 97). Harris described the habits of his *vibrans* as follows: 'Seen on currant leaves in May and June, where they are seen to run about, shaking their wings, as if highly delighted.' The name *vibrans* was synonymised with *cynipsea* by Walker (1833: 247), but this is most unlikely to be correct: Harris' notes on the habits of his species fit *fulgens* much more closely than *cynipsea*.

Sepsis concinna: Walker described this as being on grass, beneath trees, near London. No material has been found in any of the sources for Walker material known to ACP. Walker's description of his *concinna* ♀ as 'tibiae anticae nigrofuscae' is difficult to place, but it is most likely to refer to a dark *fulgens*. 2♂ from Madeira in BMNH named by Walker as *concinna* are actually *thoracica*. There is no material in the Curtis collection (NMV).

Sepsis communis: Although proposed as a 'new name' for *cynipsea* sensu Frey (1908: 503) (= *nigripes* sensu Collin, 1910: 1975) [misidentifications], strictly speaking this was a new species. The syntype series is therefore Frey's 1908 material of *cynipsea*, which consisted of both sexes, 'über ganz Finnland verbreitet'. Collin noted simply 'Frey's *cynipsea* is the same as our *nigripes*'. It is now neither practical nor worthwhile to pinpoint Frey's entire type-series, i.e. his 1908 series of *cynipsea*. ACP has recognised what is an obvious ♂ syntype in UZMH, and has labelled it and designates it herewith as lectotype. It was collected at Uskela (Ab) by E.J. Bonsdorff, and was determined by Frey as '*Sepsis cynipsea* L.'. It is in excellent condition; it is *fulgens* as currently understood.

35. ***Sepsis lateralis*** Wiedemann, 1830

(Figs 401, 436-440)

Sepsis lateralis Wiedemann, 1830: 468. Lectotype ♂, 'CHINA', in ZMUC [see designation below].

Sepsis complicata Wiedemann, 1830: 468. Holotype ♂, 'CHINA', in ZMUC [see below].

Sepsis inpunctata Macquart *in* Webb & Berthelot, 1839: 118. Holotype ♂, locality not stated [from title: CANARY IS], in MNHNP [see below].

Nemopoda algira Macquart, 1843a: 389; 1843b: 232. Holotype ♂, ALGERIA: Algiers, probably in MHNL.

Sepsis hyalipennis Macquart, 1851a: 269; 1851b: 296. Lectotype ♂, EGYPT, in UMO [see designation below].

Sepsis rufa Macquart, 1851a: 269; 1851b: 296. Holotype ♀, EGYPT, in UMO.

Meroplius melitensis Rondani, 1874: 176. Syntypes 2♂, MALTA, probably in MZLS.

Meroplius schembrii Rondani, 1874: 176. Holotype ♂, MALTA, probably in MZLS.

Sepsis fragilis Becker, 1903b: 145. Syntypes ♂, EGYPT: Lake Birket-el-Karûn, in ZMHU.

Diagnosis. – Adult *lateralis* differs from all other European *Sepsis* by the lack of a dark wing-spot. The ♂ fore femoral armature is very characteristic (Figs 436-437).

Larval morphology as diagnosed for *cynipsea*; i.e. the larvae of this species cannot be reliably separated from those of *cynipsea*, *fulgens*, *luteipes*, *neocynipsea*, *orthocnemis*, *punctum*, *thoracica*, and *violacea* (but see Meier, 1996: table 2).

Adult. – Colour: Head dark, with only the genal area brown to yellow; to mostly yellow, with only fronto-orbital plates and median occipital sclerite brown. Antenna brown, the articulation between pedicel and flagellomere yellow; to almost wholly reddish-yellow. Thorax variable: wholly dark with brown postpronotal lobes and proepisternal area; to entirely yellow except for central part of scutum (between dorsocentrals) and a median line on postnotum. Scutellum black to reddish-yellow. Abdomen wholly dark; to mainly dark with basal part (syntergite 1+2 and part of tergite 3) and apical part (much of tergite 4 and following segments) reddish to yellow. Legs wholly yellow; to mid and hind coxae basally, fore femur mainly, mid and hind femora except at base, mid tibia except on apical quarter, hind tibia wholly, and tarsomeres 4-5 dark. Wing clear, without a dark spot at tip; calypter creamy, margin yellow to creamy, fringe-hairs brown to yellow; haltere yellow.

Head: Frontal plate mostly shining, occiput weakly grey pruinose. Outer vertical just over 1/2 and postocellar 2/3 length of inner vertical. Gena half width of antennal flagellomere. Subvibrissal setae numerous, fine and short.

Thorax: Scutum, postpronotal lobes and scutellum brownish-grey to light-grey matt, and nowhere shining. Setae: anterior notopleural slightly shorter than posterior; 2 strong dorsocentrals; acrostichal and dorsocentral ground-setulae minute; postalar quite long in large specimens, half as long as supraalar. Pleural pruinosity (not easy to see in specimens with yellow pleura): anepisternum subshining, entirely shagreened; katepisternum extensively glossy, only white pruinose on upper posterior part; anepimeron shining but shagreened, pruinose on hind-margin; meron pruinose, narrowly glossy on hind-margin. Setae reduced: no katepisternal setulae. Scutellum short, twice as wide as long, with a pair of basal setulae.

Legs: Femora subshining. ♂ fore legs in Figs 436-438; ♀ fore legs simple. Mid femur without setae except for 1-3 submedian anterior setae in ♂ or 0-1 in ♀. Mid tibia with 0 anterodorsal, 1-3 posterodorsal and 1 anteroventral setae. Hind femur with 1-2 stout anterodorsals before tip in ♂, bare in ♀. Hind tibia with 1 anterodorsal, 1 posterodorsal and 1 anteroventral in ♂, but with only the anteroventral in ♀.

Wing: Narrow. Veins bare except for a few setulae in basal part of stem-vein on ventral surface. Covered with microtrichiae except in anterior basal, costal and subcostal cells, or in ♀ only parts of these cells.

Abdomen: Subshining but shagreened, the dark areas sometimes with violet reflections. Syntergite 1+2 strongly constricted behind in ♂. Without setae, except for lateral marginals and often median marginals too on tergites 3-5 in ♂ and tergites 3-4 in ♀; syntergite 1+2 with erect setulae at sides. ♂ sternites with short setulae, strong on sternite 5; genitalia in Figs 439-440.

Wing-length: 1.9-3.3 mm (♂), 1.9-2.9 mm (♀).

Remarks. – This species is extremely variable in colour and size, the ♀ tending to be smaller and darker than the ♂. Some *lateralis* ♂ have the head yellow except for a brownish frontal vitta, thorax yellow except for the central part of the scutum (between the dorsocentral rows), and legs wholly yellow.

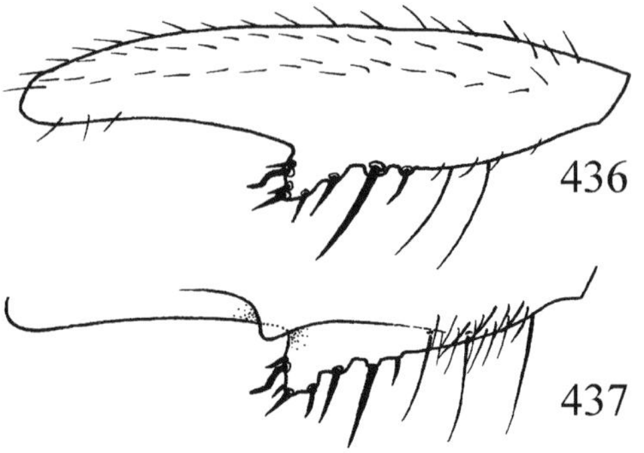

436

437

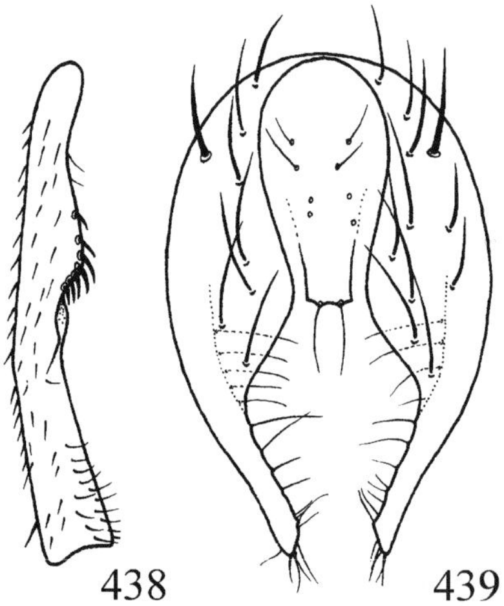

438

439

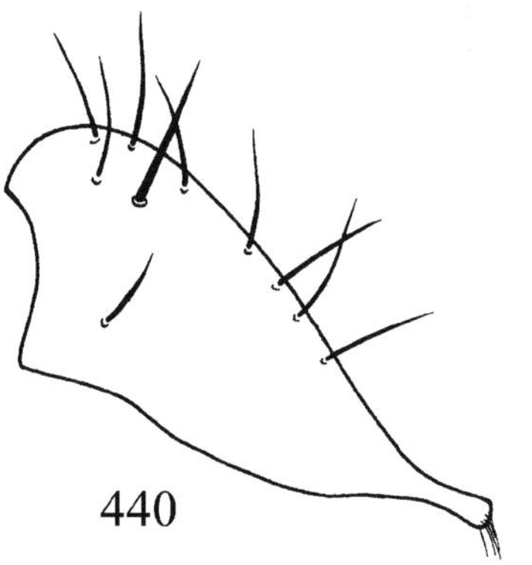

440

Figs 436-440. *Sepsis lateralis* Wiedemann, ♂. 436: fore femur, posterior view; 437: part of fore femur, anterior view; 438: fore tibia, posterior view; 439: hypopygium, dorsal view; 440: hypopygium, lateral view.

Egg. – The egg has been described by Hafez (1947: 320-325, figs 1-5).

Larva. – All the immature stages are described by Hafez (1947: 320-325, figs 1-5), and the 3rd instar larva is fully described by Meier (1996: 80-82, figs 265-273)

Distribution. – Not in Fennoscandia.

Southern Europe, the Macaronesian islands, North Africa, Middle East, eastwards to China (Zuska & Pont, 1984: 161) and Japan (Iwasa, 1980: 404; Iwasa, 1995: 793). Afrotropical, Oriental and Australasian regions.

Records in the Mediterranean area are from April to September.

Biology. – *S. lateralis* is a generalist, breeding in very different substrates (Hafez, 1939: 319: 'all sorts of dung'). It has been reared from cow dung (Hafez, 1939: 319; 1947: 307; 1948: 102; Iwasa, 1980: 412; RM, pers. obs.), chicken dung (Disney, 1973: 212), ostrich dung (RM, pers. obs.), dog dung (Hardy, 1975c: 13), and even a dead locust (Deeming, 1969: 66). The females deposit their eggs readily on cow dung, and a fresh substrate is preferred. At 25-30°C the development from egg to adult takes 6-7 days (Hafez, 1939: 325), at 26-28°C the three larval instars require 4 days and 8 hours (Hafez, 1948: 102). Hafez (1939: 319) states that in the presence of cow dung the species avoids camel and horse dung and is less abundant in pig dung. It can be extremely common on fresh cow dung. The mating behaviour is described by Hafez (1948: 101) and there is no precopulatory guarding (RM, pers. obs.).

Type-material. – *Sepsis lateralis*: Described from 'Dr. Trentepohl's und meiner Sammlung'. The ZMUC syntype ♂ (seen by ACP), collected by Trentepohl and listed by Zimsen (1954: 27), is in excellent condition. It was studied in 1966 by J. Zuska, who labelled it as lectotype. This designation has not been published. ACP herewith designates this ♂ as lectotype of *lateralis*. There are no syntypes among Wiedemann's material in NMW (R. Contreras-Lichtenberg, pers. comm.).

Sepsis complicata: Described from 'Dr. Trentepohl's Sammlung'. The holotype ♂ is in ZMUC (seen by ACP), collected by Trentepohl and listed by Zimsen (1954: 27), and is in excellent condition. It was studied in 1966 by J. Zuska, who identified it as *Sepsis lateralis*.

Sepsis inpunctata: The holotype has been seen by ACP in MNHNP, under no. 381 in box 8 of the Macquart collection. It is labelled by Macquart 'N.° [100 - deleted]. 101. / Sepsis / inpunctata'. It is a little dirty, and the thorax is largely destroyed by the pin, but otherwise it is in good condition.

Sepsis hyalipennis: There are 2♂ 4♀ syntypes in UMO (seen by ACP). One of the males was labelled as lectotype by J. Zuska in 1967, but this designation has never been published. ACP herewith designates this male as lectotype of *hyalipennis*, and has labelled the other 1♂ 4♀ as paralectotypes. The lectotype lacks the right wing and all four mid and hind legs, but the fore leg ornamentation and genitalia identify it as *lateralis*.

Sepsis rufa: The holotype in UMO (seen by ACP) consists just of a thorax on a pin; head, legs, wings, abdomen missing. It is completely unrecognisable, and the name is best kept as a junior synonym of *lateralis*.

36. *Sepsis luteipes* Melander & Spuler, 1917

(Figs 53, 391, 441-444)

Sepsis luteipes Melander & Spuler, 1917: 29. Lectotype ♂, USA: Idaho, Moscow Mt., in USNM [by designation of Ozerov, 1998a: 86].
Sepsis lamellifera Frey, 1917: 90. Nomen nudum [see below].
Sepsis lamellifera Frey, 1925: 73. Conditional name. Syntype 1♂, FINLAND: Kuopio, in UZMH [see below].
Sepsis helvetica Munari, 1985b: 52. Holotype ♂, SWITZERLAND: NE, Le Cachot, 1050 m, in MHNN.

Diagnosis. – In the group of species related to *punctum* (basal medial cell mainly or wholly devoid of microtrichiae, ♂ fore femur with a spinose tubercle at about apical quarter), adult *luteipes* can be recognised by the pattern of pruinosity on anepimeron (Fig. 391) and by the armature of the ♂ surstylus which has short setulae rather than long setae on inner margin (Fig. 443).

Larval morphology as diagnosed for *cynipsea*; i.e. the larvae of this species cannot be reliably separated from those of *cynipsea*, *fulgens*, *lateralis*, *neocynipsea*, *orthocnemis*, *punctum*, *thoracica*, and *violacea* (but see Meier, 1996: table 2).

Adult. – Colour: Head reddish-brown, only occiput, ocellar triangle and upper half of fronto-orbital plate black. Antenna brown, the articulation between pedicel and flagellomere, and most of inner surface of the latter, extensively reddish to yellow. Thorax, scutellum and abdomen black, sometimes postpronotal lobes and proepisternal area orange. Fore leg yellow except the last two tarsomeres and an indefinite dorsal brownish streak on femur; mid and hind legs with coxae and trochanters yellow, femora yellow, usually with indefinite areas of slight darkening, tibiae yellow but darkened on basal third (mid tibia) or on basal third to two-thirds (hind tibia); tarsi yellow, the apical two tarsomeres dark. Wing clear except for a dark round to squarish spot at tip of vein R_{2+3}; calypter, margin and fringe-hairs brown; haltere creamy to yellow.

Head: Frontal plate mostly shining, occiput weakly brownish-grey pruinose. Outer vertical 1/2 and postocellar 2/3 length of inner vertical. Gena 1/2 to 2/3 width of antennal flagellomere. Subvibrissal setae numerous and short.

Thorax: Scutum, postpronotal lobes and scutellum brownish-grey matt and nowhere shining. Setae: anterior notopleural rather shorter than posterior; 1 or 2 dorsocentrals, the anterior pair short; acrostichal and dorsocentral ground-setulae minute. Pleural pruinosity: anepisternum subshining, shagreened; katepisternum entirely white pruinose; anepimeron glossy, broadly brownish-grey pruinose on posterior margin and along part of lower margin (Fig. 391); meron pruinose, narrowly glossy on hind-margin. Setae reduced: no katepisternal setulae. Scutellum short, twice as wide as long, with a pair of basal setulae.

Legs: Femora subshining. ♂ fore legs in Figs 441-442; ♀ fore legs simple. Mid femur with or without a submedian anterior seta. Mid tibia with 0 anterodorsal, 0-1 posterodorsal and 1 anteroventral setae. Hind femur without setae. Hind tibia with 0-1 submedian anteroventral.

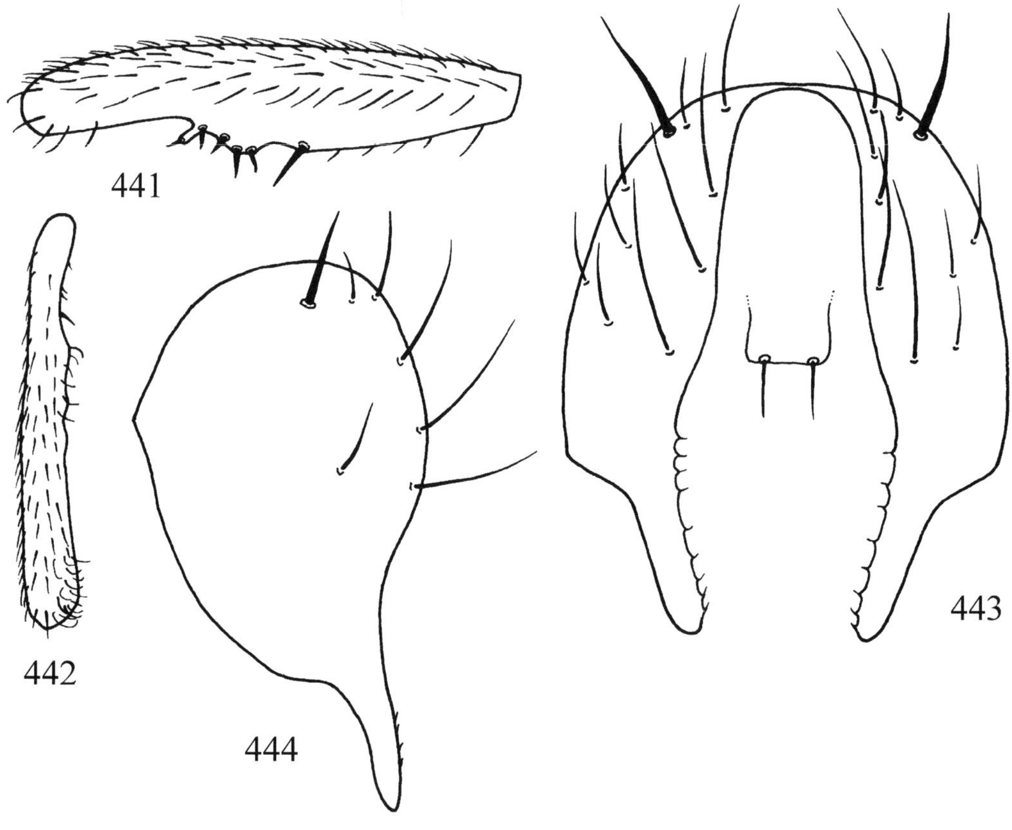

Figs 441-444. *Sepsis luteipes* Melander & Spuler, ♂. 441: fore femur, posterior view; 442: fore tibia, posterior view; 443: hypopygium, dorsal view; 444: hypopygium, lateral view.

Wings: Narrow. Veins bare except for a few setulae in basal part of stem-vein on ventral surface. Covered with microtrichiae except in basal medial cell.

Abdomen: Subshining but shagreened, with violet reflections. ♂ tergites only weakly expanded behind. Without setae, except for lateral marginals on tergites 4-5 in ♂, without setae in ♀ (as in Fig. 383); syntergite 1+2 with erect setulae at sides (as in Fig. 385). ♂ sternites with short setulae, strong on sternite 5; genitalia in Figs 443-444.

Wing-length: 2.3-2.7 mm (♂), 2.3-2.6 mm (♀).

Remarks. – *S. luteipes* was fully described and illustrated by Ozerov (2000e: 166-169, fig. 3); the ♂ genitalia were also illustrated by Ozerov (1999e: fig. 350.11).

In what appear to be females of this species (collected together with males and identified by A. L. Ozerov), the abdomen lacks strong setae on tergites 4 and 5.

Larva. – The larva has been fully described by Meier (1996: 74-76, figs 238-246, as *helvetica*).

Distribution. – A little-known species in Europe, only recently confirmed as distinct, and probably more widely distributed in Fennoscandia, and in northern and central Europe, than appears to be the case at present. Denmark: absent. Norway: no records, but probably occurs there. Sweden: TO. Finland: Ab to LkW (Frey, 1917: 90; Frey, 1925: 73, as *lamellifera*).

Great Britain (Ozerov, 1999e: 570), Germany (Meier, 1999: 190, based on F. Püchel, pers. comm.; Ozerov, 2000c: 121), Czech Republic (M. Barták, pers. comm.), Poland and Croatia (Melander & Spuler, 1917: 303, but these old records require verification), Switzerland (Munari, 1985b: 55; Haenni, 1997: 73), Italy (Munari & Rivosecchi, 1995: 5); in Russia, from the Moscow district, Stavropol district, North Ossetia, Transbaikalia, and the Far East province (Ozerov, 1999e: 570). Nearctic: sparsely from British Columbia to Quebec, south to California and the south-western states, Iowa, also Mexico (Ozerov, 2000e: 169).

Records are from mid-May to mid-September.

Biology. – Adults have been found on goose droppings (A. L. Ozerov, pers. comm.) and have been reared from cow dung (Coffey, 1966: 214).

Type-material. – *Sepsis lamellifera*: In 1917, Frey included this name in a list of species from Savolaks in Finland, without any descriptive matter ('beskrivning kommer att lämnas i annat sammanhang'). Subsequently, in 1925, he provisionally identified these specimens as the Nearctic *luteipes* Melander & Spuler, but added that if his species were to prove to be different from *luteipes* then it could be called *lamellifera*. His inclusion of it in a key, together with his notes, provide sufficient data to give this name nomenclatural availability. One of Frey's original specimens, labelled as 'Sepsis lamellifera n. sp.', from Kuopio (in Savolax), is in UZMH (P. Vilkamaa, pers. comm.).

The synonymy of *luteipes* and *helvetica* was established by Ozerov (1999e: 570).

37. *Sepsis neocynipsea* Melander & Spuler, 1917

(Figs 49, 445-448)

Sepsis neocynipsea Melander & Spuler, 1917: 28. Lectotype ♂, USA: Idaho, Moscow Mt., in USNM [by designation of Ozerov, 2000e: 172].

Diagnosis. – Adult *neocynipsea* is very similar to *cynipsea*, having katepisternum wholly pruinose and meron glossy on posterior half (as in Fig. 381), but it has ♂ fore tibia with a ventral swelling, not incision (Fig. 446), and ♀ mid femur usually with a submedian anterior seta.

Larval morphology as diagnosed for *cynipsea*; i.e. the larvae of this species cannot be reliably separated from those of *cynipsea*, *fulgens*, *lateralis*, *luteipes*, *orthocnemis*, *punctum*, *thoracica*, and *violacea* (but see Meier, 1996: table 2).

Adult. – Colour: Head mostly black; parafacial and facial carina reddish to yellow, and sometimes anterior part of frontal plate also brownish; gena yellowish in ♀. Antenna dark brown, the articulation between pedicel and flagellomere, and sometimes part of inner surface of the latter, yellow. Thorax, scutellum and abdomen entirely black. Fore coxa mostly dark to wholly yellow, mid and hind coxae brown to partly yellow; trochanters brown to yellowish-brown; femora dark with yellow knees, to dark with broad basal (except fore femur) and apical yellow rings, or even more extensively yellow especially in ♀; fore tibia mostly brown to wholly yellow; mid and hind tibiae dark brown, mid tibia often yellow at tip; tarsi mostly dark, to basal tarsomeres yellow and apical ones darker. Wing clear except for a dark elongate oval to oblong spot at tip of vein R_{2+3}; calypter and margin dirty yellow to creamy, fringe-hairs pale; haltere creamy to yellow.

Head: Frontal plate mostly shining, occiput weakly brownish-grey pruinose. Outer vertical 1/2 and postocellar 2/3 length of inner vertical. Gena 1/3 to 1/2 width of antennal flagellomere. Subvibrissal setae numerous and short.

Thorax: Scutum, postpronotal lobes and scutellum brownish-grey matt and nowhere shining. Setae: notopleurals subequal; 2 dorsocentrals; acrostichal and dorsocentral ground-setulae minute. Pleural pruinosity: anepisternum subshining, entirely shagreened; katepisternum entirely white pruinose; anepimeron glossy, thinly pruinose on upper anterior corner and on posterior margin; meron pruinose anteriorly, broadly glossy posteriorly (as in *cynipsea*, Fig. 381). Setae reduced: no katepisternal setulae. Scutellum short, twice as wide as long, with or without a pair of basal setulae.

Legs: Femora subshining. ♂ fore legs in Figs 445-446; ♀ fore legs simple. Mid femur with a strong submedian anterior seta. Mid tibia with 0 anterodorsal, 2-3 posterodorsal and 1 anteroventral setae. Hind femur without setae. Hind tibia with 1 submedian anterodorsal.

Wing: Narrow. Veins bare except for a few setulae in basal part of stem-vein on ventral surface. Covered with microtrichiae except in basal medial cell.

Abdomen: Subshining but shagreened, with violet reflections especially in smaller specimens. ♂ tergites only weakly expanded behind. Without setae, except for lateral marginals on tergites 3-5 in ♂; syntergite 1+2 with erect setulae at sides. ♂ sternites with short setulae, strong on sternite 5; genitalia in Figs 447-448.

Wing-length: 2.6-3.0 mm (♂), 2.9-3.4 mm (♀).

Remarks. – *S. neocynipsea* was fully described and illustrated by Ozerov (2000e: 170-172, fig. 5); the ♂ genitalia were also illustrated by Ozerov (1999e: fig. 351.11).

S. neocynipsea is very variable in size and colour, the larger specimens having much more orange colour on legs than the smaller ones.

Larva. – The larva has been fully described by Meier (1996: 88-90, figs 301-309).

Distribution. – Not in Fennoscandia.

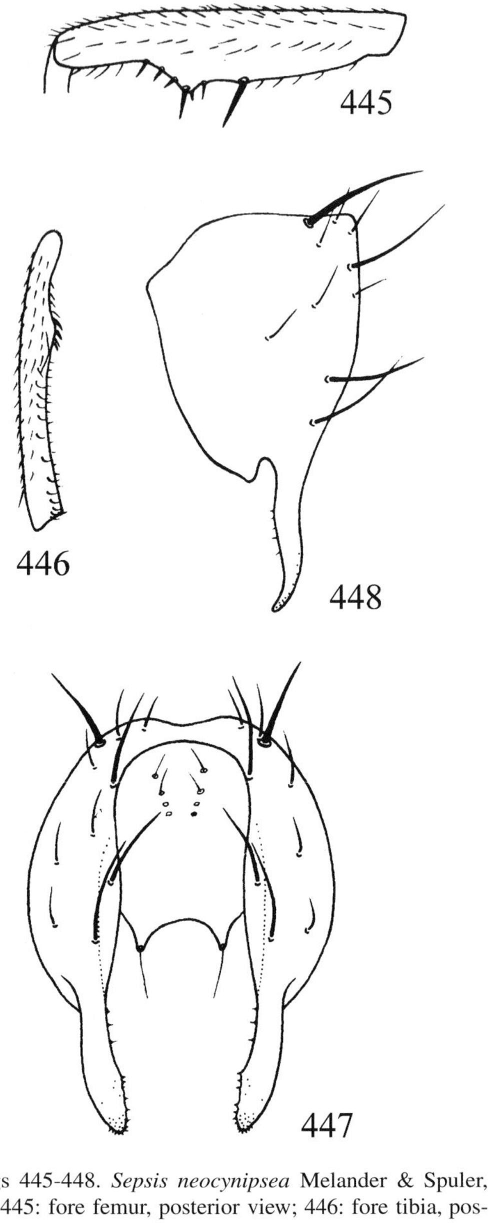

Figs 445-448. *Sepsis neocynipsea* Melander & Spuler, ♂. 445: fore femur, posterior view; 446: fore tibia, posterior view; 447: hypopygium, dorsal view; 448: hypopygium, lateral view.

British Isles and middle belt of Europe, Afghanistan, Mongolia, North China (Zuska & Pont, 1984: 162), the Far East province of Russia (Ozerov, 1989: 844; 1999e: 568), Japan (Iwasa, 1980: 405; 1995: 796). Nearctic: Alaska to Quebec, south to California and Alabama, in the south-west into Mexico as far as Mexico City

(Steyskal, 1965: 683; Mangan, 1977: 338; Ozerov, 2000e: 172); Bermuda (Woodley & Hilburn, 1994: 46). Oriental: Pakistan (Gregor & Daniel, 1971: 352; Iwasa, 1989: 55), Nepal (Iwasa, 1984a: 78; Iwasa & Thapa, 1994: 15) and Kashmir (ZMUC, det. A. L. Ozerov).

Records are from May to August.

Since it is distributed in the uplands of central Europe and in northern England and Scotland, and is a Holarctic species, the complete absence of *neocynipsea* from Fennoscandia (and from European Russia) is very curious.

Biology. – *S. neocynipsea* is the most common sepsid in lowland North America but in the Old World it is a characteristic upland species (Hennig, 1949: 15; Randall et al., 1981: 184; F. Püchel, pers. comm.). It has been found in the subalpine zone of the Tien Shan Mts up to 2555 m (Sychevskaya & Vtorov, 1969: 820) and in the Tirich Valley of Pakistan at 3650 m (Gregor & Daniel, 1971: 352). Adults have been predominantly found on cow dung (Naswaty, 1960: 99; Papp, 1976: 130), but other attractants are carrion (Reed, 1958: 238; Mangan, 1977: 339), small mammal faeces (Mangan 1977: 339), sheep dung (Coffey, 1966: 214; Randall, et al., 1981: 187), mink and pig dung (Coffey, 1966: 214; Iwasa, 1980: 407), horse dung (Coffey, 1966: 214; Papp, 1976: 130), fox dung (Iwasa, 1980: 407), human faeces (Iwasa, 1980: 407) and yak dung (Papp, 1976: 130). In Arizona, Eberhard (1999: 130) observed a female feeding on the sticky secretion of the plants *Ribes inebrians* and *Ribes cereum*. Schulz (1999: 115) has described precopulatory guarding for some populations of this species.

In North America it is one of the first sepsids to appear on freshly deposited cow pads and it can be very common. Based on the form of the ♂ surstylus and the presence of the anterodorsal seta on hind tibia, Hennig (1949: 78) placed *neocynipsea* and *cynipsea* in a monophyletic species group. Both species are particularly common visitors to fresh cow pads. However, in Europe where they are sympatric, *neocynipsea* is rare at low elevations. This may be the result of competitive exclusion. A similar phenomenon is observed with the species pair *biflexuosa* and *flavimana*. Both belong to Hennig's (1949: 72) *flavimana* species-group and breed in older cow pads. They are sympatric in Europe where *biflexuosa* is very rare, while *biflexuosa* is very common in the USA where *flavimana* does not occur.

Immature stages have frequently been found in or adults reared from cow dung (Coffey, 1966: 214; Valiela, 1969: 213; 1974: 374, Blume, 1970: 1023; Mangan, 1977: 339; Wharton & Moon, 1979: 88; Iwasa, 1980: 407; RM, pers. obs.). Larvae have also been reported from the dung of horses (Coffey, 1966: 214), pigs (Iwasa, 1980: 407), and humans (Iwasa, 1980: 407). Puparial parasitoids in North America are listed in Blume (1986: 216).

38. ***Sepsis nigripes*** Meigen, 1826

(Figs 449-454)

Sepsis nigripes Meigen, 1826: 289; Meigen plate 157 fig. 7 [Morge, 1976a]. Lectotype ♂, locality not stated [probably GERMANY: Hamburg], in NMW [see designation below].

Sepsis minima Strobl, 1893: 225. Holotype ♂, AUSTRIA: 'Styria sup.', Admont, in BAA, Strobl collection [Morge, 1984: 331; see below].

Diagnosis. – Adult *nigripes* is a very small species and is most reliably recognised in the ♂ sex, by the form of the fore tibia (Figs 451-452), the armature of the fore femur (Figs 449-450), and the characteristic, blunt-tipped surstylus (Fig. 454).

Adult. – Colour: Head mostly black; parafacial, face (or just facial carina) and gena brownish to yellow. Antenna dark brown, the articulation between pedicel and flagellomere, and part of inner surface of the latter, orange. Thorax, scutellum and abdomen entirely black. Fore coxae mostly to entirely yellow, mid and hind coxae mainly brown; trochanters brown to yellowish; femora dark brown, yellow at tips, sometimes base of hind femur yellow; fore tibia yellow; mid and hind tibiae dark brown, mid tibia sometimes yellow at tip; tarsi dark brown, tarsomeres 1-2 on fore and mid legs sometimes yellowish. Wing clear except for a dark elongate oblong spot at tip of vein R_{2+3}; calypter, margin and fringe-hairs brown, sometimes paler in ♀; haltere creamy.

Head: Frontal plate mostly shining, occiput weakly grey to brownish-grey pruinose. Outer vertical 1/2 and postocellar at least 3/4 length of inner vertical. Gena at least half width of antennal flagellomere. Subvibrissal setae numerous and short.

Thorax: Scutum, postpronotal lobes and scutellum brownish-grey matt and nowhere shining. Setae: anterior notopleural rather weaker than posterior; 2 dorsocentrals; acrostichal and dorsocentral ground-setulae minute. Pleural pruinosity: anepisternum subshining, entirely shagreened; katepisternum entirely white pruinose; anepimeron glossy, thinly pruinose on upper anterior corner and on posterior margin; meron pruinose, hardly if at all glossy on upper part of posterior margin. Scutellum short, twice as wide as long, usually with a pair of basal setulae.

Legs: Femora subshining. ♂ fore legs in Figs 449-452; ♀ fore legs simple. Mid femur without setae. Mid tibia with 0 anterodorsal, 0 posterodorsal and 0-1 anteroventral setae. Hind femur and tibia without setae.

Wing: Narrow. Veins bare except for a few setulae in basal part of stem-vein on ventral surface. Entirely covered with microtrichiae.

Abdomen: Subshining but shagreened, with violet reflections. ♂ tergites only weakly expanded behind. Without setae except lateral marginals on tergites 4-5 in ♂; syntergite 1+2 with erect setulae at sides. ♂ sternites with short setulae, strong on sternite 5; genitalia in Figs 453-454.

Wing-length: 1.4-1.9 mm (♂), 1.8-2.1 mm (♀).

Remarks. – This is a small, rare species, difficult to distinguish from *flavimana* and *biflexuosa* except by the ♂ fore legs and genitalia. The surstylus is blunt at tip, with no trace of a point when turned as in e.g. *biflexuosa*. The ♀ can only be separated with difficulty from that of *flavimana* and *biflexuosa*, and the characters given in the key are only a guide. Single females in this group of species cannot be named with certainty.

Egg and larva. – Not known.

Distribution. – A little-known and scarce species. Most records in the literature are misidentifications, mostly of *fulgens*. Denmark: NEZ. Norway: F. Sweden: ÖL and GO. Finland: Al (Pont, 1990: 154).

Sparsely through Central and Northern Europe: Great Britain (Pont, 1986a: 91), Germany (Bährmann, 1993a: 110), Poland (Hennig, 1949: 74), Austria (Strobl, 1893: 225; Franz, 1989: 153), Hungary (Duda, 1926a: 127); in Russia from the Moscow district (A. L. Ozerov, pers. comm.) through Transbaikalia to the Far East province (Ozerov, 1999e: 570), Mongolia, and Hokkaido (Iwasa, 1985: 635; 1995: 796).

Records are from June and July.

Biology. – Adults have been collected on goose droppings (A. L. Ozerov, pers. comm.). They have

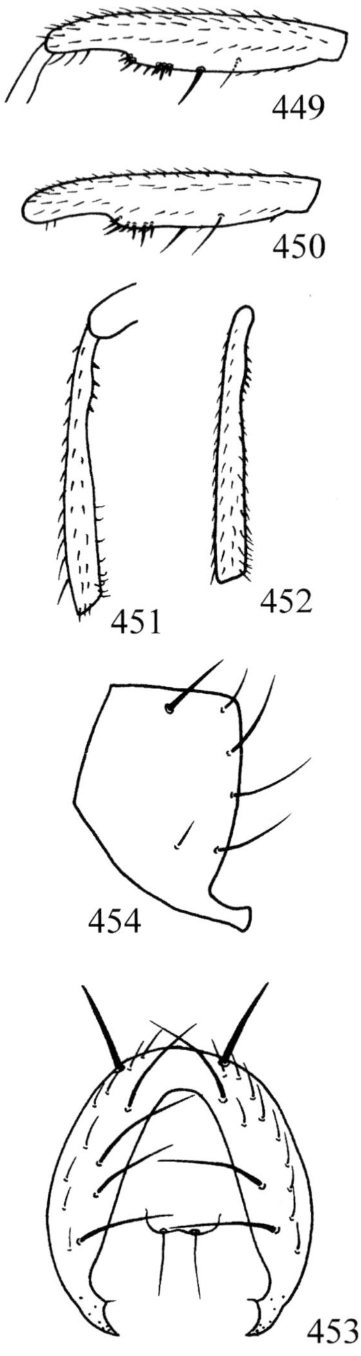

Figs 449-454. *Sepsis nigripes* Meigen, ♂. 449, 450: fore femur, posterior view; 451, 452: fore tibia, posterior view; 453: hypopygium, dorsal view; 454: hypopygium, lateral view.

been reared from cow dung (Laurence, 1952: 82; 1954: 281), but the species is a very rare member of the cow dung community.

Pridantseva's (1967: 830) record of *nigripes* associated with camel dung in South Tuva (Russia, on the Mongolian border) is based on incorrectly identified specimens of *Sepsis neglecta* Ozerov (A. L. Ozerov, pers. comm.).

Type-material. – *Sepsis nigripes*: Meigen described *nigripes* from the ♂ sex only, without giving any locality: 'nur einige Male das Männchen'. ACP has studied 1♂ 1♀ in NMW and 2♀ in MNHNP, which were also all seen by Becker (1902: 229) and Duda (1926a: 111, 127). Frey (1925: 71) also saw the NMW ♂. Becker (l.c.), with surprising percipience, concluded that the NMW ♂ represented a good species close to *ruficornis* (i.e. *flavimana*). Duda (l.c.) and Frey (1925: 71) both studied this ♂ and came to the same conclusion. ACP has labelled it and designates it herewith as lectotype. The NMW ♀ is *duplicata*.

In MNHNP, Meigen collection no. 2240, ACP has seen 1♀, labelled erroneously 'nigripes ♂' by Meigen, and 1♀, labelled by him as 'nigripes ♀' and therefore not a syntype. Becker (1902: 229) also pointed out that both specimens are ♀. The '♂' is actually an immature ♀ of *fulgens*; the ♀ is also *fulgens*.

Sepsis minima: The holotype is in BAA, in the Strobl collection, box 5, no. 2834 (Morge, 1984: 331). Morge (1974a: 357; 1974b: 412) had previously reported on 1♂ 1♀ syntypes, collected by Strobl at 'Admont unter Obstbäumen 17.7.1891', but Strobl only described the male sex and so this ♂ must be the holotype.

39. *Sepsis niveipennis* Becker, 1903
(Figs 400, 403, 455-458)

Sepsis niveipennis Becker, 1903b: 143. Lectotype ♂, EGYPT: Assiut, in ZMHU [by designation of Zuska, 1968: 474].

Diagnosis. – Adult *niveipennis* can be recognised by the fused basal medial and basal radial cells (as in *Saltella*, Fig. 13).

Larval pseudocephalon with fused cephalic lobes (Fig. 400), but anal division not elongated (Fig. 403) as in the remaining species with fused cephalic lobes (*biflexuosa*, *duplicata*, *flavimana*).

Adult. – Colour: Highly variable in colour. Head wholly yellow to wholly black. Antenna yellow to dark brown, with the articulation between pedicel and flagellomere reddish. Thorax and scutellum wholly yellow to wholly black. Abdomen mostly yellow, with some darkening on hind margins of the tergites, to wholly black. Legs wholly yellow to mainly black with only fore coxa, fore tibia and some tarsomeres yellow. Wing clear except for a diffuse dark round spot at tip of vein R_{2+3}, the wing-apex otherwise milky-white; calypter, margin and fringe-hairs white; haltere creamy to yellow.

Head: Frontal plate mostly shining, occiput weakly brownish-grey to yellowish-white pruinose. Outer vertical 1/2 and postocellar 2/3 length of inner vertical. Gena 1/2 to 2/3 width of antennal flagellomere. Subvibrissal setae numerous and short.

Thorax: Scutum, postpronotal lobes and scutellum matt, only shining around postalar callus. Setae: anterior notopleural much shorter than posterior; 1 dorsocentral in small specimens, 2 in large; acrostichal and dorsocentral ground-setulae minute. Pleural pruinosity: anepisternum shining; most of katepisternum shining, with a white pruinose patch in upper posterior corner; anepimeron shining; meron mostly shining. Setae reduced: no katepisternal setulae. Scutellum short, twice as wide as long, with a pair of basal setulae.

Legs: Femora subshining. ♂ fore legs in Figs 455-456; ♀ fore legs simple, femur with a conspicuous posteroventral seta at 2/3 of length. Mid femur without setae, but in large specimens with several submedian anterior setae. Mid tibia with 0 anterodorsal, 1-2 posterodorsal (up to 4 in large ♂) and 1 anteroventral setae. Hind femur without setae. Hind tibia with 1 submedian anterodorsal, 1 anteroventral, and, in large specimens, 1 posterodorsal.

Wings: Narrow. Veins bare except for a few setulae in basal part of stem-vein on ventral surface. Basal medial and basal radial cells fused (as in *Saltella*, Fig. 13). Covered with microtrichiae except in basal medial cell and on alula, which is narrow and hyaline.

Abdomen: Subshining. Strongly constricted between syntergite 1+2 and tergite 3, tip of syntergite 1+2 and base of tergite 3 equally narrow. ♂ tergites only weakly expanded behind. Without setae, except for lateral marginals on tergites 4-5 in ♂, absent in ♀; syntergite 1+2 with strong and

455

456

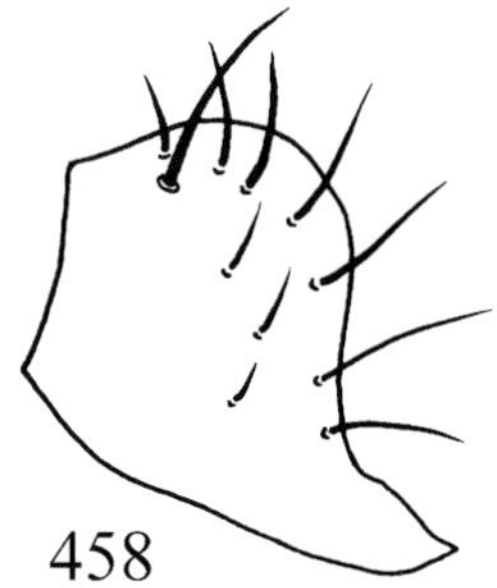

458

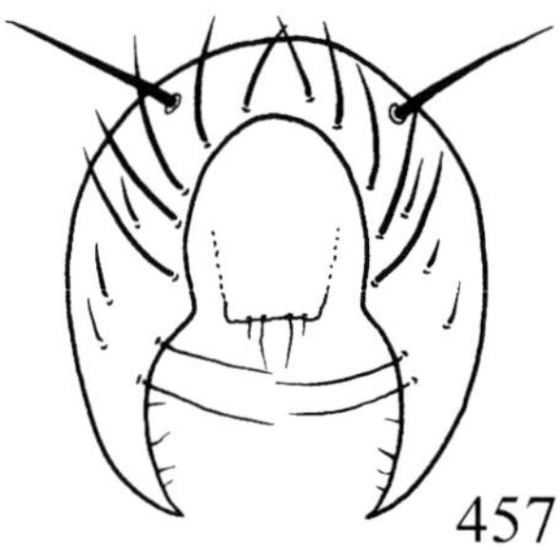

457

Figs 455-458. *Sepsis niveipennis* Becker, ♂. 455: fore femur, posterior view; 456: fore tibia, posterior view; 457: hypopygium, dorsal view; 458: hypopygium, lateral view.

erect setulae at sides. ♂ sternites with short setulae, strong on sternite 5; genitalia in Figs 457-458.

Wing-length: 1.8-2.2 mm (♂), 1.6-1.7 mm (♀).

Remark. – As in other species of *Sepsis*, the males may be large and yellow or small and black, whilst the females are always small and black.

Larva. – The 3rd instar larva has been fully described by Meier (1996: 14-16, figs 1-9).

Distribution. – Not in Fennoscandia.

In Europe only known from Cyprus (Zuska & Pont, 1984: 155); North Africa and the Middle East (Zuska & Pont, 1984: 155), eastwards through Turkmenistan (Ozerov, 1986a: 639) to Japan (Iwasa, 1984b: 305; 1995: 792). Afrotropical, Oriental and Australasian regions.

Biology. – Adults of *niveipennis* have been collected mainly from cow dung (Zuska, 1968: 475-476; Iwasa, 1984b: 306; Modassir, 1993b: 243). In Australia, where it has probably been introduced, it dominates the cow dung communities in arid areas (Colless, 1980: 74), where its abundance is positively correlated with temperature and thus increases from spring to autumn (Bishop et al., 1998: 94-95). It generally favours open pastures and is exceptionally heat- and drought-tolerant (Bishop et al., 1998: 98). Snowball (1944: 234) and Hafez (1947: 307; 1948: 100) reared specimens from cow dung (see also Ferrar, 1987: 342), but also reported that the species is associated with carrion. In dung the larval development took 4 days at a temperature of 26-28°C in Egypt, while Snowball (1944: 234) reports a larval life of 10 days (6-7 days at 20°C). Adults can emerge as quickly as 7 days after dung deposition (Bishop et al., 1998: table 2), but at 17°C emergence occurs between day 19 and 34 (Bishop et al., 1998: 95-96). Adults readily deposit eggs on fresh cow dung, and the total development time from egg to adult can be as little as 6 days (RM, pers. obs.). Despite its abundance on fresh cow pads, no pre- or postcopulatory guarding has been observed (RM, pers. obs.). Adults have been found carrying 2 unnamed species of Parasitidae mites. Snowball (1944: 234) reared an unidentified alysiine (Hymenoptera) from a puparium of *niveipennis*.

40. *Sepsis orthocnemis* Frey, 1908

(Figs 24, 388, 393, 459-462)

Sepsis orthocnemis Frey, 1908: 583. Lectotype ♂, FINLAND: Ta, Kangasala, in UZMH [see designation below].

Diagnosis. – Adult *orthocnemis* can be recognised by the pattern of pruinosity on anepimeron, where a diagonal line separates the pruinose and shining areas (Fig. 388), and in the ♂ by the presence of two ventrally-directed and usually diverging spinules in apical third of fore femur (Fig. 459).

Larval morphology as diagnosed for *cynipsea*; i.e. the larvae of this species cannot be reliably separated from those of *cynipsea*, *fulgens*, *lateralis*, *luteipes*, *neocynipsea*, *punctum*, *thoracica*, and *violacea* (but see Meier, 1996: table 2).

Adult. – Colour: Head mostly black; parafacial, facial carina and gena yellowish-brown to brownish. Antenna dark brown, the articulation between pedicel and flagellomere, and part of inner surface of the latter, orange-yellow. Thorax, scutellum and abdomen entirely black. Fore coxa yellow except at base, mid and hind coxae mostly brown; trochanters yellow; femora black except on knees, especially fore knees; fore tibia wholly or mostly yellow, mid and hind tibiae black except at tips; tarsi with tarsomeres 1-2 yellow, the rest dark brown. Wing clear except for a dark rounded to squarish spot at the tip of vein R_{2+3}; calypter, margin and fringe-hairs dark brown; haltere creamy to yellow.

Head: Frontal plate shining, occiput weakly grey pruinose. Outer vertical 1/2 and postocellar 2/3 length of inner vertical. Gena half width of antennal flagellomere. Subvibrissal setae numerous and short.

Thorax: Scutum, postpronotal lobes and scutellum brownish-grey matt and nowhere shining. Setae: anterior notopleural a little shorter than posterior; 2 dorsocentrals; acrostichal and dorsocentral ground-setulae minute. Pleural pruinosity: anepisternum subshining, entirely shagreened; katepisternum entirely white pruinose; anepimeron glossy, thinly pruinose on upper anterior part and more obviously and broadly pruinose on lower posterior half (Fig. 388); meron pruinose, narrowly glossy on hind-margin. Setae reduced: no katepisternal setulae. Scutellum short, twice as wide as long, with a pair of basal setulae.

Legs: Femora subshining. ♂ fore legs in Figs 459-460; ♀ fore legs simple. Mid femur without setae. Mid tibia with 0 anterodorsal, 0-1 posterodorsal and 1 anteroventral. Hind femur and tibia without setae.

Wing: Narrow. Veins bare except for a few setulae in basal part of stem-vein on ventral surface. Entirely covered with microtrichiae.

Abdomen: Subshining but shagreened, with violet reflections. ♂ tergites only weakly expanded behind. Without setae except for lateral marginals on tergites 4 and 5 in ♂; syntergite 1+2 with erect setulae at sides. ♂ sternites with short setulae, not much stronger on sternite 5; genitalia in Figs 461-462; also illustrated by Ozerov (1999e: fig. 351.5).

Wing-length: 2.2-2.8 mm (♂), 2.4-2.9 mm (♀).

Remark. – *S. orthocnemis* is rather constant in colour and structure.

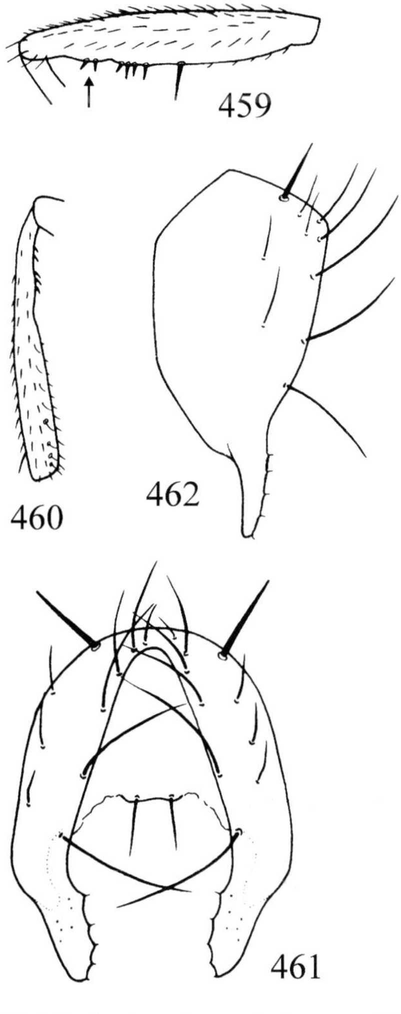

Figs 459-462. *Sepsis orthocnemis* Frey, ♂. 459: fore femur, posterior view; 460: fore tibia, posterior view; 461: hypopygium, dorsal view; 462: hypopygium, lateral view.

Egg. – The egg has been described by Meier (1995b: 427-431).

Larva. – The 3rd instar larva has been fully described by Meier (1996: 91-93, figs 310-318).

Distribution. – Common and widespread in Fennoscandia. Denmark: all provinces (e.g. Hammer, 1941: passim; Lyneborg, 1965: 231). Norway: Ø and A to F (e.g. Hennig, 1949: 78). Sweden: SK to TO (e.g. Ringdahl, 1954b: 231) Finland: Al and Ab to Li (Frey, 1908: 583; Hedström, 1964: 81). Also in the adjacent parts of Russia.

Throughout Europe, North Africa (Algeria), through the Middle East to Afghanistan (Zuska & Pont, 1984: 162), and through Russia to the Far East province (Ozerov, 1989: 844; 1999e: 570). Oriental: Pakistan (Iwasa, 1989: 55).

Records are from mid May to early September in the north, and March to October in the south. The species is most abundant in early spring and autumn.

Biology. – Adults are rather eurytopic with a certain preference for drier habitats (e.g. Bährmann 1993a: 104-105). They occur mainly in pastures (Minder, 1963: 387), on cow dung (Papp, 1976: 128), around animal sheds and byres (Veselkin, 1966: 289), human faeces (Gregor, 1966: 81; RM, pers. obs.), sheep dung (Randall et al., 1981: 187), and horse dung (Pont, 1987: 270). Schulz (1989) collected this species on cow pads that were at least one day old. They have also been found in coastal dunes (Krogerus, 1932: 116) and North Moravian peat-bogs (Roháček & Zuska, 1983: 13). This was the most abundant sepsid species in old gravel pits near Köln, Germany (Püchel, 1992: 420). Adults have also been found on carrion (Duffield, 1938: 165), and on flowers (Tiensuu, 1948: 213), e.g. *Potentilla fruticosa* (Wahlgren, 1915: 54), *Saxifraga hirculus* (Warncke et al., 1993: 145), *Heracleum* and *Polygonum* (Van der Goot, 1986a: 5, 1986b: 33-34), *Rheum* (Randall et al., 1981: 186). A mass occurrence on *Polygonum cuspidatum* was recorded by Van der Goot (1986b: 33).

Overwintering adults have been found in undergrowth (Minder, 1963: 387; Pont, 1979: 21). Its ecology in grasslands in Thuringia has been discussed by Bährmann (1993a: 110), and its general ecology and biology by Hammer (1941: passim). It is a potential carrier of the phoretic stage of the coprophagous nematode *Diplogaster coprophila* Sudhaus & Rehfeld, although no nematodes were found on flies caught in the field (Kiontke, 1996: 356).

S. orthocnemis breeds in cow dung (Hammer, 1941; Laurence, 1954: 250; Papp, 1971a: table 2; 1971b: 432; 1975: tables 1-2; 1976: 122, table 3; Lavčiev & Tsankova, 1980: 4; Papp & Garzó, 1985: 156; Schweiger, 1988; Schulz, 1989). However, larvae have occasionally also been found in sheep dung (Randall et al., 1981: 187; Papp, 1985: 372, 374). Development times are reported by several authors: up to 24 days (Hammer, 1941: 273), 22-24 days (Laurence, 1954: 252), 25 days in midsummer (Pont, 1979: 4), and 25 days (RM, pers.obs.: 12-14 days in the laboratory at 24°C).

Type-material. – Frey described *orthocnemis* from both sexes, with the comment 'wahrscheinlich ist diese Art über ganz Finnland verbreitet'. It is now neither practical nor worthwhile to pinpoint his entire type-series, and for present purposes ACP regards as syntypes only those Finnish specimens in UZMH that are labelled by Frey himself as 'types'. A ♂ from Ta, Kangasala, R. Frey, seen by ACP, has been labelled and is herewith designated as lectotype; it is in excellent condition. Paralectotypes labelled by ACP are as follows (all in UZMH): 1♂, Ta, Birkkala, R. Frey; 1♂, Ib, Maaninga, Lundström; 1♂, Ks, Kuusamo, Mäklin; 1♂, Al, Saltvik, R. Frey; 1♀, Al, Sund, R. Frey. All these specimens are conspecific and belong to *orthocnemis* as currently understood.

41. ***Sepsis pseudomonostigma*** Ursu, 1969

(Figs 390, 463-466)

Sepsis pseudomonostigma Ursu, 1969a: 319, figs 2e-g, 3a-c, 4a-c. Holotype ♂, ROMANIA: Ilfov, Băneasa, in MGAB.

Diagnosis. – In the group of species related to *punctum* (basal medial cell mainly or wholly devoid of microtrichiae, ♂ fore femur with a spinose tubercle at about apical quarter), adult *pseudomonostigma* can be recognised by the wholly pruinose anepimeron (Fig. 390) and by the unusual triangular form of the ♂ surstylus (Figs 465-466).

Adult. – Colour: Head extensively reddish-yellow to yellow, only the occiput, ocellar triangle and upper half of fronto-orbital plate black. Antenna brown, the articulation between pedicel and flagellomere, and part of inner surface of the latter, reddish to yellow. Thorax, scutellum and abdomen entirely black. Legs yellow except for tarsomeres 3-5 on fore leg, mid and hind tibiae brown in basal half. Wing clear except for a dark round to squarish spot at tip of vein R_{2+3}; calypter, margin and fringe-hairs brownish; haltere creamy to yellow.

Head: Frontal plate mostly shining, occiput weakly brownish-grey pruinose. Outer vertical 1/2 and postocellar 2/3 length of inner vertical. Gena 1/2 to 2/3 width of antennal flagellomere. Subvibrissal setae numerous and short.

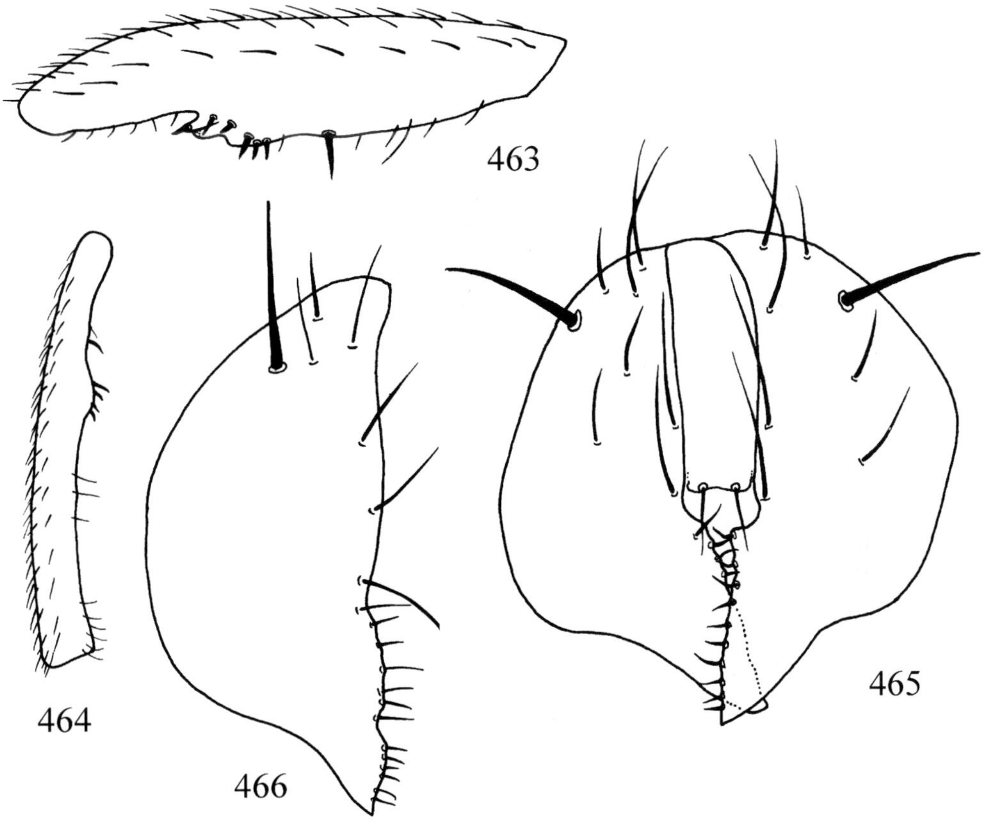

Figs 463-466. *Sepsis pseudomonostigma* Ursu, ♂. 463: fore femur, posterior view; 464: fore tibia, posterior view; 465: hypopygium, dorsal view; 466: hypopygium, lateral view.

Thorax: Scutum, postpronotal lobes and scutellum brownish-grey matt and nowhere shining. Setae: anterior notopleural rather shorter than posterior; 2 dorsocentrals; acrostichal and dorsocentral ground-setulae minute. Pleural pruinosity: anepisternum subshining, entirely shagreened; katepisternum entirely white pruinose; anepimeron completely but thinly pruinose (Fig. 390); meron pruinose, very narrowly glossy on hind-margin. Setae reduced: no katepisternal setulae. Scutellum short, twice as wide as long, with a pair of basal setulae.

Legs: Femora subshining. ♂ fore legs in Figs 463-464; ♀ fore legs simple. Mid femur without, or rarely with, a submedian anterior seta. Mid tibia with 0 anterodorsal, 2 (♂) or ?0 (♀) posterodorsal, and 1 anteroventral setae. Hind femur without setae. Hind tibia without submedian setae.

Wings: Narrow. Veins bare except for a few setulae in basal part of stem-vein on ventral surface. Covered with microtrichiae except in basal medial cell.

Abdomen: Subshining but shagreened, with violet reflections. ♂ tergites only weakly expanded behind. Without setae, except for lateral marginals on tergites 3-5 in ♂ and tergites 4-5 in ♀ (where they are weak); syntergite 1+2 with erect setulae at sides (as in Fig. 385). ♂ sternites with short setulae, strong on sternite 5; genitalia in Figs 465-466.

Wing-length: 2.0-2.2 mm (♂), 2.3-2.6 mm (♀).

Remarks. – This is a small species, similar to *violacea* or a small *punctum*, but it differs from them by the thin pruinosity covering the entire anepimeron and by the very distinctive ♂ genitalia (Figs 465-466).

The taxonomy and species characteristics have been discussed by Munari & Vienna (1982: 10 ff., figs 1-2, 6c, 7-8). The ♂ fore legs and genitalia have also been illustrated by Ozerov (1986b: figs 2-3).

Egg and larva. – Not known.

Distribution. – Not in Fennoscandia.

A relatively little-known and scarce, mainly south and south-east European species: Lithuania (A. L. Ozerov, pers. comm.); Czech Republic (Zuska & Barták, 1997: 74; Barták et al., 1997: 226), Italy (Munari & Vienna, 1982: 16; Munari, 1991a: 26), Sardinia (Munari & Vienna, 1982: 16), Sicily (Munari & Vienna, 1979a: 34), Romania (Ursu, 1984: 228), Turkey (Munari & Vienna, 1982: 16), and Turkmenistan (Ozerov, 1986b: 14).

Records are from April to October.

Biology. – Little is known of the ecology and biology. Barták et al. (1997: 227) speculate that it overwinters as an adult, because this stage has been found in April and October in Bohemia.

Type-material. – The holotype is no. 181520 in MGAB (Weinberg, 1982: 293). It has also been studied by Munari & Vienna (1982: 16).

42. *Sepsis punctum* (Fabricius, 1794)

(Figs 389, 394, 467-470)

Musca punctum Fabricius, 1794: 351. Holotype ♂, 'in Galliae' [FRANCE], in MNHNP [see below].

Musca stigma Panzer, 1798: 21. Holotype/syntypes ♂, AUSTRIA, not in NMW and presumed destroyed [see below].

Sepsis cornuta Meigen, 1826: 288, plate 47 fig. 9; Meigen plate 157 fig. 4 [Morge, 1976a]. Holotype ♀, locality not stated [probably GERMANY: Stolberg], in MNHNP [see below]. **Syn. n.**

Sepsis ornata Meigen, 1826: 290; Meigen plate 281 fig. 6 [Morge, 1976c]. Holotype ♀, locality not stated [probably GERMANY, Stolberg], in MNHNP [see below].

Sepsis pectoralis Macquart, 1835: 478. Syntypes ♂♀, FRANCE: Bordeaux, probably in MHNL.

Sepsis rufocincta Hoffmeister, 1844: 13. Syntypes, sex and locality not stated [GERMANY, Hessen: Kassel district], not located and presumed destroyed [see below]. **Syn. n.**

Sepsis punctum var. *violacea* subvar. *zernyi* Duda, 1926a: 57, 116. Syntypes ♂ ♂, SICILY, Monreale, AUSTRIA, Seebenstein, and TUNISIA, Tunis, in NMW and MNM.

Sepsis punctum var. *quadrisetosa* Duda, 1926a: 57, 116. Syntypes ♂♀, over 130 specimens, many localities in EUROPE and ASIA, in NMW, ZMHU and MNM.

Sepsis icaria Séguy, 1932: 186, figs 111-115. Holotype ♂, FRANCE: Gard, Aigoual, 12-1400 m, in MNHNP [seen by ACP].

Sepsis punctum var. *meridionalis* Séguy, 1932: 190, plate 9 fig. 95. Syntypes 14♂, FRANCE: Var, Callian, SPAIN, MOROCCO, ALGERIA, in MNHNP [see below].

Diagnosis. – Adult *punctum* can be distinguished from its closest relatives (basal medial cell mainly or wholly devoid of microtrichiae, ♂ fore femur with a spinose tubercle at about apical quarter), by the extremely reduced pruinosity on anepimeron (Fig. 389), single dorsocentral seta, and by the form of the ♂ surstylus which is relatively broad and has long setae along the inner margin (Figs 469-470).

Larval morphology as diagnosed for *cynipsea*; i.e. the larvae of this species cannot be reliably separated from those of *cynipsea*, *fulgens*, *lateralis*, *luteipes*, *neocynipsea*, *orthocnemis*, *thoracica*, and *violacea* (but see Meier, 1996: table 2).

Adult. – Colour: Head mostly black; parafacial, facial carina and gena brownish to yellow. Antenna dark brown, the articulation between pedicel and flagellomere, and sometimes part of inner surface of the latter, reddish to yellow. Thorax, scutellum and abdomen entirely black, but abdominal syntergite 1+2 orange in large specimens, especially in ♂. Fore coxa yellow except at base, mid and hind coxae partly to mostly brown; trochanters yellow; femora mostly dark, only narrowly yellow at bases and on knees, to entirely yellow, usually paler in ♀ than in ♂; fore tibia mostly or wholly yellow; mid and hind tibiae mostly dark with paler tips, to mostly yellow with darker bases, also paler in ♀ than in ♂; tarsi with first 2-3 tarsomeres yellow, the others dark. Wing clear except for a dark round to squarish spot at tip of vein R_{2+3}; calypter, margin and fringe-hairs dark brown; haltere creamy to yellow.

Head: Frontal plate mostly shining, occiput weakly brownish-grey pruinose. Outer vertical 1/2 and postocellar 2/3 length of inner vertical. Gena 1/2 to 2/3 width of antennal flagellomere. Subvibrissal setae numerous and short.

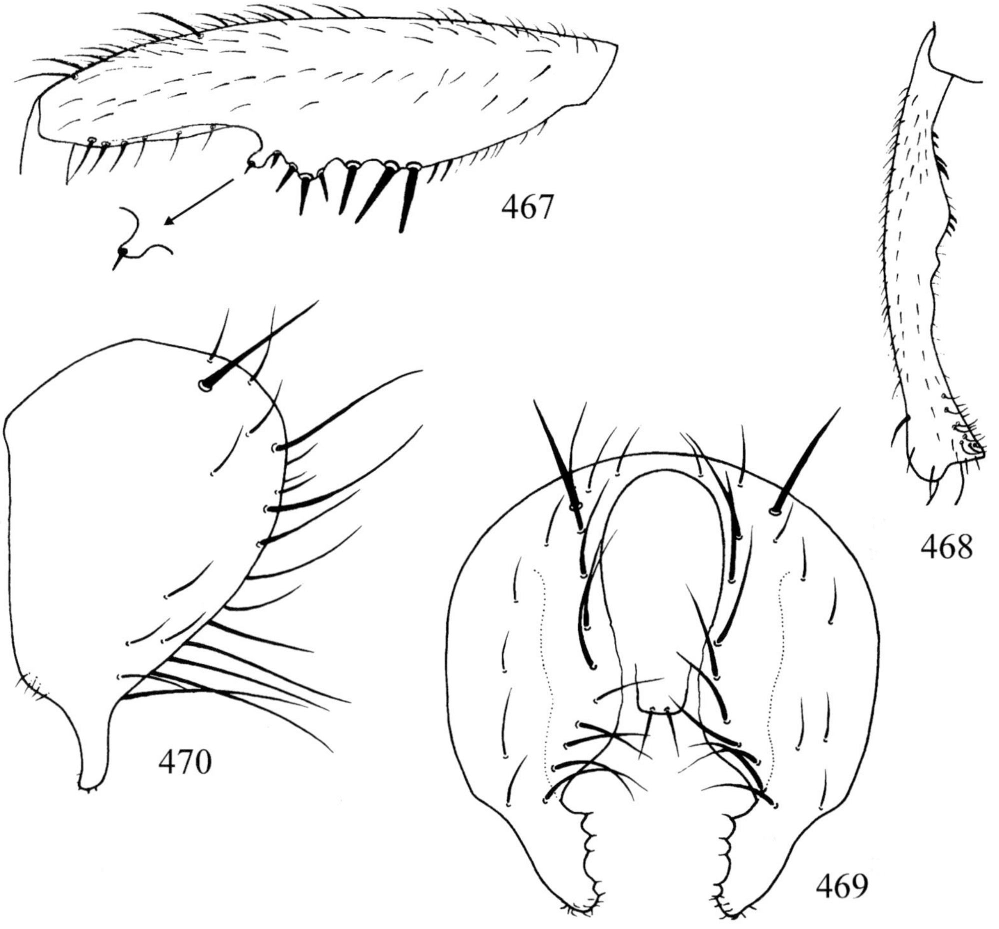

Figs 467-470. *Sepsis punctum* (Fabricius), ♂. 467: fore femur, posterior view (detail: ventral protuberance); 468: fore tibia, posterior view; 469: hypopygium, dorsal view; 470: hypopygium, lateral view.

Thorax: Scutum, postpronotal lobes and scutellum brownish-grey matt and nowhere shining. Setae: anterior notopleural rather shorter than posterior; 1 or 2 dorsocentrals, usually 1; acrostichal and dorsocentral ground-setulae minute, but distinct in large specimens. Pleural pruinosity: anepisternum subshining, entirely shagreened; katepisternum entirely white pruinose; anepimeron glossy, with a little thin pruinosity on upper anterior corner and more conspicuous pruinosity along posterior margin (Fig. 389); meron pruinose, narrowly glossy on hind-margin. Setae reduced: no katepisternal setulae. Scutellum short, twice as wide as long, with a pair of basal setulae.

Legs: Femora subshining. ♂ fore legs in Figs 467-468; ♀ fore legs simple. Mid femur without setae, but in large specimens with several submedian anterior setae. Mid tibia with 0 anterodorsal, 1-2 posterodorsal (up to 4 in large ♂) and 1 anteroventral setae. Hind femur without setae. Hind tibia with 1 submedian anterodorsal.

Wings: Narrow. Veins bare except for a few setulae in basal part of stem-vein on ventral surface. Covered with microtrichiae except in basal medial cell.

Abdomen: Subshining but shagreened, with violet reflections or, in large ♂, with greenish reflections. ♂ tergites only weakly expanded behind. Without setae, except for lateral marginals on tergites 3-5 in ♂ and tergites 4-5 in ♀; syntergite 1+2 with erect setulae at sides (as in Fig. 385). ♂ sternites with short setulae, strong on sternite 5;

genitalia in Figs 469-470.

Wing-length: 2.4-4.3 mm (♂), 2.6-4.0 mm (♀).

Remarks. – *S. punctum* was fully described and illustrated by Ozerov (2000e: 172-174, fig. 6); the ♂ genitalia were also illustrated by Ozerov (1999e: fig. 351.10).

A very variable species in size, colour and setae, and often difficult to separate from *violacea* and *luteipes*, from which it differs most consistently and reliably by the anepimeral pruinosity. The taxonomy has been discussed by Munari & Vienna (1982: 10 ff., figs 3-4, 6a, 9). It varies from having legs and syntergite 1+2 wholly yellow, to legs except knees and abdomen black; larger specimens are more yellow, and also have stronger setae on head, thorax, abdomen and legs (Naswaty, 1960: 113-114). Minder (1963: 380) noted that specimens collected in drier habitats were small and dark, with the abdomen and most of the legs black.

Egg. – Schumann (1962: 425-426, figs 37-44) has included a description of the egg. The egg has been most recently described by Meier (1995b: 427-431).

Larva. – The larva was illustrated by Peterson (1960: fig. D29, i-j, as *pectoralis*). All three larval instars were described by Schumann (1962: 425-426, figs 37-44). The 3rd instar is also described by Mangan (1977: figs 3, 10) and in detail by Meier (1996: 93-96, figs 319-327).

Distribution. – Common and widespread in Fennoscandia. Denmark: most provinces. Norway: Ø and A to F (few published records since Siebke, 1877: 152). Sweden: SK to TO (e.g. Zetterstedt, 1847: 2288; Ringdahl, 1931: 25; Ringdahl, 1951: 142; Ringdahl, 1954b: 231). Finland: Al and Ab to Li (e.g. Frey, 1908: 582; Hedström, 1964: 81; Hedström & Nuorteva, 1971: 123). Also known from the adjacent areas of Russia, north to Arkhangelsk (Frey, 1918: 22).

Throughout Europe, the Macaronesian islands, North Africa, Middle East, eastwards through Russia (Zuska & Pont, 1984: 162) to the Far East province (Ozerov, 1999e: 570). Records from South Korea (Iwasa et al., 1994: 205; Suh & Kwon, 1994: 62) and Japan (Iwasa, 1980: 406; 1995: 794) are misidentifications of *Sepsis monostigma* Thomson (A. L. Ozerov, pers. comm.). Nearctic: Alaska to Quebec, south to California and Alabama, also Mexico (Steyskal, 1965: 683; Mangan, 1977: 338; Ozerov, 2000e: 174). Oriental: Pakistan (Iwasa, 1989: 56) and Taiwan (Iwasa, 1987: 6).

It was reported as declining in abundance in the Netherlands, unlike *violacea* (Van der Goot, 1987a: 82).

Records are from mid May to August in the north, late March to early October in the south.

Biology. – This is mainly a meadow species, though Naswaty (1960: 102-103) found a preference for damp woodland (see also Minder, 1963: 387; Zerbe, 1995: 118). The ecology in grasslands in Thuringia has been discussed by Bährmann (1993a: 104-105, 110) who also concluded that the species is mesophilic. Adults are found around animal sheds and byres (Veselkin, 1966: 789), in coastal sand dunes (Krogerus, 1932: 116), and up to 1740 m in the Tien Shan Mts (Sychevskaya & Vtorov, 1969: 820). They have also been reported from animal feeding pens, small mammal faeces, carrion, rotting vegetation (Mangan, 1977: 339), horse dung (Pont, 1987c: 270), fox dung (Iwasa, 1980: 407), human excrement (Gregor, 1966: 81; Minder, 1963: 387; Iwasa, 1980: 406), and red deer excrement (Roháček & Zuska, 1983: 13).

Adults are recorded as visiting various flowers: Alfken (1924: 470, 474, 475) on *Angelica silvestris*, *Cochlearia officinalis*, *Daucus carota*; Morley (1938: xxvii) on *Smyrnium olusatrum*; Bährmann (1993a: 105) on *Cornus mas*. Adult flies are known to overwinter (Minder, 1963: 387; ACP, pers. obs.), and have been found in late autumn with a well-developed fat-body (Minder, 1963: 387). However, puparia can also hibernate (Lobanov, 1962a: 37).

Adults have been found on cow dung carrying the phoretic nematode *Nygolaimodes borborophilus* (De Man, 1876) Meyl, 1960 (Chizhov, 1996: 185): the nematode is 1.9 mm in length and lives on the surface of the fly, at the rate of one nematode per fly. *S. punctum* is also a potential carrier for the phoretic stage of the coprophagous nematode *Diplogaster coprophila* Sudhaus & Rehfeld, although no nematodes were found on flies caught in the field (Kiontke, 1996: 356).

S. punctum belongs to the group of *Sepsis* species that appear soon after the deposition of the dung (Schulz, 1989; Zerbe, 1995: 121). Parker (1972a: 201-202) first reported precopulatory guarding in this species. He noticed that male attacks on precopulatory pairs and take-overs of the females are much more common in this species than in *cynipsea*. Males arrive earlier at the substrate than females and are much more common on fresh cow pads (Zerbe, 1995: 121). Males attempt to establish territories on the cow pad, and small males are forced into the surrounding vegetation. On a fresh cow pad, only small females succeed in depositing eggs without being mounted by a male, whilst large females are immediately mounted by large males. Precopula pairs are subject to attacks by males looking for a mate. Of 14 pairs that Zerbe (1995: 125) observed, seven pairs were attacked once and four pairs were attacked several times. The almost invariable outcome is for large males to displace small males which may have initially succeeded in mounting a female. Take-over fights can be very violent. Body size varies dramatically, by a factor of 10 when measured by body weight (Zerbe, 1993: 23) and is linked to the abundance of food during larval development. After oviposition, copulation almost always takes place in the surrounding vegetation, but preliminary evidence indicates that additional matings may occur between oviposition bouts, although the sperm of one mating is sufficient to fertilize 500-800 eggs (Schulz, 1999: 20, 31). Small and large males follow different strategies for mate searching: small males wait in the surrounding vegetation for females that leave the pad without being accompanied by a male (Zerbe, 1993: 52; 1995: 129); large males are found on the fresh dung. Since Schulz (1989) found under laboratory conditions that small males live longer than large males, the competitive disadvantage of small adults at fresh cow pads may be partly overcome by a greater longevity. Interestingly, precopulatory guarding as described above is restricted to the European populations and apparently lacking in American populations (Schulz, 1999: fig. 4.2), which led to the study of the genetic basis of the this behaviour (see the general biology section of this volume, page 12).

Compared with other *Sepsis* species, *punctum* is a generalist. It breeds in pig manure (Perris, 1876: 230; Hammer, 1941), rabbit droppings (Werner, 1997: 47), dog dung (RM, pers. obs.), sheep dung (Papp, 1985: 372-377), brown bear dung (Nishijima & Iwasa, 1984: 323-324), liquid manure (Werner, 1997: 47), carrion (Hanski, 1976: 114; Mangan, 1977: 339), and even in rotting woodland vegetation (Mangan, 1977: 339). But the main breeding substrate is probably cow dung (Schumann, 1962: 425; Minder, 1963: 387; Papp, 1975: table 2; Wharton & Moon, 1979: 88; Lavčiev & Tsankova, 1980: 4; Schweiger, 1988; Schulz, 1989; Zerbe, 1995: 116). There are several reports about the development time from egg to adult. Minder (1963: 387) gave 15-18 days for the Moscow area, which is in agreement with the 16 days that Schulz (1989) observed for laboratory cultures. At 24°C development is considerably faster, with the first adults emerging after 11 days. Females can be highly fecund. Under laboratory conditions an average of 82 eggs are produced every 0.8 days, and females become receptive 3-5 days after eclosion (Schulz, 1989). On average the females live 5 weeks and produce 17 clutches (Schulz, 1999: 30).

Type-material. – *Musca punctum*: Fabricius did not give the sex or number of specimens for this species, but gave the locality as 'Habitat in Galliae locis umbrosis, humosis Dom. Bosc.' Zimsen (1964: 494) stated that the type is in the Bosc collection in MNHNP, and the type-specimen was illustrated by Coquebert (1804: 109, plate 24 fig. 14). What appears to be the holotype ♂ was studied by ACP in MNHNP, in the General collection, and is labelled in Bosc's handwriting 'T. [?] punctum / n.s. Fab.'.

Musca stigma: Panzer's illustration shows a male. He writes: 'Habitat in Austria. M. R. Megerle.' Megerle's collection in NMW is known to have been destroyed (Pont, 1986b: 204).

Sepsis cornuta: Meigen described *cornuta* from the ♀ sex without any statement of locality. The specimen was clearly aberrant, perhaps immature: 'diese Art ... zeichnet sich durch die vorne in ein kurzes Horn verlängerte Stirne aus.' The 'horn' was probably the remains of the dry ptilinum. ACP has studied the single ♀ in MNHNP, no. 2231 of the Meigen collection, which is the holotype; there is no material in NMW (see also Becker, 1902: 228). It is labelled 'cornuta ♀' by Meigen. The head, abdomen and several legs are missing. The characters that remain are those of *punctum*; the specimen cannot be *cynipsea*, with which *cornuta* has customarily been synonymised, because the meron is wholly pruinose.

Sepsis ornata: Meigen did not give the number of specimens or locality for his *ornata*, writing simply 'nur das Weibchen'. ACP has studied the holotype in MNHNP, Meigen collection no. 2237 (see also Becker, 1902: 229). It is labelled 'ornata ♀' by Meigen. It belongs to *punctum* as currently defined.

Sepsis rufocincta: Nothing is known about the fate of Hoffmeister's collection, and enquiries by ACP to Professor R. Bährmann and through him to the curator of the Kassel Museum have been unsuccessful; the collection must now be assumed to have been destroyed. Hoffmeister compared his new species with *punctum*, but there is nothing in his description (wing-spot present, abdomen golden-green with 1st segment and antennae and legs red, hind tibia brown at base) to suggest that this is anything but a typical large *punctum*.

Séguy's var. *meridionalis* is the large form of *punctum*, with wholly yellow legs and red syntergite 1+2. ACP has seen 14♂ in MNHNP which appear to be syntypes: France, Var, Callian, 1♂; Morocco, Tanger, various dates, 11♂; Morocco, Rabat, 1♂; Algeria, Boufarik, 1♂.

42a. ***Sepsis setulosa*** (Duda, 1926)

Sepsidimorpha setulosa Duda, 1926a: 148. Holotype ♂, GERMANY: Allgäuer Alpen, Hinterstein, in ZMHU [Ozerov, 1997a: 484].

We are very doubtful about the status of this species. According to the holotype ♂ (ZMHU), which ACP has studied, it has no close relationship with *duplicata*. External characters and the shape of the surstylus, rather, suggest that it is closely related to *flavimana*, as Duda himself observed. Katepisternum and meron pruinose, hind-marginal streak on meron complete; anepimeron glossy, except on hind-margin; syntergite 1+2 setulose at sides; tergite 3 with a lateral marginal seta; gena very narrow; hind tibia without anteroventral seta. In the key given here it runs to *flavimana*, and is most probably a dwarf form of that highly variable species. The holotype has, as is clearly shown in Duda's figure, 5 short posteroventral setae on fore femur, and near base of fore tibia, where the ventral swelling of *flavimana* would usually be, there are 2 strong setulae. The surstylus matches that of the variable *flavimana*. A. L. Ozerov (pers. comm.) originally thought that this is a form of *flavimana*, but in a recent paper (Ozerov, 1997a: 484) he considers it to be a good species of *Sepsis*.

For the present, this remains a 'species dubia', and it is still known only from the holotype. Coe's (1968: 138) record from Slovenia (Postojne) (seen by ACP in BMNH) is only a dwarf specimen of *fulgens* as it has the surstylus and syntergite 1+2 of that species.

43. ***Sepsis thoracica*** (Robineau-Desvoidy, 1830)

(Figs 31, 382, 471-474)

Micropeza thoracica Robineau-Desvoidy, 1830: 742. Syntypes, sex not stated, locality not stated, 'cette espèce se trouve plus particulièrement dans les lieux humides' [FRANCE: probably Saint-Sauveur], not in MNHNP and presumed destroyed.

Sepsis tridens Becker, 1903b: 145. Holotype ♂, EGYPT: Lake Birket-el-Karûn, in ZMHU [Ozerov, 1997a: 485].

Sepsis consanguinea Villeneuve, 1920: 355. Syntypes 7♂ 1♀, 'répandu dans toute la France et connu aussi d'Angleterre' [FRANCE, ENGLAND], in IRSNB [see below].

Sepsis goetghebueri Frey, 1925: 71. 'New name' for *nigripes* sensu Goetghebuer & Bastin (1925: 129, fig. 8), not Meigen [misidentification]. Lectotype ♂, BELGIUM: Virton, in IRSNB [see designation below].

Sepsis idmais Séguy, 1932: 187, figs 116-118. Holotype ♂, FRANCE: Var, Callian, in MNHNP [seen by ACP].

Sepsis ino Séguy, 1932: 188, figs 119-121. Holotype ♂, FRANCE: Morbihan, La Trinité-sur-Mer, in MNHNP [seen by ACP].

Diagnosis. – Among the species with meron glossy in posterior half and pruinose in anterior half, and katepisternum also partly shining and without pruinosity (Fig. 382), adult *thoracica* differs by having a dark spot at the tip of vein R_{2+3} and lacking a milky-white patch beyond this.

Larval morphology as diagnosed for *cynipsea*; i.e. the larvae of this species cannot be reliably separated from those of *cynipsea*, *fulgens*, *lateralis*, *luteipes*, *neocynipsea*, *orthocnemis*, *punctum*, and *violacea* (but see Meier, 1996: table 2).

Adult. – Colour: Head mostly black, with parafacial, facial carina and gena yellow; to almost wholly yellow, with only vertical and occipital areas black. Antenna dark brown, the articulation between pedicel and flagellomere, and sometimes part of inner surface of the latter, yellow; to entirely yellow. Thorax, scutellum and abdomen entirely black; to pleura, sides of scutum, abdominal segment 1+2, ventral part of tergites, and genital segments yellow. Legs black, with coxae brown to yellowish, trochanters, base of mid and hind femora, fore tibia mostly, and tarsomeres 1 yellow; to entirely yellow except for the last 2-3 tarsomeres and part of mid and hind tibiae. Wing clear except for an oval to squarish spot at tip of vein R_{2+3}; calypter and fringe-hairs creamy, margin yellowish; haltere creamy to orange-yellow.

Head: Frontal plate mostly shining, occiput weakly brownish-grey pruinose. Outer vertical 1/2 and postocellar 2/3 length of inner vertical. Gena 1/2 to 2/3 width of antennal flagellomere. Subvibrissal setae numerous and short.

Thorax: Scutum, postpronotal lobes and scutellum brownish-grey matt and nowhere shining. Setae: notopleurals subequal; 2 dorsocentrals; acrostichal and dorsocentral ground-setulae long and conspicuous in large specimens. Pleural pruinosity: anepisternum subshining, entirely shagreened; katepisternum white pruinose, but with a large glossy area on anterior lower part (Fig. 382); anepimeron glossy black, only thinly and weakly pruinose on hind-margin; meron pruinose anteriorly, broadly glossy posteriorly (Fig. 382). Setae reduced: no katepisternal setulae. Scutellum short, twice as wide as long, usually with a pair of basal setulae.

Legs: Femora subshining. ♂ fore legs in Figs 471-472; ♀ fore legs simple. Mid femur without setae, but often with a submedian anterior seta in ♂. Mid tibia with 0 anterodorsal, 1-2 short posterodorsals (but up to 4 in large ♂) and 1 anteroventral setae. Hind femur without setae. Hind tibia with 1 submedian anterodorsal, often with a posterodorsal in ♂ and a second anterodorsal in large ♂.

Wing: Narrow. Veins bare except for a few setulae in basal part of stem-vein on ventral surface. Covered with microtrichiae except in basal medial cell.

Abdomen: Subshining but shagreened, with violet reflections. ♂ tergites only weakly expanded behind. Without setae, except for lateral marginals, or even some median marginals too, on tergites 3-5 in ♂; syntergite 1+2 with erect setulae at sides. ♂ sternites with short setulae, strong on sternite 5; genitalia in Figs 473-474; also illustrated by Ozerov (1999e: fig. 351.6).

Wing-length: 2.0-3.5 mm (♂), 2.4-3.2 mm (♀).

Remarks. – Extremely variable in size and colour, the larger specimens with much more orange colour on head, pleura, abdomen and legs.

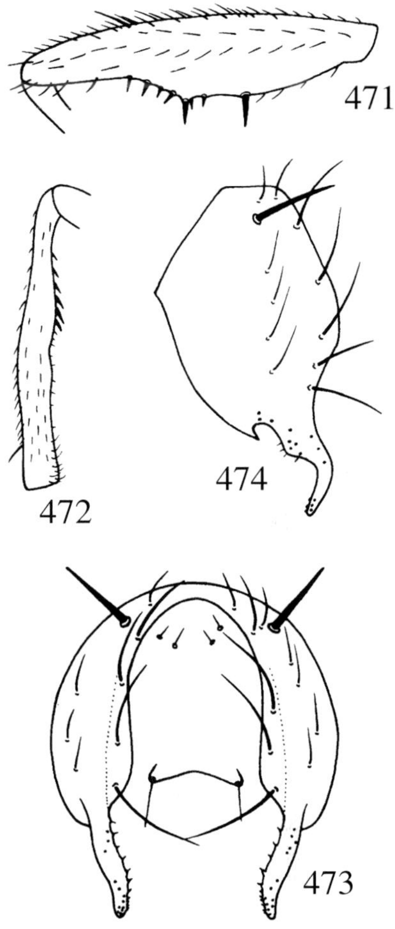

Figs 471-474. *Sepsis thoracica* (Robineau-Desvoidy), ♂. 471: fore femur, posterior view; 472: fore tibia, posterior view; 473: hypopygium, dorsal view; 474: hypopygium, lateral view.

Naswaty (1960: 113) described a ♂ with 4 vertical setae, pleura yellow, syntergite 1+2 yellow-brown, tergites with many setae additional to the usual strong ones, mid tibia with 1 anteroventral and 4 anterodorsals, hind tibia brownish and with 2 anteroventrals and 1 posterodorsal.

Egg. – The egg has been described by Meier (1995b: 427-431).

Larva. – The larva has been illustrated by Hennig (1949: 20, figs 26, 35, 38, 57, 58), Ursu (1973: plate 1 fig. 6), and has been fully described by Meier (1996: 98-100, figs 337-345).

Distribution. – Uncommon in Fennoscandia, and at the northern edge of its range in southern Fennoscandia. Denmark: WJ, LFM, NEZ. Norway: absent. Sweden: SK, HA, NB. Finland: Ab (Pont, 1990: 154).

Common, especially in central and southern Europe, the Macaronesian islands and North Africa, Middle East, eastwards to China (Zuska & Pont, 1984: 163), the Far East province of Russia (Ozerov, 1989: 844), South Korea (Iwasa et al., 1994: 207; Suh & Kwon, 1994: 62), and Japan (Iwasa, 1995: 793). Afrotropical and Oriental regions. Hawaii (Hardy, 1975a, b).

Records are from July to September; in southern Europe from May onwards, even to mid November.

Biology. – Adults are common in pastures and have been found on the dung of cattle and buffalo (Papp, 1976: 128, 130), horse and donkey (Papp, 1976: table 2, 128, 130; Iwasa, 1980: 410), yak (Papp, 1976: 128, 130), sheep (Papp, 1976: 128, 130), pig (Papp, 1974b: table 1; 1976: 128, 130), including manure heaps (Sychevskaya, 1957a: 110), and sometimes on human faeces.

In Europe the species does not exhibit any pre- or postcopulatory guarding. In Zimbabwe, RM commonly collected a species that is considered conspecific with *thoracica*. At mid-elevations it was very common on cow dung and was the earliest visitor to freshly deposited cow pads. Here the males displayed precopulatory guarding. Single females on the cow pad were exceedingly rare. Copulation took place in the surrounding vegetation. Take-over attempts were rarely observed, and the precopulatory guarding very closely resembled that of *cynipsea*.

Adults of *thoracica* have been found to carry the mite *Bonomoia sphaerocerae* Vitzthum. The following Eucoilidae (Hymenoptera) have been reared from puparia (Ionescu & Ursu, 1971: 526-527): *Cothonaspis gracilis* Hartig, *C. nigricornis* Kieffer, *Kleidotoma brevicornis* Thomson, *K. certa* Belizin, *K. geniculata* Hartig and *K. scutellaris* Thomson.

The main breeding substrate is apparently cattle dung. Larvae have been reported in this substrate or adults been reared from it by Hafez (1947: 307), Hennig (1949: 20), Laurence (1954: 252), Papp (1975: tables 1-2; 1976: 122, table 3), Papp & Garzó (1985: 156, 163), Schweiger (1988), and Kirk (1992: 510, 514). Other breeding records include buffalo dung (Hafez, 1947: 307; 1948: 102) and sheep dung (Papp, 1985a: 372, 374). The development time from egg to adult can be very short in buffalo dung in Egypt, only 4 days, 5 hours at 26-28°C (Hafez, 1948: 102). Under laboratory conditions at 24°C development requires 7-8 days. In cow dung, the field development time is 15-22 days (Laurence 1954: 252). The species can overwinter as a larva (Lavčiev & Tsankova, 1982: 61).

Type-material. – *Sepsis goetghebueri*: Although proposed as a 'new name' for *nigripes* sensu Goetghebuer & Bastin (1925: 129) [misidentification], strictly speaking this was a new species. The syntype series is therefore the material listed by Goetghebuer & Bastin from Belgium: Sleidinge, Hingene (Fl.), Deurne (Camp.), Modave (Calc.), Virton (Jur.), June-August. ACP has seen 4♂ syntypes in IRSNB. A ♂ from Virton, dated 16.ix.1920 but collected by Goetghebuer and clearly one of his '*nigripes*', has been labelled by ACP and is herewith designated as lectotype. 3♂ from Modave have been labelled by ACP as paralectotypes.

Sepsis consanguinea: ACP has seen 7♂ 1♀ syntypes in the Villeneuve collection in IRSNB, named as *consanguinea* by Villeneuve, all from France: Luc s. Mer (2♂), St Lunaire (2♂), Royan (1♂ 1♀), St Léger (1♂), and Rambouillet (1♂).

44. *Sepsis violacea* Meigen, 1826
(Figs 384-385, 392, 397-398, 475-478)

Sepsis violacea Meigen, 1826: 289; Meigen plate 157 fig. 2 [Morge, 1976a]. Lectotype ♂, locality not stated [probably GERMANY: Stolberg], in MNHNP [see designation below].

Sepsis fuscipes Roser, 1840: 60. Syntypes 3♂, locality not stated [GERMANY, probably Württemberg], in SMN [Duda, 1926a: 115].

Sepsis uncta Becker, 1913: 648, plate 14 fig. 39. Holotype ♂, IRAN: Khorâsân, 'Gerri-schatur', in ASSP [Ozerov, 1986b: 12].

Sepsis ciliforceps Duda, 1926a: 57, 116, figs 24a-b. Syntypes ♂♀, AUSTRIA, DENMARK, HUNGARY, TUNISIA, POLAND, RUSSIA, SIBERIA, TYROL, in ZMHU, MNM, ZMUC.

Sepsis ciliforceps var. *schineri* Duda, 1926a: 57, 117. Syntypes ♂♀, locality not stated, probably AUSTRIA, in NMW.

Diagnosis. – In the group of species related to *punctum* (basal medial cell mainly or wholly devoid of microtrichiae, ♂ fore femur with a spinose tubercle at about apical quarter), adult *violacea* can be recognised by the pattern of pruinosity on anepimeron (Fig. 392) and by the form of the ♂ surstylus which is relatively slender and has long setae along inner margin (Figs 477-478).

Larval pseudocephalon with two separate cephalic lobes (as in Fig. 401); anal division not elongated and constricted behind anal pad (Fig. 397); with hairs instead of warts around ventral tubercles (Fig. 397); posterior spiracle with all three spiracular openings of approximately the same length (Fig. 398); all peristigmatic tufts with more than two hairs.

Adult. – Colour: Head mostly black; parafacial, facial carina and gena brownish to yellow. Antenna dark brown, the articulation between pedicel and flagellomere, and usually part of inner surface of the latter, reddish to yellow. Thorax, scutellum and abdomen entirely black. Fore coxa yellow except at base, mid and hind coxae mostly brown; trochanters yellow; femora broadly yellow, with indefinite median areas of dark colour, to dark with narrow yellow bases (except fore femur) and yellow knees; fore tibia yellow; mid and hind tibiae mostly dark, mid tibia weakly yellow at tip or both tibiae broadly yellow at tip; tarsi with tarsomeres 1-2 and even 3 yellow, the others dark. Wing clear except for a small dark round or squarish spot at the tip of vein R_{2+3}; calypter, margin and fringe-hairs brown; haltere creamy to yellow.

Head: Frontal plate mostly shining, occiput wholly brownish-grey pruinose. Outer vertical 1/2 and postocellar 3/4 length of inner vertical. Gena 1/3 to 2/3 width of antennal flagellomere. Subvibrissal setae numerous and short.

Thorax: Scutum, postpronotal lobes and scutellum brownish-grey matt and nowhere shining. Setae: notopleurals subequal; 2 dorsocentrals; acrostichal and dorsocentral ground-setulae minute. Pleural pruinosity: anepisternum subshining, entirely shagreened; katepisternum entirely white pruinose; anepimeron extensively pruinose, with only a diagonal central oval or elongate area remaining glossy (Fig. 392); meron pruinose, narrowly glossy on hind-margin. Setae reduced: no katepisternal setulae. Scutellum short, twice as wide as long, with a pair of basal setulae.

Legs: Femora subshining. ♂ fore legs in Figs 475-476; ♀ fore legs simple. Mid femur with 1-2 anterior setae in ♂, otherwise without setae. Mid tibia with 0 anterodorsal, 1 posterodorsal and 1 anteroventral setae. Hind femur without setae. Hind tibia with 1 short submedian anterodorsal in ♂.

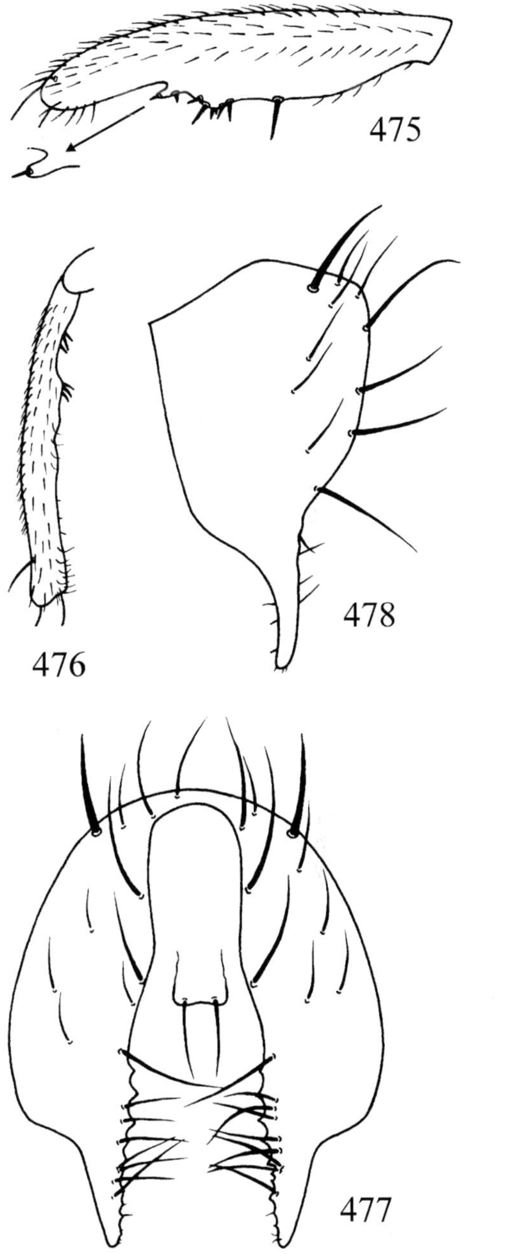

Figs 475-478. *Sepsis violacea* Meigen, ♂. 475: fore femur, posterior view (detail: ventral protuberance); 476: fore tibia, posterior view; 477: hypopygium, dorsal view; 478: hypopygium, lateral view.

Wing: Narrow. Veins bare except for a few setulae in basal part of stem-vein on ventral surface.

Covered with microtrichiae except in basal medial cell.

Abdomen: Subshining but shagreened, with violet reflections. ♂ tergites only weakly expanded behind. Without setae, except for lateral marginals on tergites 3-5 in both sexes; syntergite 1+2 with erect setulae on sides (Fig. 385). ♂ sternites with short setulae, strong on sternite 5; genitalia in Figs 477-478; also illustrated by Ozerov (1999e: fig. 351.13).

Wing-length: 2.6-3.4 mm (♂), 2.8-3.6 mm (♀).

Remarks. – Usually smaller and darker than *punctum*, but also rather variable and often and easily confused with a small *punctum* with 2 dorsocentral setae and *luteipes*; however, it can be reliably separated by the pattern of the anepimeral pruinosity. The taxonomy and specific characters have been discussed by Munari & Vienna (1982: 10 ff., figs 5, 6b, 10).

Egg. – The functional morphology of the egg is discussed in Hinton (1960: 47-50, fig 1). The egg has been described by Meier (1995b: 427-431).

Larva. – There is a description of the puparium in Mohr (1943: 287), and an old description of the larva by Porchinskiy (1910: 86-88, figs 90, 92-94). The larva is fully described by Meier (1996: 100-102, figs 346-354).

Distribution. – Widespread and common in Fennoscandia. Denmark: sparse (few published records, e.g. Stæger, 1844: 26, correctly identified). Norway: from Ø and A to TR (few records, e.g. Siebke, 1877: 153). Sweden: SK to TO (e.g. Zetterstedt, 1847: 2289; Ringdahl, 1951: 142; 1954b: 231; 1960: 18). Finland: from Al and Ab to Li (e.g. Frey, 1908: 582; Hedström, 1964: 81). Also in the adjacent parts of Russia, except for Lapland.

Throughout Europe, North Africa, Middle East, eastwards to Mongolia (Zuska & Pont, 1984: 163) and to the Far East province of Russia (Ozerov, 1999e: 570) and Japan (Iwasa, 1980: 407; 1995: 795). Oriental: Pakistan (Iwasa, 1989: 56).

It was reported as increasing in abundance in the Netherlands, unlike *punctum* (Van der Goot, 1987a: 82).

Records are from May to mid September in the north, from late March to mid September in the south.

Biology. – This is a grassland, meadow species, often around dung heaps and sewage (Duda, 1926a: 19), but seldom together with *punctum*. Its ecology in grasslands in Thuringia has been discussed by Bährmann (1993a: 104-105, 110). It has also been recorded from Moravian peat bogs (Roháček & Zuska, 1983: 14). Adults of *violacea* and *punctum* are morphologically very similar, but they differ remarkably in larval morphology, behaviour and ecology (Schweiger, 1988). Some of the older reports on the ecology of *violacea* should be used with care since they may actually refer to *punctum*, or even to *luteipes*. This is particularly the case for Hinton's (1960: 49) note that this species has a precopula. In laboratory cultures of this species, there was never any indication of such behaviour.

Zuska & Laštovka (1969: 209) considered this to be the most synanthropic species of European *Sepsis*, but Minder (1963: 386) reported that although common near buildings it is rarely found in houses, although it can be common on the walls of outhouses (see also Evenhuis in Van der Goot, 1985b: 179; 1987a: 82). Naswaty (1960: 102-103) found *violacea* to have greater ecological tolerance than *punctum*, and found it more commonly in meadows and less commonly around cow dung. The adults occasionally visit flowers (Silén, 1906: 132; Bährmann, 1993: 105, *Cornus mas*). Adults have been found carrying the mites *Bonomoia sphaerocerae* Vitzthum, *Macrocheles plumiventris* Hull, and unidentified species of Parasitiformes. The species has been found in very low numbers in swarms of *fulgens*, but tends to form aggregations around ammonia sources (Pont; 1987a: 314-315).

Larvae have been found in cow dung and dunghills (Papp, 1975: tables 1-2; Lavčiev & Tsankova, 1980: 4), cow manure inside stables (Zuska, 1960: 22; Van der Goot, 1987a: 82), rabbit dung (Werner, 1997: 97), pig manure (Papp, 1974b: 144), wild brown bear dung (Nishijima & Iwasa, 1984: 323-324), liquid manure (Werner, 1997: 97), and human faeces (Sychevskaya, 1968: 51, 52; 1972c: 539). It breeds in cattle dung (Mohr, 1943: 287; Van der Goot, 1987a: 82; RM, pers. obs.: in the laboratory), pig dung (Papp, 1974b: 144), human excrement (see Mohr, 1943: 287), and decaying vegetable matter (see Mohr, 1943: 287). The development time under laboratory conditions is 7-10 days at 24°C. *S. violacea* is known to pupariate in the cow pad. It hibernates as a puparium in North America (Mohr, 1943: 287), but in Europe overwintering puparia were found at Ivanovo (Lobanov, 1962a: 37), and overwintering adults have also been found in Uzbekistan (Sychevskaya, 1981: 89) and England (ACP, pers. obs.).

Type-material. – *Sepsis violacea*: Meigen described *violacea* without any indication of sex, number of specimens or locality, beyond the comment 'sie ist sehr gemein'. ACP has studied 1♂ in MNHNP, Meigen collection no. 2234 (see also Becker, 1902: 229; Collin, 1910: 175). It is labelled 'violacea ♂' by Meigen. ACP has labelled it and designates it herewith as lectotype. Duda (1926a: 114, 115, 118) saw specimens in NMW, including 4♀ which he identified as *ciliforceps* (= *violacea*), but ACP located only 1♂, labelled by Meigen as 'Sepsis / viola- / cea', which has been labelled as paralectotype. Both syntypes are *violacea* as currently understood.

As is evident from Frey's (1908) key and notes, his early *violacea* consisted mainly of *punctum* (together with some *violacea* and *fulgens*).

Species from peripheral areas

Sepsis alanica Ozerov, 1999

Sepsis alanica Ozerov, 1999a: 57, 7 figs. Holotype ♂, RUSSIA: North Ossetia, Caucasus Mts, mountain ridge Bakhty Lapparyrag, 10 km SE Alagir, in ZMUM.

Palaearctic species, known only from North Ossetia, and close to *flavimana* and *nigripes*.

Archisepsis discolor (Bigot, 1857)

Sepsis discolor Bigot, 1857a: 346; 1857b: 823. Holotype ♀, CUBA, in MNHNP [see below].
Sepsis mequignoni Séguy, 1936: 17, fig. a. Holotype ♂, AZORES: San Miguel, Sete Cidades, in MNHNP [see below].

Type-material. – *S. discolor*: described from an unspecified number of unsexed specimens. The holotype ♀ has been seen by ACP in MNHNP, no. 2564 in box 85 of the Macquart collection (Guérin-Méneville's Diptera from Cuba).

S. mequignoni: The holotype has been seen by ACP in MNHNP, collected viii.1930 by L. Chopard. One fore leg is mounted on slide no. 562.

Neotropical species; recorded from the Azores Islands (as *mequignoni* Séguy) and certainly an introduction. This synonymy was established by Ozerov (1993d: 248; 1994a: 81).

Sepsis freidbergi Ozerov, 1987

Sepsis freidbergi Ozerov, 1987b: 79, 3 figs. Holotype ♂, ISRAEL: Sinai, Qzaima, in TAU.

Palaearctic species, known only from Israel, and close to *punctum*.

Sepsis geniculata Bigot, 1891

Sepsis geniculata Bigot, 1891: 278. Holotype ♂, CANARY ISLANDS, in UMO [see below].

Type-material. – Described from 1♂, collector C. Alluaud. ACP has seen the holotype ♂, which is Diptera no. 467 in the UMO Type Collection. With a small handwritten label 'Canaria / 17.', and a Bigot provisional label: 'Sepsis / Geniculata. ♂. / n.sp. Inedict. / Quincy. Jan [?] / 1891. J.Bigot / Canaria / Alluaud.'. It was lent to T. Becker on 20.x.1906.

A. L. Ozerov (pers. comm.) considers *geniculata* to be a good species (rather than a junior synonym of *punctum* as it has always been treated) and *geniculata* (2 dorsocentrals) and *punctum* (1 dorsocentral) to be a species-pair.

Sepsis macrochaetophora Duda, 1926

Sepsis macrochaetophora Duda, 1926a: 53. Lectotype ♂, ABYSSINIA: Marako, in ZMHU [by designation of Ozerov, 1997a: 482].
Sepsis nephodes Séguy, 1936: 18, fig. b. Holotype ♂, AZORES: Terceira, Monte Brasil, in MNHNP [see below].

Type-material. – *S. nephodes*: ACP has seen the holotype in MNHNP, collected in 1930 by L. Chopard. One fore leg is mounted on slide no. 704.

Afrotropical species; recorded from the Azores Islands (as *nephodes* Séguy) and certainly an introduction. This synonymy was established by Ozerov (1993d: 248).

Sepsis mediana Iwasa, 1989

Sepsis mediana Iwasa, 1989: 54, figs 3-6. Holotype ♂, PAKISTAN: Lalazar, Naran, in OUAVM.

Oriental species (Pakistan); recorded from Uzbekistan and Turkmenistan by Ozerov (1999a: 57), and close to *flavimana* and *nigripes*.

Catalogue

				DENMARK											SWEDEN																													
		Germany	Gr. Britain	SJ	EJ	WJ	NWJ	NEJ	F	LFM	SZ	NWZ	NEZ	B	SK	BL	HA	SM	ÖL	GO	GS	ÖG	VG	BO	DS	NÄ	SÖ	UP	VS	VR	DR	GÄ	HS	ME	HR	JÄ	ÅN	VB	NB	ÅS	LY	PI	LU	TO
Orygma luctuosum Meigen, 1830	1	●	●					●					●		●		●							●																				
Ortalischema albitarse (Zetterstedt, 1847)	2																	●		●						●		●			●					●								
Zuskamira inexpectata Pont, 1987	3																											●	●		●													
Themira annulipes (Meigen, 1826)	4	●	●	●	●		●	●	●	●		●	●	●	●	●	●	●	●	●		●	●	●		●	●	●	●		●					●			●		●			●
Themira arctica (Becker, 1915)	5																																								●		●	●
Themira biloba Andersson, 1975	6	●	●										●		●																													
Themira germanica Duda, 1926	7	●	●												●																?								●		●		●	●
Themira gracilis (Zetterstedt, 1847)	8	●	●										?		●		●	●										●	●		●					●								
Themira leachi (Meigen, 1826)	9	●	●		●	●	●	●		●		●	●		●							●					●				●					●		●	●		●			●
Themira lucida (Stæger, 1844)	10	●	●	●	●	●	●	●	●	●			●	●	●	●	●	●	●	●		●						●								●			●	●				●
Themira malformans Melander & Spuler, 1917	11																																								●			●
Themira minor (Haliday, 1833)	12	●	●		●	●	●	●	●	●			●	●	●		●	●	●	●		●						●			●													

		NORWAY																		
		Ø + AK	HE s+n	O s+n	B ø+v	VE	TE y+i	AA y+i	VA y+i	R y+i	HO y+i	SF y+i	MR y+i	ST y+i	NT y+i	Ns y+i	Nn ø+v	TR y+i	F v+i	F n+ø
Orygma luctuosum Meigen, 1830	1	●								●				●	●		●	●	●	
Ortalischema albitarse (Zetterstedt, 1847)	2														●					
Zuskamira inexpectata Pont, 1987	3																			
Themira annulipes (Meigen, 1826)	4	●			●	●			●	●	●		●					●		
Themira arctica (Becker, 1915)	5																	●		
Themira biloba Andersson, 1975	6																			
Themira germanica Duda, 1926	7																	●		
Themira gracilis (Zetterstedt, 1847)	8									●								●		
Themira leachi (Meigen, 1826)	9	●	●		●					●	●	●		●	●	●	●	●	●	●
Themira lucida (Stæger, 1844)	10									●							●	●		
Themira malformans Melander & Spuler, 1917	11			●													●	●		
Themira minor (Haliday, 1833)	12	●																		

		FINLAND																				RUSSIA		
		Al	Ab	N	Ka	St	Ta	Sa	Oa	Tb	Sb	Kb	Om	Ok	Ob S	Ob N	Ks	Lk W	Lk E	Le	Li	Vib	Kr	Lr
Orygma luctuosum Meigen, 1830	1																							●
Ortalischema albitarse (Zetterstedt, 1847)	2							●															●	
Zuskamira inexpectata Pont, 1987	3																							
Themira annulipes (Meigen, 1826)	4	●	●	●	●		●	●	●		●	●		●	●	●			●			●	●	
Themira arctica (Becker, 1915)	5																	●		●	●			●
Themira biloba Andersson, 1975	6			●	●																			
Themira germanica Duda, 1926	7		?	●		?	●																	
Themira gracilis (Zetterstedt, 1847)	8		●				●	●															●	
Themira leachi (Meigen, 1826)	9	●	●	●	●	●	●	●	●	●		●	●	●		●	●	●		●	●	●		●
Themira lucida (Stæger, 1844)	10	●	●	●			●	●	●	●		●			●		●				●	●		●
Themira malformans Melander & Spuler, 1917	11																	●			●			●
Themira minor (Haliday, 1833)	12	●	●	●				●					●		●							●	●	?

				DENMARK											SWEDEN																													
		Germany	Gr. Britain	SJ	EJ	WJ	NWJ	NEJ	F	LFM	SZ	NWZ	NEZ	B	SK	BL	HA	SM	ÖL	GO	GS	ÖG	VG	BO	DS	NÄ	SÖ	UP	VS	VR	DR	GÄ	HS	ME	HR	JÄ	ÅN	VB	NB	ÅS	LY	PI	LU	TO
Themira nigricornis (Meigen, 1826)	13	●	●		●				●				●		●		●	●									●	●			●						●		●					●
Themira paludosa Elberg, 1963	14																																											
Themira pusilla (Zetterstedt, 1847)	15		●												●													●			?					●			?	●			●	●
Themira putris (Linnaeus, 1758)	16	●	●	●	●	●	●	●	●	●	●		●	●	●		●	●	●	●		●		●		●		●	●		●					●			●				●	●
Themira ringdahli Pont, sp. n.	17																																											●
Themira superba (Haliday, 1833)	18	●	●	●	●	●	●	●		●		●	●		●			●	●	●		●																			●			
Saltella sphondylii (Schrank, 1803)	19	●	●	●	●	●	●	●	●	●	●	●	●	●	●				●	●				●								●												
Nemopoda nitidula (Fallén, 1820)	20	●	●	●	●	●	●	●	●	●	●	●	●	●	●	●	●	●	●	●		●	●	●	?	●	●	●	●	●	●		?			●	?		●	●	●		●	●
Nemopoda pectinulata Loew, 1873	21	●	●		●			●		●	●		●	●	●			●				●				●	●	●		●	●					●	?		●	●	●			●
Nemopoda speiseri (Duda, 1926)	22	●																●									●																	
Meroplius minutus (Wiedemann, 1830)	23	●	●	●	●				●				●		●			●						●		●		●			●	●												
Sepsis biflexuosa Strobl, 1893	24	●	●		●	?	?	●					●		●				●	●																								

		NORWAY																			FINLAND																				RUSSIA		
		Ø + AK	HE s+n	O s+n	B ø+v	VE	TE y+i	AA y+i	VA y+i	R y+i	HO y+i	SF y+i	MR y+i	ST y+i	NT y+i	Ns y+i	Nn ø+v	TR y+i	F v+i	F n+ø	Al	Ab	N	Ka	St	Ta	Sa	Oa	Tb	Sb	Kb	Om	Ok	Ob S	Ob N	Ks	Lk W	Lk E	Le	Li	Vib	Kr	Lr
Themira nigricornis (Meigen, 1826)	13	●				●								●	●							●	●			●	●	●			●		●	●		●					●	●	
Themira paludosa Elberg, 1963	14																					●	●		●					●						●						●	
Themira pusilla (Zetterstedt, 1847)	15			●														?												●			●		?	●	●	●	●	●		●	●
Themira putris (Linnaeus, 1758)	16	●		●						●	●	●	●		●	●	●	●	●	●	●	●	●			●	●	●	●	●	●	●	●	●	●	●	●		●	●	●	●	●
Themira ringdahli Pont, sp. n.	17																																										
Themira superba (Haliday, 1833)	18			●						●	?								●		●	●	●			●	●	●			●	●		●	●		●				●	●	
Saltella sphondylii (Schrank, 1803)	19									●																●	●				●										●		
Nemopoda nitidula (Fallén, 1820)	20	●	?	?	?	?		?	●	●	●		●			●	●	●	●	●	●	●	●	●	●	●	●	●	●		●	?	●	●	●	●	●		●	●	●	●	●
Nemopoda pectinulata Loew, 1873	21	●	?	●	●				●		●	●		●		?	●	●				●	●	●	●	●	●	●	●	●	●	●	●	●		●	●	?	●	●	●	●	●
Nemopoda speiseri (Duda, 1926)	22																									●					●												
Meroplius minutus (Wiedemann, 1830)	23	●						●														●	●	●		●		●	●		●										●	●	
Sepsis biflexuosa Strobl, 1893	24																	●				●														●							

				DENMARK												SWEDEN																													
		Germany	Gr. Britain	SJ	EJ	WJ	NWJ	NEJ	F	LFM	SZ	NWZ	NEZ	B	SK	BL	HA	SM	ÖL	GO	GS	ÖG	VG	BO	DS	NÄ	SÖ	UP	VS	VR	DR	GÄ	HS	ME	HR	JÄ	ÅN	VB	NB	ÅS	LY	PI	LU	TO
Sepsis cynipsea (Linnaeus, 1758)	25	●	●	●	●	●	●	●	●	●	●	●	●	●	●	●	●	●	●	●	●	●	●	●		●	●	●	●	●	●			●		●			●	●	●		●	●
Sepsis duplicata Haliday, 1838	26	●	●	●	●		●	●						●	●	●	●	●	●	●			●			●	●				●													●
Sepis flavimana Meigen, 1826	27	●	●	●	●	●	●	●	●	●			●	●	●	?	●	●	●	●		?	●	●	?	●	●	●	●		●					●	?		●	●	●			●
Sepsis fulgens Meigen, 1826	28	●	●	●	●	●	●	●	●	●	●	●	●	●	●	●	●	●	●	●	●	●	●	●	●	●	●	●	●	●	●					●	●				●			
Sepsis luteipes Melander & Spuler, 1917	29	●	●																																									●
Sepsis nigripes Meigen, 1826	30	●	●										●						●	●																								
Sepsis orthocnemis Frey, 1908	31	●	●	●	●	●	●	●	●	●	●	●	●	●	●	●	●	●	●	●		●	●	●		●	●	●	●	●	●					●			●		●		●	●
Sepsis punctum (Fabricius, 1794)	32	●	●	●	●			●	●	●	●		●	●	●	●	●	●	●	●		●	●	●		●	●	●	●	●	●	●				●			●		●		●	●
Sepsis thoracica (Robineau-Desvoidy, 1830)	33	●	●			●				●			●		●		●																						●					
Sepsis violacea Meigen, 1826	34	●	●	●				●		●				●	●	●	●	●	●	●		●	●	●	●	●	●	●	●	●	●					●	●	●	●	●	●		●	●

		NORWAY																			FINLAND																				RUSSIA		
		Ø + AK	HE s+n	O s+n	B ø+v	VE	TE y+i	AA y+i	VA y+i	R y+i	HO y+i	SF y+i	MR y+i	ST y+i	NT y+i	Ns y+i	Nn ø+v	TR y+i	F v+i	F n+ø	Al	Ab	N	Ka	St	Ta	Sa	Oa	Tb	Sb	Kb	Om	Ok	Ob S	Ob N	Ks	Lk W	Lk E	Le	Li	Vib	Kr	Lr
Sepsis cynipsea (Linnaeus, 1758)	25	●		●	●		●	●	●	●	●	●			●	●	●	●	●	●	●	●	●	●	●	●	●	●	●		●		●	●	●	●	●	●	●	●	●	●	●
Sepsis duplicata Haliday, 1838	26																				●	●	●			●	●			●	●											●	
Sepis flavimana Meigen, 1826	27	●		?	●			●	●	●	●		●		●	●	●	●	●	●	●	●	●	●	●	●	●	●	●	●				●	●	●	●		●	●	●	●	●
Sepsis fulgens Meigen, 1826	28	●		●		●		●	●	●	●	●			●	●	●	●			●	●	●	●	●	●	●	●	●		●						●				●	●	
Sepsis luteipes Melander & Spuler, 1917	29																					●	●			●			●	●						●	●						
Sepsis nigripes Meigen, 1826	30																		●		●																						
Sepsis orthocnemis Frey, 1908	31	●		●	●			●		●	●	●			●	●	●	●	●	●	●	●	●	●	●	●	●	●	●	●	●	●	●	●	●	●	●		●	●	●	●	●
Sepsis punctum (Fabricius, 1794)	32	●		●	●	●	●		●	●	●	●	●	●			●		●	●	●	●	●	●	●	●	●	●	●	●	●	●	●	●	●	●	●			●	●	●	●
Sepsis thoracica (Robineau-Desvoidy, 1830)	33																					●																					
Sepsis violacea Meigen, 1826	34	●	●	●	●		●			●	●	●		●	●	●	●	●			●	●	●	●	●	●	●	●	●	●	●		●	●			●	●		●	●	●	

Plates

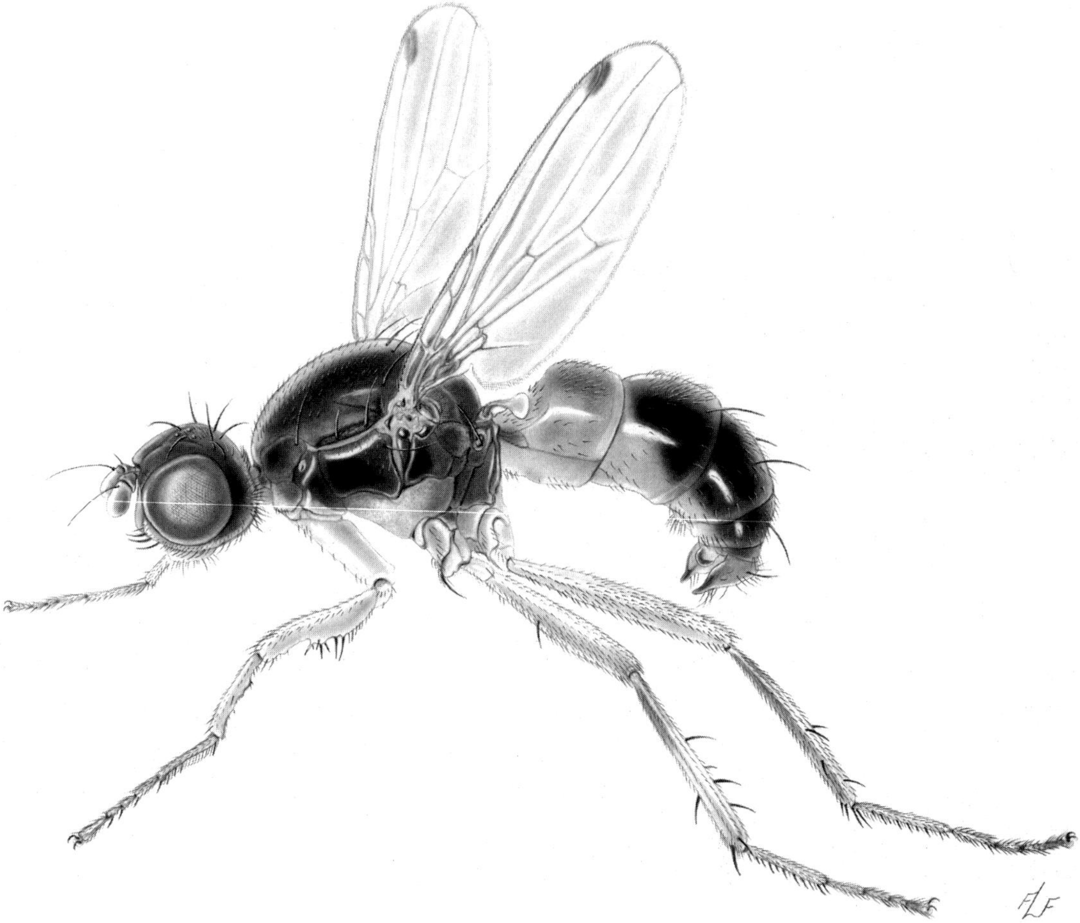

Plate 1. *Sepsis punctum* (Fabricius).

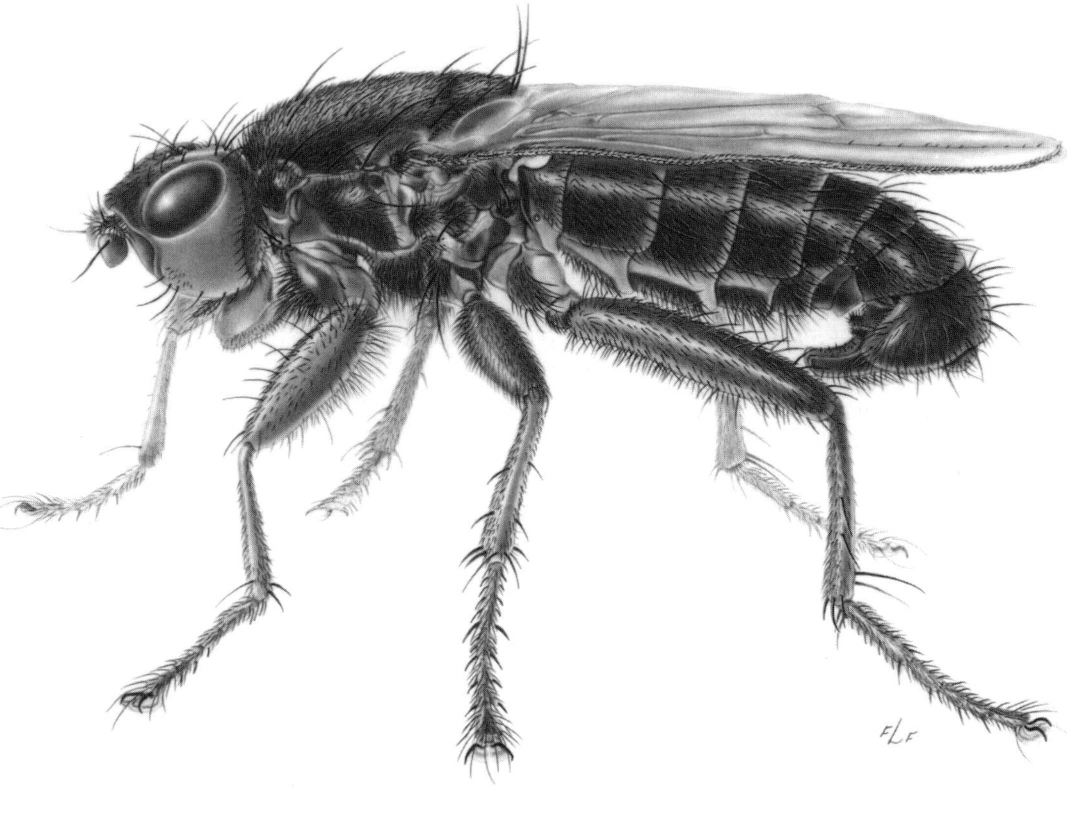

Plate 2. *Orygma luctuosum* Meigen.

Literature

Ahlqvist, H., 1935: Dipterenfunde auf Rågöarna und bei Baltischport (NW-Estland). – Memo Soc. Fauna Flora fenn. 10: 212-215.

Alfken, J. D., 1924: Die Insekten des Memmert. Zum Problem der Besiedelung einer neuentstehenden Insel. – Abh. naturw. Ver. Bremen 25 (3): 358-481.

Allen, G. R. & Simmons, L. W., 1996: Coercive mating, fluctuating asymmetry and male mating success in the dung fly *Sepsis cynipsea*. – Anim. Behav. 52: 737-741, 1 table.

Andersson, H., 1967: Faunistic, ecological and taxonomic notes on Icelandic Diptera. – Opusc. ent. 32 (1-2): 101-120, 6 figs.

Andersson, H., 1975: A new species of *Themira* R.-D. from South Sweden (Diptera: Sepsidae). – Entomologica scand. 6 (1): 57-60, 12 figs.

Ardö, P., 1957: Studies in the marine shore dune ecosystem with special reference to the Dipterous fauna. – Opusc. ent. Suppl. 14: 255 pp., 82 figs, 11 tables.

Ashmole, N. P., Nelson, J. M., Shaw, M. R. & Garside, A., 1983: Insects and spiders on snowfields in the Cairngorms, Scotland. – J. nat. Hist. 17 (4): 599-613.

Aubertin, D., 1932: Notes on certain species of the genus *Orthellia*, with a description of one new species. – Ann. Mag. nat. Hist. (10) 11 [1933]: 139-144.

Backlund, H. O., 1945: Wrack fauna of Sweden and Finland, ecology and chorology. – Opusc. ent. Suppl. 5: 237 pp., 29 figs, 6 plates, 11 tables.

Báez, M., 1982: Revision of the Sepsidae (Diptera) from the Canary Islands. – Vie Milieu 32 (1): 69-72, 18 figs.

Bährmann, R., 1993a: Zum Vorkommen der Sepsidae in unterschiedlichen Rasenökosystemen Thüringens. – Beitr. Ent. 43 (1): 97-113, 3 figs, 13 tables.

Bährmann, R., 1993b: 19. Sepsidae. – Pp. 62-63 *in* Menzel, F. & Bährmann, R.: Zweiflügler (Diptera) Ostdeutschlands. Kritische Liste ausgewählter Familien. – Selbstverlag DEI, Eberswalde-Finow (Nova Supplementa Entomologica 5), 92 pp.

Bährmann, R., 1996: Winteraktive Zweiflügler (Insecta, Diptera Brachycera) in Xerothermrasen Thüringens. – Studia dipterologica 3 (2): 259-274, 10 figs, 3 tables.

Barták, M., Zeman, V. & Zuska, J., 1997: Sepsidae, Piophilidae, Chloropidae. – Folia Facultatis Scientiarum Naturalium Universitatis Masarykianae Brunensis Biologia, 95: 226-227.

Becker, T., 1902: Die Meigen'schen Typen der sogen. Muscidae acalypterae (Muscaria holometopa) in Paris und Wien. [Part.] – Z. syst. Hymenopt. Dipterol. 2 (4): 209-256.

Becker, T., 1903a: Die Typen der v. Roser'schen Dipteren-Sammlung in Stuttgart. – Jh. Ver. vaterl. Naturk. Württ. 59: 52-66.

Becker, T., 1903b: Aegyptische Dipteren. (Fortsetzung und Schluss.) – Mitt. zool. Mus. Berl. 2 (3): 67-195, plates 4b-4e (figs 10-91).

Becker, T., 1907: Zur Kenntniss der Dipteren von Central-Asien. I. Cyclorrhapha schizophora holometopa und Orthorrhapha brachycera. – Ezheg. zool. Muz. 12 (3): 253-317, 2 plates with 18 figs.

Becker, T., 1908: Dipteren der Kanarischen Inseln. – Mitt. zool. Mus. Berl. 4 (1): 1-180, plates 1-4 (figs 1-59).

Becker, T., 1913: Persische Dipteren von den Expeditionen des Herrn N. Zarudny 1898 und 1901. – Ezheg. zool. Muz. 17 (3-4) [1912]: 503-654, figs 20-27, plates 12-13.

Becker, T., 1915a: Cyclorrhapha schizophora. Holometopa. – Pp. 63-67, figs 5-7 *in* Becker, T., Dziedzicki, H., Schnabl, J., & Villeneuve, J.: Résultats scientifiques de l'éxpédition des frères Kuznecov (Kouznetzov) à l'Oural Arctique en 1909, sous la direction de H. Backlund. Diptera. – Zap. imp. Akad. Nauk (VIII), cl. phys.-math. 28 (7):1- 67 (+ 4), 7 figs, plates 1-3 (11-13). [Publication date ii.1915.]

Becker, T., 1915b: Fauna Faeröensis. Ergebnisse einer Reise nach den Faeröer, ausgeführt im Jahre 1912 von Alfons Dampf und Kurt v. Rosen. VI. Dipteren. 3. Orthorrhapha brachycera, Cyclorrhapha aschiza und schizophora (exkl. Anthomyiinae). – Zool. Jb. (Abt. Syst.) 39 (1): 121-134, 2 figs. [Publication date 8.x.1915.]

Bigot, J. M. F., 1857a: Dipteros. – Pp. 328-49, plate 20 *in* Sagra, R. de la: Historia fisica politica y natural de la Isla de Cuba. Segunda parte. Historia natural. Tomo VII. Crustaceos, aragnides é insectos. – A. Bertrand, Paris, xxxii + 371 pp.

Bigot, J. M. F., 1857b: Ordre des Diptères, Latr. – Pp. 783-829, plate 20 *in* Guérin-Méneville, F. E.: Histoire physique, politique et naturelle de l'Ile de Cuba. Animaux articulés à pieds articulés. – A. Bertrand, Paris, lxxxvii + 868 pp.

Bigot, J. M. F., 1891: Voyage de M. Ch. Alluaud aux Iles Canaries (Novembre 1889 - Juin 1890). Diptères. – Bull. Soc. zool. Fr. 16: 275-279.

Bishop, A. L., McKenzie, H. J., Barchia, I. M. & Harris, A. M., 1998: Occurrence and effect of temperature regimes on four species of fly (Diptera) found with *Culicoides brevitarsis* Kieffer (Ceratopogonidae) in bovine dung. – Gen. Appl. Ent. 28: 93-99, 5 tables, 2 figs.

Bishop, A. L., McKenzie, H. J., Barchia, I. M., Murison, R. & Spohr, L. J., 1996: Positions of juvenile stages of *Culicoides brevitarsis* Kieffer (Diptera: Ceratopogonidae) and of four other flies in bovine dung. – Aust. J. Ent. 35: 209-212, 3 tables.

Blanckenhorn, W. U., 1997: Altitudinal life history variation in the dung flies *Scathophaga stercoraria* and *Sepsis cynipsea*. – Oecologia 109: 342-352, 4 tables, 6 figs.

Blanckenhorn, W. U., 1998: Altitudinal differentiation in the diapause response of two species of dung flies. – Ecol. Ent. 23: 1-8, 1 table, 3 figs.

Blanckenhorn, W. U., Morf, C., Mühlhäuser, C. & Reusch, T., 1999: Spatiotemporal variation in selection on body size in the dung fly *Sepsis cynipsea*. – J. Evol. Biol. 12: 563-576, 7 tables, 2 figs.

Blanckenhorn, W. U., Reusch, T. & Mühlhäuser, C., 1998: Fluctuating asymmetry, body size and sexual selection in the dung fly *Sepsis cynipsea* - testing the good genes assumptions and predictions. – J. Evol. Biol. 11: 735-753, 6 tables, 1 fig.

Blume, R. R., 1970: Insects associated with bovine droppings in Kerr and Bexar counties, Texas. – J. econ. Ent. 63: 1023-1024, 1 table.

Blume, R. R., 1986: Parasites of Diptera associated with bovine droppings on a pasture in east central Texas. – SWest Ent. 11 (3): 215-222, 1 table, 5 figs.

Boness, M., 1975: Arthropoden im Hochwassergenist von Flüssen. – Bonn. zool. Beitr. 26 (4): 383-401.

Bouché, P. F., 1834: Naturgeschichte der Insekten, besonders in Hinsicht ihrer ersten Zustände als Larven und Puppen. Erste Lieferung. – Nicolai, Berlin, vi + 216 pp., 10 plates.

Brander, T., 1965: Lounais-Hämeen Kärpäsistä, Brachycera (Diptera). – Lounais-Hämeen Luonto 16: 51-69.

Brindle, A., 1965: Taxonomic notes on the larvae of British Diptera. No. 20 - The Sepsidae. – Entomologist 98: 137-140, 11 figs.

Bristowe, W. S., 1979: The mysterious swarms of sepsid flies and their unpalatability to spiders. – Proc. Trans. Br. Ent. nat. Hist. Soc., 12 (1-2): 16-19.

Buxton, P. A., 1954: British Diptera associated with fungi. 2. - Diptera bred from Myxomycetes. – Proc. R. ent. Soc. Lond. (A) 29 (10-12): 163-171, 1 table.

Carles-Tolrá, M., 1993: New and interesting records of Diptera Acalyptrata from Spain. Part III: Lauxaniidae, Chamaemyiidae, Coelopidae, Dryomyzidae, Sciomyzidae and Sepsidae. – Fragm. ent. 25 (1): 21-41.

Chandler, P. J. (Ed.), 1998: Checklists of Insects of the British Isles (New Series), Part 1: Diptera (incorporating a list of Irish Diptera). – Handbk Ident. Br. Insects 12: xx + 234 pp.

Chernov, Yu. I., 1965: A complex of synanthropic Diptera in the tundra zone of the USSR. [Russian, with English summary.] – Ent. Obozr. 44 (1): 74-83, 1 table. [English translation in Ent. Rev., Wash. 44 (1): 39-44.]

Chizhov, V. N., 1996: Phoresia of *Nigolaimoides* [sic] *borborophilus* (Nematoda: Dorylaimida) by scatobiotic dipterans. – Russian J. Nematol. 4(2): 185-186, 1 fig.

Chizhov, V. N. & Sturhan, D., 1998: *Pratinema sepsis* gen. n., sp. n. (Tylenchida: Allantonematidae), a parasite of coprophilous flies of the genus *Sepsis* (Diptera: Sepsidae). – Russian J. Nematol. 6: 5-8, 1 fig.

Coe, R. L., 1968: A further collection of Diptera from Yugoslavia, with localities and notes. Part three. – Glasn. Muz. Beogr. (B) 23: 131-149.

Coffey, M. D., 1966: Studies on the association of flies (Diptera) with dung in southeastern Washington. – Ann. ent. Soc. Am. 59: 207-218, 1 fig., 3 tables.

Cole, F. R., 1969: The flies of Western North America. – University of California Press, Berkeley and Los Angeles, xi + 693 pp., 360 figs, 1 col. pl.

Colless, D. H., 1980: Biogeography of Australian Sepsidae (Diptera). – Aust. J. Zool. 28: 65-78, 13 figs, 4 tables.

Collin, J. E., 1910: Additions and corrections to the British list of Muscidae Acalyptratae. [Part.] – Entomologist's mon. Mag. 46 [= (2) 21]: 173-178.

Coquebert, A. J., 1804: Illustratio iconographica insectorum quae in Musaeis parisinis observavit et in lucem edidit Joh. Christ. Fabricius, praemissis ejusdem descriptionibus; accedunt species plurimae, vel minus aut nondum cognitae. Decas tertia. – 'Anno XII', Petri Didot, Parisiis [= Paris], [iv +] pp. 91-142, plates 21-30.

Coquillett, D. W., 1910: The type-species of the North American genera of Diptera. – Proc. U.S. natn. Mus., 37: 499-647.

Curtis, J., 1829: British Entomology; being illustrations and descriptions of the genera of insects found in Great Britain and Ireland: containing coloured figures from nature of the most rare and beautiful species, and in many instances of the plants upon which they are found. Volume 6. – 'Printed for the author', London, plates 242-289.

Curtis, J., 1831: A guide to an arrangement of British insects; being a catalogue of all the named species hitherto discovered in Great Britain and Ireland. [First edition.] [Part.] – London, columns 161-256.

Curtis, J., 1837: A guide to an arrangement of British insects; being a catalogue of all the named species hitherto discovered in Great Britain and Ireland. Second edition, greatly enlarged. – London, vi + 282 numbered columns (141 pp.) + 283-284.

Czerny, L., 1903: 'Dreikönigs-Fliegen.' (Ausbeute an Dipteren am 6.-8. Jänner 1903.) – Verh. zool.-bot. Ges. Wien 53 (3-4): 238-240.

Dahl, R., 1968: Studies on the Diptera Brachycera fauna of the sea shores in North Norway. – Norsk. ent. Tidsskr. 15 (1): 19-27, 3 figs, 1 table.

Dampf, A., 1924: Zur Kenntnis der estländischen Moorfauna. (II. Beitrag.) – Protok. Obshch. Estest., Yur'ev

[= Sber. Naturf.-Ges. Univ. Dorpat] 31 (1-2): 17-71, 10 figs.

Dampf, A. (Ed.), 1927: Zur Kenntnis der estländischen Hochmoorfauna. (IV. Beitrag.) – Protok. Obshch. Estest., Yur'ev [= Sber. Naturf.-Ges. Univ. Dorpat] 33 (2) [1926]: 57-98.

Debry, J.-M., 1978: Introduction au dénombrement et à l'étude écologique d'une communauté de diptères coprophages sur du lisier de porcs. – Bull. Écol. 9 (2): 129-140, 3 figs, 4 tables.

Deeming, J. C., 1969: Diptera from Nepal. Sphaeroceridae. – Bull. Br. Mus. (nat. Hist.) Ent., 23: 53-74, 32 figs.

Dinter, A. & Paarmann, W., 1989: Untersuchungen zur Populationsökologie der Arthropoden einer Schlammdeponie für Zuckerrübenerde. – Verh. Ges. Ökol. (Essen 1988) 18: 769-776, 5 figs, 2 tables.

Disney, R. H. L., 1973: A note on some filth-inhabiting flies of Cameroon. – Entomologist's mon. Mag. 108: 212-213.

Doumerc, A., 1833: Mémoire sur le *Psalidomyia fucicola*, nouvelle espèce de diptère vivant sur les bords de la mer et formant un nouveau genre dans la famille des Athéricères, tribu des Muscides (Latr.). – Annls Soc. ent. Fr. 2 (1): 89-93, plate 6 figs A 1-8.

Draeger, U., 1997: Ökofaunistische und verhaltensbiologische Untersuchungen klärschlammbewohnender, koprophager Sepsidae (Diptera). Dissertation thesis, Freie Universität Berlin.

Draeger, U. & Scheurer, S., 1999: The carbohydrate diet of *Themira putris* adults (Diptera: Sepsidae) living in a sewage treatment plant in Central Europe. – Studia dipterologica 6 (1): 67-72, 2 tables.

Drapiez, P. A. J., 1820: Description de six espèces d'insectes nouveaux. – Annls gén. Sci. phys. Brux. 4: 349-354, plate 48.

Duda, O., 1926a: Monographie der Sepsiden (Dipt.). I. – Annln naturh. Mus. Wien 39 [1925]: 1-153, 2 figs, 7 plates, 1 table. [Publication date 16.i.1926]

Duda, O., 1926b: Monographie der Sepsiden (Dipt.). II. – Annln naturh. Mus. Wien 40: 1-110, 9 plates.

Duffield, J. E., 1937: Notes on some animal communities of Norwegian Lapland. An account of the dung and carrion communities and of those insects found in human dwellings. – J. anim. Ecol. 6 (1): 160-168, plate 6, 1 table.

Dufour, L., 1851: Recherches anatomiques et physiologiques sur les Diptères, accompagnées de considérations rélatives à l'histoire naturelle de ces insectes. – Mém. prés. div. Sav. Acad. Sci. Inst. Fr. 11: 171-360, plates 1-11.

Eberhard, W. G., 1999: Mating systems of sepsid flies and sexual behavior away from oviposition sites by *Sepsis neocynipsea* (Diptera: Sepsidae). – J. Kans. ent. Soc. 72 (1): 129-130.

Eberhard, W. G. & Huber, B. A., 1998: Copulation and sperm transfer in *Archisepsis* flies (Diptera, Sepsidae) and the evolution of their intromittent genitalia. – Studia dipterologica 5 (2): 217-248, 60 figs.

Eberhard, W. G. & Pereira, F., 1996: Functional morphology of male genitalic surstyli in the dungflies *Archisepsis diversiformis* and *A. ecalcarata* (Diptera: Sepsidae). – J. Kans. ent. Soc. 69, Suppl.: 43-60, 14 figs.

Egglishaw, H. J., 1960: Studies on the family Coelopidae (Diptera). – Trans. R. ent. Soc. Lond. 112 (6): 109-140, 24 figs, 11 tables.

Egglishaw, H. J., 1965: Observations on the fauna of wrack beds. – Trans. Soc. Br. Ent. 16 (8): 189-216.

Elberg, K., 1963: *Themira paludosa* sp. n. (Diptera, Sepsidae) aus Estland. [Russian, with German summary.] – Ent. Obozr. 42 (4): 909-911, 3 figs. [English translation in Ent. Rev., Wash. 42 (4): 491.]

Elberg, K., 1966: On some rare species of flies (Diptera Brachycera) from the Baltic districts. [Russian, with English summary.] – Eesti NSV Tead. Akad. Toim. (Biol. Seer.) 1966 (3): 397-401.

Enderlein, G., 1934: Entomologische Ergebnisse der Deutsch-Russischen Alai-Pamir-Expedition 1928 (III). 1. Diptera. – Dt. ent. Z. 1933 (2-3): 129-146, 3 figs.

Evenhuis, N. L., 1990: Dating of the livraisons and volumes of d'Orbigny's *Dictionnaire Universel d'Histoire Naturelle*. – Occ. Pap. Bernice P. Bishop Mus. 30: 219-225, 2 tables.

Evenhuis, N. L. & Thompson, F. C., 1990: Type designations of genus-group names of Diptera given in d'Orbigny's *Dictionnaire Universel d'Histoire Naturelle*. – Occ. Pap. Bernice P. Bishop Mus. 30: 226-258, 1 fig., 2 tables.

Fabricius, J. C., 1794: Entomologia systematica emendata et aucta. Secundum classes, ordines, genera, species adjectis synonimis, locis, observationibus, descriptionibus. 4. – C. G. Proft, Hafniae [= Copenhagen], 472 pp.

Fabritius, K., 1981: Über das natürliche Vorkommen und den Wirtskreis von Parasiten synanthroper Fliegen. – Z. angew. Zool. 68 (2): 139-149, 2 tables.

Fallén, C. F., 1810: Specim. entomolog. novam Diptera disponendi methodum exhibens. – Litteris Berlingianis, Lundae [= Lund], 26 pp., 1 plate.

Fallén, C. F., 1820a: Heteromyzides sveciae. – Literis [sic] Berlingianis, Lundae [= Lund], 10 pp.

Fallén, C. F., 1820b: Ortalides Sveciae. [Part.] – Litteris Berlingianis, Lundae [= Lund] [2]: 13-24.

Ferrar, P., 1987: A guide to the breeding habits and immature stages of Diptera Cyclorrhapha. Parts 1, 2. – Entomonograph 8: 907 pp. E.J. Brill/Scandinavian Science Press, Leiden & Copenhagen.

Foster, W. A., 1967: Co-operation by male protection of ovipositing female in the Diptera. – Nature 214: 1035-1036.

Franz, H., 1989: Diptera Cyclorapha [sic]. *In* Franz, H. et al.: Die Nordost-Alpen im Spiegel ihrer Landtierwelt. Eine Gebietsmonographie umfassend: Fauna,

Faunengeschichte, Lebensgemeinschaften und Beeinflussung der Tierwelt durch den Menschen. Band VI (2). – Universitätsverlag Wagner, Innsbruck, 445 pp.

Frey, R., 1906: Tvenne för vår fauna nya dipterer. – Meddn Soc. Fauna Flora fenn. 32: 67.

Frey, R., 1908: Ueber die in Finnland gefundenen Arten des Formenkreises der Gattung *Sepsis* Fall. (Dipt.). – Dt. ent. Z. 1908 (5): 577-588, 6 figs.

Frey, R., 1917: Entomologiska anteckningar från norra Savolaks. – Meddn Soc. Fauna Flora fenn. 43: 84-97.

Frey, R., 1918: Beitrag zur Kenntnis der Dipterenfauna des nördl. europäischen Russlands. II. Dipteren aus Archangelsk. – Acta Soc. Fauna Flora fenn. 46 (2): 32 pp.

Frey, R., 1921: Studien über den Bau des Mundes der niederen Diptera Schizophora nebst Bemerkungen über die Systematik dieser Dipterengruppe. – Acta Soc. Fauna Flora fenn. 48 (3): 246 pp., 10 plates.

Frey, R., 1925: Zur Systematik der Diptera Haplostomata. II. Fam. Sepsidae. – Notul. ent. 5 (3): 69-76, 10 figs. [Publication date 10.x.1925]

Frey, R., 1950: Dipterfaunan vid Tana älv i Utsjoki sommaren 1949. – Notul. ent. 30 (1-2): 5-18, 1 fig.

Frey, R., 1953: [Note on Coelopidae at meeting of 17 February 1953.] – Notul. ent. 33 (3-4): 122.

Gallis, R., 1994: *Xenosepsis fukuharai* Iwasa, espèce nouvelle pour la France (Diptera, Sepsidae). – Entomologiste 50 (2): 146.

Gaunitz, S., 1951: Koprofiler från Åtvidaberg. – Ent. Tidskr. 72 (1-2): 76.

Gistel [Gistl], J. von N. F. X., 1837: Kritische Revisionen und Ergänzungen zu Schrank's 'Enumeratio Insectorum Austriae, Fauna boica, u.s.w.' Aus dem bisher noch ungedruckten literarischen Nachlasse des sel. Schrank mitgethgeilt von Gistl. – Gistl's Faunus, Zeitschrift für Zoologie und vergleichende Anatomie (2) 1 (1): 5-19.

Gistel [Gistl], J. von N. F. X., 1848: Naturgeschichte des Thierreichs für höhere Schulen. – Hoffmann, Stuttgart, xvi + 216 (+ 4) pp., 32 col. plates, text ills.

Goetghebuer, M. & Bastin, F., 1925: Contribution à l'étude des Sepsidae de Belgique. – Bull. Annls Soc. ent. Belg. 65 (4): 123-137, 1 fig., 2 plates. [Publication date 30.iv.1925]

Goot, V. S. van der, 1985a: De Sepsidae van Nederland, 1856-1984 (Diptera). 1. Literatuur van de Nederlandse fauna en de determinaties van voor 1939. – Ent. Ber., Amst. 45 (10): 139-144.

Goot, V. S. van der, 1985b: De Sepsidae van Nederland, 1856-1984 (Diptera). 2. Mijn eigen determinaties, enkele gevolgtrekkingen daaruit en enige vergelijkende aantekeningen van de Belgische fauna. – Ent. Ber., Amst. 45 (12): 177-182, 1 fig., 1 table.

Goot, V. S. van der, 1986a: De Sepsidae van Nederland, 1856-1984 (Diptera). 3. Catalogus van de Nederlandse Sepsidae. – Ent. Ber., Amst. 46 (1): 1-6.

Goot, V. S. van der, 1986b: Een opeenhoping van wenkvliegjes op een vreemde plaats (Diptera: Sepsidae). – Ent. Ber., Amst. 46 (3): 33-34.

Goot, V. S. van der, 1987a: *Sepsis violacea* Meigen (Dipt. Sepsidae) on the increase in the Netherlands since the sixties. – Entomologist's mon. Mag. 123: 82.

Goot, V. S. van der, 1987b: *Meroplius minutus* (Wiedemann) (Dipt. Sepsidae) extinct in the Low Countries. – Entomologist's mon. Mag. 123: 82.

Goot, V. S. van der, 1989: *Themira biloba* nieuw voor de Nederlandse fauna (Diptera: Sepsidae). – Ent. Ber., Amst. 49 (3): 36.

Gregor, F., 1966: Redeskription von *Themira gracilis* (Zett.) und Bemerkungen zur Ökologie und Synanthropie einiger *Themira*-Arten (Diptera, Sepsidae). – Acta ent. bohemoslovaca 63 (1): 76-83, 13 figs, 1 photo, 1 table.

Gregor, F. & Daniel, M., 1971: A contribution to the knowledge of fauna of synanthropic flies in East Hindu Kush. – Folia parasit. 18 (4): 347-356, 4 figs, 2 tables.

Gregor, F. & Povolný, D., 1959a: Eine Ausbeute von synanthropen Fliegen aus Slowenien. – Cslká Parasit. 6 (1): 97-112, 26 figs, 3 tables.

Gregor, F. & Povolný, D., 1959b: Beitrag zur Kenntnis synanthroper Fliegen Bulgariens. – Pr. brn. Zakl. čsl. Akad. Věd. 31 (7): 377-384, 1 table, 2 plates.

Gregor, F. & Povolný, D., 1960a: Beitrag zur Kenntnis synanthroper Fliegen Albaniens. – Cslká Parasit. 7: 115-131, 4 plates, 2 tables.

Gregor, F. & Povolný, D., 1960b: Beitrag zur Kenntnis der synanthropen Fliegen Ungarns. – Cas. čsl. Spol. ent. 57 (2): 158-177, 1 map, 2 figs, 4 plates, 2 tables.

Gregor, F. & Povolný, D., 1961: Resultate stationärer Untersuchungen von synanthropen Fliegen in der Umgebung einer Ortschaft in der Ostslowakei. – Zool. Listy 10 (1): 17-44, 36 figs, 7 tables, 1 photo.

Gregor, F. & Povolný, D., 1964: Eine Ausbeute von synanthropen Fliegen aus Tirol. – Zool. Listy 13 (3): 229-248, 3 figs, plates 7-14.

Griffiths, G. C. D., 1972: The phylogenetic classification of Diptera Cyclorrhapha, with special reference to the structure of the male postabdomen. – Series ent. 8: 340 pp., 154 figs, 2 plates, 2 tables.

Grootaert, P., 1991: Sepsidae. – P. 135 *in* Grootaert, P., de Bruyn L. & de Meyer, M.: Catalogue of the Diptera of Belgium. – Docums Trav. Inst. r. Sci. Nat. Belg. 70: 338 pp.

Grzegorzek, A., 1873: Übersicht der bis jetzt in der Sandezer Gegend West-Galiziens gesammelten Dipteren. – Verh. zool.-bot. Ges. Wien 23: 25-36.

Hackman, W., 1980: A check list of the Finnish Diptera. II. Cyclorrhapha. – Notul. ent. 60 (3): 117-162.

Haenni, J.-P., 1997: Sepsidae (Diptera) nouveaux pour la faune de Suisse. – Bull. romand Ent. 15: 69-78.

Haenni, J.-P., 1998: 64. Sepsidae. – Pp. 249-250 *in* Merz, B., Bächli, G., Haenni, J.-P. & Gonseth, Y.:

Fauna Helvetica, 1. Diptera - Checklist. – Centre Suisse de Cartographie de la Faune, and Schweizerische Entomologische Gesellschaft, Neuchâtel, 369 pp., figs.

Hafez, M., 1939: The life history of *Sepsis impunctata* [sic] Macq. (Diptera: Sepsidae). – Bull. Soc. Fouad 1er Ent. 23: 319-325, 5 figs.

Hafez, M., 1947: Further additions to the dipterous fauna of dung in Egypt, with some biological observations. – Bull. Soc. Fouad 1er Ent. 31: 307-312.

Hafez, M., 1948: Ecological and biological observations on some coprophagous Sepsidae (Diptera). – Proc. R. ent. Soc. Lond. (B) 23: 99-104, 1 fig.

Haliday, A. H., 1833: Catalogue of the Diptera occurring about Holywood in Downshire. – Entomol. Mag. 1 (2): 147-180.

Haliday, A. H., 1838: New British insects indicated in Mr. Curtis's Guide. [Part.] – Ann. nat. Hist. 2 (9): 183-190.

Haliday, A. H., 1851: Sendschreiben an C. A. Dohrn über die Dipteren der in London befindlichen Linnéischen Sammlung. Aus dem Englischen übersetzt von Anna Dohrn. – Stettin. ent. Ztg 12: 131-145.

Hammer, O., 1941: Biological and ecological investigations on flies associated with pasturing cattle and their excrement. – Vidensk. Meddr dansk naturh. Foren. 105: 141-394, 50 figs, 21 tables. [Also reprinted with pagination 1-257.]

Hanski, I., 1976: Breeding experiments with carrion flies (Diptera) in natural conditions. – Suom. hyönt. Aikak. 42 (3): 113-121, 1 fig., 2 tables.

Hardy, D. E., 1975a: Notes and exhibitions. *Sepsis thoracica* Robineau-Desvoidy. – Proc. Hawaii. ent. Soc. 22 (1): 11.

Hardy, D. E., 1975b: Notes and exhibitions. *Sepsis thoracica* (Robineau-Desvoidy). – Proc. Hawaii. ent. Soc. 22 (1): 12-13.

Hardy, D. E., 1975c: Notes and exhibitions. *Sepsis lateralis* Wiedemann. – Proc. Hawaii. ent. Soc. 22 (1): 13.

Harris, M., 1780a: An exposition of English insects, with curious observations and remarks, wherein each insect is particularly described; its parts and properties considered; the different sexes distinguished, and the natural history faithfully related. The whole illustrated with copper plates, drawn, engraved, and coloured by the author. [Decad 3.] – 'Printed for the author', London, pp. 73-99, plates 21-30.

Harris, M., 1780b: An exposition of English insects, with curious observations and remarks, wherein each insect is particularly described; its parts and properties considered; the different sexes distinguished, and the natural history faithfully related. The whole illustrated with copper plates, drawn, engraved, and coloured by the author. [Decad 4.] – 'Printed for the author', London, pp. 100-138, plates 31-40.

Harris, M., 1780c: An exposition of English insects, with curious observations and remarks, wherein each insect is particularly described; its parts and properties considered; the different sexes distinguished, and the natural history faithfully related. The whole illustrated with copper plates, drawn, engraved, and coloured by the author. [Decad 5.] – 'Printed for the author', London, pp. 139-166, plates 41-50 + 1 plate.

Hedström, L., 1964: Lounais-Hämeen Kärpäsiä, Brachycera (Diptera). – Lounais-Hämeen Luonto 15: 77-89.

Hedström, L. & Nuorteva, P., 1971: Zonal distribution of flies on the hill Ailigas in subarctic northern Finland. – Suom. hyönt. Aikak. 37 (2): 121-125, 1 fig., 1 table.

Hendel, F., 1902: Dipterologische Anmerkungen. – Wien. ent. Ztg 21 (10): 265.

Hennig, W., 1937: Coelopidae. – *In* Lindner, E.: Fliegen palaearkt. Reg. 52: 39 pp., 40 figs. Schweizerbart, Stuttgart.

Hennig, W., 1949: Sepsidae. – *In* Lindner, E.: Fliegen palaearkt. Reg. 39a: 91 pp., 82 figs, 10 plates. Schweizerbart, Stuttgart.

Hennig, W., 1952: Die Larvenformen der Dipteren. Eine Übersicht über die bisher bekannten Jugendstadien der zweiflügeligen Insekten. 3. Teil. – Akademie Verlag, Berlin, vii + 628 pp., 338 figs, 21 plates.

Hennig, W., 1965: Die Acalyptratae des Baltischen Bernsteins und ihre Bedeutung für die Erforschung der phylogenetischen Entwicklung dieser Dipteren-Gruppe. – Stuttg. Beitr. Naturk. 145: 215 pp., 319 figs.

Hinton, H. E., 1960: The structure and function of the respiratory horns of the eggs of some flies. – Phil. Trans. R. Soc. Lond. (B) 243: 45-73, 14 figs, 6 tables, plate 3.

Hinton, H. E., 1961: How some insects, especially the egg stages, avoid drowning when it rains. – Proc. Trans. S. Lond. ent. nat. Hist. Soc. 1960: 138-154, 4 figs, plates 5-8.

Hinton, H. E., 1981: Biology of insect eggs in three volumes. Volume 2. – Pergamon Press, Oxford, etc: xxiii + 1-473, figs 1-135, 155 plates.

Hoffmeister, H., 1844: Einige Nachträge zu Meigen's Zweiflüglern. – Jber. Ver. Naturk. Cassel 8: 11-14.

Ionescu, M. A. & Ursu, A., 1971: Contribuţii la cunoaşterea paraziţilor la dipterele sinantrope. [Romanian, with English abstract and French summary: Contribution à la connaissance des parasites chez les diptères synanthropes.] – Studii Cerc. Biol. (Ser. Zool.) 23 (6): 525-529, 1 table.

Iwasa, M., 1980: Studies on the Sepsidae from Japan. I. Revision of the genus *Sepsis* Fallén, with a key to the Japanese genera. – Kontyû 48 (3): 402-413, 14 figs.

Iwasa, M., 1981: Studies on the Sepsidae from Japan. II. Notes on the genus *Themira* Robineau-Desvoidy, with descriptions of two new species. – Kontyû 49 (1): 45-53, 12 figs.

Iwasa, M., 1982a: A new species and some records of Sepsidae from India (Diptera). – Kontyû 50: 604-609, 5 figs.

Iwasa, M., 1982b: A new Oriental species of the genus *Sepsis* from Taiwan and Indonesia (Diptera: Sepsidae). – Pacif. Ins. 24 (3-4): 232-234, 5 figs.

Iwasa, M., 1984a: The Sepsidae from Nepal, with descriptions of eight new species (Diptera). – Kontyû 52 (1): 72-93, 40 figs, 2 tables.

Iwasa, M., 1984b: Studies on the Sepsidae from Japan (Diptera). III. On the eleven species of eight genera excluding the genera *Sepsis* Fallén and *Themira* R.-D., with description of a new species. – Kontyû 52 (2): 296-308, 14 figs.

Iwasa, M., 1985: Supplementary notes on the Sepsidae from Japan (Diptera). – Kontyû 53 (4): 632-638, 16 figs.

Iwasa, M., 1986: Notes on the genus *Toxopoda* in the Oriental region and Papua New Guinea, with descriptions of two new species (Diptera, Sepsidae). – Kontyû 54 (4): 654-660, 9 figs.

Iwasa, M., 1987: The Sepsidae from Taiwan (Diptera). – Sieboldia, Suppl.: 1-9, 18 figs.

Iwasa, M., 1989: Taxonomic study of the Sepsidae (Diptera) from Pakistan. – Jap. J. sanit. Zool. 40, Suppl.: 49-60, 12 figs.

Iwasa, M., 1994: Two new species of the genus *Meroplius* Rondani (Diptera, Sepsidae) from the Oriental region. – Jap. J. Ent. 62 (1): 161-166, 11 figs.

Iwasa, M., 1995: Revisional notes on the Japanese Sepsidae (Diptera). – Jap. J. Ent. 63 (4): 781-797, 15 figs.

Iwasa, M., 1996: The Sepsidae (Diptera) from Madagascar, with descriptions of four new species. – Int. J. Dipterol. Res. 7 (2): 135-140, 13 figs.

Iwasa, M., Hori, K. & Aoki, N., 1995: Fly fauna of bird nests in Hokkaido, Japan (Diptera). – Can. Ent. 127 (5): 613-621, 1 fig., 2 tables.

Iwasa, M. & Jayasekera, N., 1994: The Sepsidae from Sri Lanka, with description of a new species (Diptera). – Jap. J. sanit. Zool. 45 (1): 57-62, 6 figs.

Iwasa, M., Kanmiya, K. & Lee, C. E., 1994: Notes on the Sepsidae (Diptera) from South Korea. – Esakia 34: 203-208, 4 figs.

Iwasa, M. & Tewari, R. R., 1990: Two new species of the genus *Dicranosepsis* Duda from India (Diptera, Sepsidae). – Jap. J. Ent. 58 (4): 794-798, 9 figs.

Iwasa, M. & Tewari, R. R., 1994: Notes on the Sepsidae (Diptera) from India, with description of a new species. – Jap. J. sanit. Zool. 45, Suppl.: 23-29, 4 figs.

Iwasa, M. & Thapa, V. K., 1994: Nepalese Sepsidae (Diptera): a new species and records of thirty-two species. – Jap. J. sanit. Zool. 45, Suppl.: 9-22, 6 figs.

Iwasa, M., Zuska, J. & Ozerov, A. L., 1991: The Sepsidae from Bangladesh, with description of a new species (Diptera). – Jap. J. sanit. Zool. 42 (3): 229-234, 5 figs.

Jansson, A., 1925: Die Insekten-, Myriopoden [sic]- und Isopodenfauna der Gotska Sandön. – Länstidningen, Örebro, iii + 182 pp., 8 figs.

Kandybina, M. N., 1984: Contribution to the study of the fauna of flies (Diptera) of the Vyborg reserve (Bol'shoy Berezovyi Island). [In Russian.] – Trudy zool. Inst. Leningr. 123: 12-30.

Karl, O., 1936: Die Fliegenfauna Pommerns. Diptera Brachycera. [Part.] – Stettin. ent. Ztg 97 (1): 108-136.

Karps, A., 1986: Flies of the family Sepsidae in Latvia. [Latvian, with English summary.] – Latv. Ent. 29: 5-7.

Keilin, D., 1919: On the life-history and larval anatomy of *Melinda cognata* Meigen (Diptera Calliphorinae) parasitic in the snail *Helicella* (*Heliomanes*) *virgata* da Costa, with an account of the other Diptera living upon molluscs. – Parasitology, 11 (3-4): 430-455, 6 figs, plates 22-25.

Kieffer, J.-J., 1898: Description de deux diptères fucivores recueillis aux Petites-Dalles (Seine-infér.). – Annls Soc. ent. Fr. 67 (1-2): 100-104.

Kiontke, K., 1989: Über die Bedeutung von Sepsiden (Diptera) und Sphaeroceriden (Diptera) für den Transport kuhfladenbewohnender Nematoden unter besonderer Berücksichtigung der Phoresie von *Diplogaster coprophilus* Sudhaus & Rehfeld 1989 (Nematoda). – Diplomarbeit des Fachbereiches Biologie der Freien Universität Berlin.

Kiontke, K., 1996: The phoretic association of *Diplogaster coprophila* Sudhaus & Rehfeld, 1990 (Diplogastridae) from cow dung with its carriers, in particular flies of the family Sepsidae. – Nematologica 42 (3): 354-366, 5 figs.

Kirk, A. A., 1992: The effect of the dung pad fauna on the emergence of *Musca tempestiva* (Dipt.: Muscidae) from dung pads in southern France. – Entomophaga 37 (4): 507-514, 2 tables.

Kotrba, M. 1996: Sperm transfer by spermatophore in Diptera: new results from the Diopsidae. – Zool. J. Linn. Soc. 117: 305-323, 1 table, 4 figs.

Kowarz, F., 1873: Beitrag zur Dipteren-Fauna Ungarns. – Verh. zool.-bot. Ges. Wien 23: 453-464.

Krogerus, R., 1932: Über die Ökologie und Verbreitung der Arthropoden der Triebsandgebiete an den Küsten Finnlands. – Acta zool. fenn. 12: 308 pp., 39 figs, 29 maps, 31 tables, 12 diagrams.

Krogerus, R., 1960: Ökologische Studien über nordische Moorarthropoden. Artenbestand, ökologische Faktoren, Korrelation der Arten. – Commentat. biol. 21 (3): 238 pp., 5 figs, 31 tables, 29 diagrams.

Krusteva, H. T. & Beschovski, V. L., 1998: Studies on the species composition of Diptera order internal stem pests and damages caused by them on spring barley. – Acta zool. bulg. 50 (1): 43-50, 4 tables.

Laboulbène, A., 1896: Observations physiologiques sur deux insectes diptères: *Oestrus (Gastrophilus) equi* Fabricius, et *Sepsis cynipsea* Linné. – Bull. Soc. ent. Fr. 1896 (4): 110-112.

Laurence, B. R., 1952: Two coprophilous Diptera (Borboridae and Sepsidae) new to Britain. – Entomologist's mon. Mag. 88: 81-82, 7 figs.

Laurence, B. R., 1954: The larval inhabitants of cow pats. – J. anim. Ecol. 23 (2): 234-260, 7 figs, 20 tables.

Lavčiev, V. I. & Tsankova, R. N., 1980: Ecological studies on coprobiont dipterans in a dunghill near Sofia. [Bulgarian, with Russian and English summaries.] – Ekologiya, Sofiya 7: 3-14, 4 figs, 4 tables.

Lavčiev, V. I. & Tsankova, R. N., 1982: On the hibernation of some coprobiont flies in Bulgaria. [Bulgarian, with Russian and English summaries.] – Ekologiya, Sofiya 10: 59-64, 3 tables.

Lindberg, H. & Saris, N.-E., 1952: Insektfaunan i Pisavaara Naturpark (Finland, Prov. Ob.). – Acta Soc. Fauna Flora fenn. 69 (2): 82 pp., 1 map, 10 photos.

Lindroth, C. H., Andersson, H., Bödvarsson, H. & Richter, S. H., 1973: Surtsey, Iceland. The development of a new fauna, 1963-1970. Terrestrial invertebrates. – Entomologica scand. Suppl. 5: 280 pp., 50 figs, 45 tables.

Linnaeus, C., 1758: Systema naturae per regna tria naturae, secundum classes, ordines, genera, species, cum caracteribus, differentiis, synonymis, locis. 10th edition. I. – Laurentii Salvii, Holmiae [= Stockholm], 824 pp.

Linnaeus, C., 1761: Fauna Svecica sistens animalia Sveciae regni. 2nd edition [revised]. – Laurentii Salvii, Stockholmiae, [48 +] 578 pp., 2 plates.

Lioy, P., 1864: I Ditteri distribuiti secondo un nuovo metodo di classificazione naturale. [Part.] – Atti R. Ist. veneto Sci. (3), 9 (9): 1087-1126.

Lobanov, A. M., 1962a: Hibernation sites for the preimaginal stages of synanthropic flies. [In Russian.] – Nauch. Dokl. vyssh. Shk. (biol. Nauki) 1962 (2): 36-39, 1 table.

Lobanov, A. M., 1962b: Sepsidae (Diptera) as synanthropic forms. [In Russian.] – Byull. Mosk. Obshch. Ispyt. Prir. (Biol.) 67 (1): 125-128, 1 fig., 5 tables.

Lobanov, A. M., 1966: Materials to the knowledge of hibernating sites and overwintering stages of synanthropic and associated short-horned flies. – Pp. 53-59, 2 tables *in* Khelevina, N. V.: Questions of regional ecology and parasitology 2. – Ivanovo.

Loew, H., 1841: Über die Gattung *Saltella* überhaupt und über *Saltella scutellaris* besonders. – Stettin. ent. Ztg 2 (12): 182-193, plate 1 figs 10, 18.

Loew, H., 1870: Über die bisher auf der Galizischen Seite des Tatragebirges beobachteten Dipteren. – Jb. gel. Ges. Krakau 42: 1-18.

Loew, H., 1873: Beschreibungen europäischer Dipteren. Dritter Band. Systematische Beschreibung der bekannten europäischen zweiflügeligen Insecten. Von Johann Wilhelm Meigen. Zehnter Theil oder vierter Supplementband. – Schmidt, Halle, viii + 320 pp.

Lyneborg, L., 1965: Hansted-Reservatets entomologi. (Entomology of the Hansted Reservation, Thy, Jutland.) 9. Diptera, Brachycera & Cyclorrhapha – Fluer. – Ent. Meddr 30 (3): 201-262.

Lyneborg, L., 1968: Diptera Brachycera et Cyclorrhapha. – Zoology Faroes 42a: 79 pp., 3 figs, 1 table.

Macquart, J., 1835: Histoire naturelle des insectes. Diptères. Tome deuxième. [2.] – Roret, Paris, 703 or 710 pp., plates 13-24.

Macquart, J., [1839]: Diptères. – Pp. 97-119, plate 4 figs 2-11 *in* Webb, P. B. & Berthelot, S.: Histoire Naturelle des Iles Canaries. Tome deuxième. Deuxième partie. Contenant la zoologie. [Entomologie.] – '1836-1844', Béthune, Paris, 119 pp., 7 plates.

Macquart, J., 1843a: Diptères exotiques nouveaux ou peu connus. [Tome deuxième. - 3.e partie.] – Mém. Soc. Sci. Agric. Arts Lille 1842: 160-460, 36 plates.

Macquart, J., 1843b: Diptères exotiques nouveaux ou peu connus. Tome deuxième. - 3.e partie. – Roret, Paris, pp. 5-304, 36 plates. [Reprint of 1843a.]

Macquart, J., 1851a: Diptères exotiques nouveaux ou peu connus. Suite du 4.e supplément publié dans les mémoires de 1849. – Mém. Soc. Sci. Agric. Arts Lille 1850: 134-294, plates 15-28.

Macquart, J., 1851b: Diptères exotiques nouveaux ou peu connus. Suite du 4.e supplément publié dans les mémoires de 1849. – Roret, Paris, pp. 161-336, 14 plates. [Reprint of 1851a.]

Madsen, H., 1936: Investigations on the shore fauna of East Greenland with a survey of the shores of other Arctic regions. Treaarsexpeditionen til Christian den X's Land 1931-1934 under ledelse af Lauge Koch. – Meddr Grønland 100 (8): 79 pp., 17 figs, 3 tables.

Malloch, J. R., 1925: Notes on Australian Diptera. No. vii. – Proc. Linn. Soc. N.S.W. 50 (4): 311-340, 23 figs. [Publication date 15.xii.1925].

Mangan, R. L., 1976: *Themira athabasca* n. sp. (Diptera: Sepsidae) with a revised key to North American *Themira* and notes on the sexual morphology of sympatric species. – Ann. ent. Soc. Am. 69 (6): 1024-1028, 2 figs, 1 table.

Mangan, R. L., 1977: A key and selected notes for the identification of larvae of Sepsidae (Diptera) from the temperate regions of North America. – Proc. ent. Soc. Wash. 79 (3): 338-342, 10 figs.

McAlpine, J. F., 1981: Morphology and terminology - adults. – Pp. 9-63, 146 figs *in* McAlpine, J. F., Peterson, B. V., Shewell, G. E., Teskey, H. J., Vockeroth, J. R. & Wood, D. M.: Manual of Nearctic Diptera. Volume 1. – Res. Branch Agric. Canada Monogr. 27: 674 pp., figs.

McAlpine, J. F., 1989: Phylogeny and classification of the Muscomorpha. – Pp. 1397-1518, 9 figs, 10 tables *in* McAlpine, J. F. & Wood, D. M.: Manual of Nearctic Diptera. Volume 3. – Res. Branch Agric. Canada Monogr. 32: vi + 1333-1581 pp., figs.

McLachlan, A. J. & Allen, D. F., 1987: Male mating success in Diptera: advantages of small size. – Oikos 48: 11-14, 1 table, 2 figs.

Meier, R., 1995a: Cladistic analysis of the Sepsidae (Cyclorrhapha: Diptera) based on a comparative scanning electron microscopic study of larvae. – Syst. Ent. 20: 99-128, 79 figs, 1 table.

Meier, R., 1995b: A comparative SEM study of the eggs of the Sepsidae (Diptera) with a cladistic analysis based on egg, larval and adult characters. – Entomologica scand. 26: 425-438, 27 figs, 1 table.

Meier, R., 1996: Larval morphology of the Sepsidae (Diptera: Sciomyzoidea), with a cladistic analysis using adult and larval characters. – Bull. Am. Mus. nat. Hist. 228: 147 pp., 478 figs, 2 tables.

Meier, R., 1999: Sepsidae. – Pp. 190 *in* Schumann, H., Bährmann, R. & Stark, A.: Checkliste der Dipteren Deutschlands. – Studia dipterologica Supplement 2: 354 pp.

Meier, R. & Pont, A. C., 2000: A.10. Family Sepsidae. – Pp. 367-386, 43 figs *in* Papp, L. & Darvas, B.: Contributions to a Manual of Palaearctic Diptera (with special reference to flies of economic importance). Appendix. – Science Herald, Budapest, 604 pp., figs.

Meigen, J. W., 1826: Systematische Beschreibung der bekannten europäischen zweiflügeligen Insekten. Fünfter Theil. [5]. – Schulz, Hamm, xii + 412 pp., plates 42-54.

Meigen, J. W., 1830: Systematische Beschreibung der bekannten europäischen zweiflügeligen Insekten. Sechster Theil. [6]. – Schulz, Hamm, xi + 401 pp., plates 55-66.

Meigen, J. W., 1838: Systematische Beschreibung der bekannten europäischen zweiflügeligen Insekten. Siebenter Theil oder Supplementband. [7]. – Schulz, Hamm, xii + 434 (+ 1) pp., plates 67-74.

Meijere, J. C. H. de, 1910: Die Dipteren der arktischen Inseln. – Fauna arct. 5 (1): 13-72.

Melander, A. L. & Spuler, A., 1917: The dipterous families Sepsidae and Piophilidae. – Bull. Wash. agric. Exp. Stn 143: 97 (+ 5) pp., 28 figs.

Mihályi, F., 1967a: The danger-index of the synanthropic flies. – Acta zool. hung. 13 (3-4): 373-377, 1 fig., 3 tables.

Mihályi, F., 1967b: Seasonal distribution of the synanthropic flies in Hungary. – Annls hist.-nat. Mus. natn. hung. 59: 327-344, 19 figs, 12 tables.

Mik, J., 1886: Dipterologische Miscellen. II. – Wien. ent. Ztg 5 (8): 276-279.

Minder, I. F., 1963: Sepsidae (Diptera) of the Moscow region. [In Russian.] – Ent. Obozr. 42 (2): 379-389, 24 figs. [English translation in Ent. Rev., Wash. 42 (2): 210-215.]

Modassir, Y., 1993a: Colour preference and sensitivity to pesticides in *Sepsis nitens*. – Geobios 20: 81-84, 2 tables, 1 fig.

Modassir, Y., 1993b: Effect of environmental factors on the seasonal abundance of sepsid flies. – Geobios 20: 241-246, 4 tables, 3 figs.

Modassir, Y., 1995: Effect of temperature on hatching and larval duration in *Sepsis nitens* (Sepsidae: Diptera). – J. Bombay nat. Hist. Soc. 92 (2): 280-282, 1 table.

Moeller, J., 1965: Ökologische Untersuchungen über die terrestrische Arthropodenfauna im Anwurf mariner Algen. – Z. Morph. Ökol. Tiere 55: 530-586, 17 figs, 19 tables.

Mohr, C. O., 1943: Cattle droppings as ecological units. – Ecol. Monogr. 13: 275-298, 4 figs.

Morge, G., 1974a: Diptera Collectionis P. Gabriel Strobl - V (Die 'Typensammlung'). – Beitr. Ent. 24, Sonderheft: 161-431, 20 figs, 13 plates.

Morge, G., 1974b: Diptera Collectionis P. Gabriel Strobl - VI (Supplementa ad Partes I-V). – Beitr. Ent. 24 (5-8): 383-416.

Morge, G., 1976a: Dipteren-Farbtafeln nach den bisher nicht veröffentlichten Original-Handzeichnungen Meigens: 'Johann Wilhelm Meigen: Abbildung der europaeischen zweiflügeligen Insecten, nach der Natur'. Pars II: Farbtafeln LXXXI-CLX. – Beitr. Ent. 26 (1): 441, plates 81-160.

Morge, G., 1976b: Diptera Collectionis P. Gabriel Strobl - VIII (Verzeichnis der Dipteren-Arten der Kollektion Strobl). – Beitr. Ent. 26 (1): 339-439.

Morge, G., 1976c: Dipteren-Farbtafeln nach den bisher nicht veröffentlichten Original-Handzeichnungen Meigens: 'Johann Wilhelm Meigen: Abbildung der europaeischen zweiflügeligen Insecten, nach der Natur'. Pars III: Farbtafeln CLXI-CCCV. – Beitr. Ent. 26 (2): 543, plates 161-305.

Morge, G., 1978: Diptera Collectionis P. Gabriel Strobl - XII (Verzeichnis der Dipteren-Arten der Kollektion Strobl nach dessen Insekten-Hauptkatalog: 'Catalog des naturhistorischen Cabinets von Admont. I.', p. 669-741). – Beitr. Ent. 28 (1): 69-168.

Morge, G., 1984: Diptera Collectionis P. Gabriel Strobl - XIII. (Typen-Designierung der Exemplare der sogenannten 'Typensammlung'). – Beitr. Ent. 34 (2): 319-335.

Morley, C., 1938: [Report on a field-trip to Walberswick, Suffolk, 31 March 1938.] – Trans. Suffolk Nat. Soc. 4 (1): xxvii.

Müller, O. F., 1764: Favna insectorvm Fridrichsdalina, sive methodica descriptio insectorvm agri Fridrichsdalensis, cvm characteribvs genericis et specificis, nominibvs trivialibvs, locis natalibvs, iconibvs allegatis, novisqve, plvribvs speciebvs additis. – I. F. Gleditschii, Hafniae et Lipsiae [= Copenhagen and Leipzig], xxiv + 96 pp.

Munari, L., 1982: Una nuova *Dicranosepsis* Duda della Sierra Leone (Diptera, Sepsidae). Notulae Sepsidologicae VI. – Problemi attuali di Scienza e di Cultura, Sez. Missioni ed Esplorazioni VIII 255: 47-51, 6 figs.

Munari, L., 1984: Nuovi dati corologici su alcune specie di Phaeomyiidae, Sciomyzidae e Sepsidae interessanti l'Italia (Diptera Cyclorrhapha). – Lav. Soc. venez. Sci. nat. 9 (1): 43-50.

Munari, L., 1985a: Considerazioni zoogeografiche sui Sepsidae Europei con particolare riguardo alle specie Italiane (Diptera, Cyclorrhapha) Notulae Sepsidologicae VIII. – Lav. Soc. venez. Sci. nat. 10: 43-50, 1 table.

Munari, L., 1985b: Una nuova specie di *Sepsis* del gruppo '*punctum*' (Diptera, Sepsidae). Notulae Sepsidologicae IX. – Lav. Soc. venez. Sci. nat. 10: 51-57, 11 figs.

Munari, L., 1986: Ricerche ditterologiche nella palude relitta di Onara (Veneto). - II. Sciomyzidae, Sepsidae (Diptera, Cyclorrhapha). – Lav. Soc. venez. Sci. nat. 11: 61-63.

Munari, L., 1987a: Ricerche ditterologiche alle sorgenti del Fiume Sile (Veneto). I. Introduzione, Sciomyzidae e Sepsidae (Diptera, Cyclorrhapha). – Lav. Soc. venez. Sci. nat. 12: 35-38, 1 fig.

Munari, L., 1987b: Studi sulla ditterofauna della Lessinia (Veneto). I. Sulla presenza in Italia di *Hilara interstincta* Fall. (Diptera, Empididae) con alcune osservazioni preliminari sul suo habitat e sulla sua ecologiia. – Lav. Soc. venez. Sci. nat. 12: 65-69, 2 figs.

Munari, L., 1987c: Studi sulla ditterofauna della Lessinia (Veneto). II. I Sepsidae della Lessinia Centrale: osservasioni faunistiche ed ecologiche. (Notulae Sepsidologicae XI) (Diptera, Cyclorrhapha). – Lav. Soc. venez. Sci. nat. 12: 71-86, 11 figs.

Munari, L., 1991a: Ricerche ditterologiche nelle cave di Gaggio di Marcon (Venezia). III. Sciomyzidae, Sepsidae, Sphaeroceridae (Diptera, Acalyptratae). – Lav. Soc. venez. Sci. nat. 16: 23-33, 1 fig.

Munari, L., 1991b: Studi sulla ditterofauna della Lessinia (Veneto). V. Sepsidae e Sphaeroceridae: nuovi dati faunistici (Diptera, Acalyptratae). – Lav. Soc. venez. Sci. nat. 16: 45-60.

Munari, L., 1993: New records of a snail killing fly and coprophilous flies from Italy, with some additional notes on the ecology of *Pullimosina heteroneura* (Haliday, 1836) (Diptera: Sciomyzidae, Sepsidae, Sphaeroceridae). – Lav. Soc. venez. Sci. nat. 18: 11-17, 2 figs.

Munari, L., 1994: Sepsidae from Sierra Leone: new records and descriptions of two new species (Diptera: Sepsidae). – Problemi attuali di Scienza e di Cultura, Sez. Missioni ed Esplorazioni XIII, 267: 231-242, 10 figs.

Munari, L., 1997: Diptera Sepsidae. – P. 277, 1 fig. *in* Zapparoli, M.: Gli Insetti di Roma. – Comune di Roma, Dip. X Area Risorsa Suolo e Tutela Ambiente, Quaderni dell'Ambiente 6: 360 pp.

Munari, L., 1998: Sphaeroceridae and Sepsidae from soil traps in Central Italy (Diptera). – Fragm. ent. 30 (1): 177-189, 13 figs.

Munari, L. & Rivosecchi, L., 1995: Diptera Sciomyzoidea. *In* Minelli, A., Ruffo, S. & La Posta, S.: Checklist delle specie della fauna italiana 74: 7 pp. – Calderini, Bologna.

Munari, L. & Roháček, J., 1990: Diptera from north Sardinia: Sepsidae, Sphaeroceridae, Ephydridae. – Lav. Soc. venez. Sci. nat. 15: 73-86.

Munari, L. & Vienna, P., 1977: Ricerche nelle torbiere d'Iseo (Brescia). III. I Sepsidae della torbiera d'Iseo. Sepsidae of the peat-bog of Iseo (Italy) (Diptera, Brachycera). – Annali Mus. civ. Stor. nat. Brescia 13 [1976]: 43-46.

Munari, L. & Vienna, P., 1979a: I Sepsidae conservati nelle collezioni del Museo Civico di Storia Naturale di Venezia (Diptera, Brachycera). – Lav. Soc. venez. Sci. nat. 4: 25-36.

Munari, L. & Vienna, P., 1979b: *Sepsis fissa* Becker: nuovi data biometrici e corologici (Insecta Diptera, Sepsidae). – Naturalista sicil. (4) 3 (1-2): 23-26, 1 fig., 1 table.

Munari, L. & Vienna, P., 1979c: Notulae sepsidologicae IV. I Sepsidae delle collezioni della Beffa e Zangheri conservati nel Museo Civico di Storia Naturale di Verona (Diptera, Brachycera). – Boll. Mus. civ. Stor. nat. Verona 6: 630-633.

Munari, L. & Vienna, P., 1982: Le *Sepsis* italiane del gruppo '*punctum*' (Diptera, Sepsidae). Notulae Sepsidologicae V. – Lav. Soc. venez. Sci. nat. 7: 7-16, 10 figs.

Naswaty, R., 1960: Beiträge zur Ökologie und Systematik der Sepsidae und Drosophilidae Frankens. – Sber. phys.-med. Soz. Erlangen, 80 [1959]: 74-120, 6 tables.

Nielsen, P., Ringdahl, O. & Tuxen, S. L., 1954: Diptera 1 (exclusive of Ceratopogonidae and Chironomidae). – Zoology Iceland 3, 46a: 189 pp., 49 figs. – Munksgaard, Copenhagen & Reykjavik.

Nishijima, Y. & Iwasa, M., 1984: Additional data on the flies occurring from wild brown bear dung in Hokkaido. 1. – Jap. J. sanit. Zool. 35 (3): 323-324, 1 table.

Nowakowski, J. T., 1991: Sepsidae. – Pp. 180-181 *in* Razowski, J.: Wykaz zwierząt Polski. Tom II. Checklist of Animals of Poland. Volume 2. – Ossolineum Press, Wrocław, 342 pp.

Ólafsson, E., 1991: A checklist of Icelandic insects. [Icelandic, with English summary.] – Fjölrit Náttúrufrædistofnunar 17: 69 pp. – Reykjavík.

d'Orbigny, C. V. D. (Ed.), 1843: Dictionnaire universel d'histoire naturelle résumant et complétant tous les faits présentés par les encyclopédies, les anciens dictionnaires scientifiques, les oeuvres complètes de Buffon, et les meilleurs traités spéciaux sur les diverses branches des sciences naturelles donnant la description des êtres et des divers phénomènes de la nature, l'étymologie et la définition des noms scientifiques, les principales applications des corps organiques et inorganiques à l'agriculture, à la médecine, aux arts industriels, etc.; dirigé par M. Charles d'Orbigny, et enrichi d'un magnifique atlas de 288 planches gravées sur acier. Tome troisième [part]. – C. Renard, Paris, 744 pp.

d'Orbigny, C. V. D. (Ed.), 1846: Dictionnaire universel d'histoire naturelle résumant et complétant tous les faits présentés par les encyclopédies, les anciens dictionnaires scientifiques, les oeuvres complètes de Buffon, et les meilleurs traités spéciaux sur les diverses branches des sciences naturelles donnant la description des êtres et des divers phénomènes de la nature, l'étymologie et la définition des noms scientifiques, les principales applications des corps organiques et inorganiques à l'agriculture, à la médecine, aux arts industriels, etc.; dirigé par M. Charles d'Orbigny, et enrichi d'un magnifique atlas de 288 planches gravées sur acier. Tome huitième [part]. – C. Renard, Paris, 766 pp.

O'Toole, C., 1978: Association with other animals and micro-organisms. Ants, bees and wasps (Aculeate Hymenoptera). – Pp. 157-164 *in* Stubbs, A. E. & Chandler, P. J.: A dipterist's handbook. – Amat. Ent., 15: 255 pp.

Ozerov, A. L., 1983: Notes on Sepsidae (Diptera) of the Far East. [In Russian.] – Ent. Obozr. 62 (3): 635-637, 5 figs. [English translation in Ent. Rev., Wash. 62 (3): 177-180.]

Ozerov, A. L., 1985a: On the taxonomy of the genus *Ortalischema* (Diptera, Sepsidae). [Russian, with English summary.] – Zool. Zh. 64 (8): 1267-1269, 3 figs.

Ozerov, A. L., 1985b: New and little-known species of Sepsidae (Diptera) from the Far East. [Russian, with English summary.] – Ent. Obozr. 64 (4): 839-844, 19 figs. [English translation in Ent. Rev., Wash. 65 (2): 154-158.]

Ozerov, A. L., 1986a: To the knowledge of Sepsidae (Diptera) of the fauna of the USSR. [Russian, with English summary.] – Ent. Obozr. 65 (3): 639-644, 18 figs. [English translation in Ent. Rev., Wash. 66 (2): 134-139.]

Ozerov, A. L., 1986b: Revision of the type-specimens of species of the genus *Sepsis* (Diptera, Sepsidae) with the description of three new species from Central Asia and Kazakhstan. [Russian, with English summary.] – Vest. Zool. 1986 (6): 12-15, 8 figs.

Ozerov, A. L., 1986c: Larvae of the Palaearctic species of the genus *Saltella* R.-D. (Diptera, Sepsidae). [Russian, with English summary.] – Nauch. Dokl. vyssh. Shk. Biol. Nauki 1986 (7): 42-47, 2 figs.

Ozerov, A. L., 1986d: A review of Palaearctic species of the '*leachi*' group in the genus *Themira* R.-D. (Diptera, Sepsidae) with description of a new species, *Themira przewalskii* sp. n. [Russian, with English summary.] – Byull. Mosk. Obshch. Ispyt. Prir. (Otdel Biol.) 91(2): 51-54, 6 figs.

Ozerov, A. L., 1987a: Larvae of the Palaearctic species of the genus *Nemopoda* R.-D. (Diptera, Sepsidae). [In Russian.] – Pp. 107-111, 12 figs *in* Lehr, P. A. & Kanyukova, E. V.: The taxonomy of insects of Siberia and Far East of the USSR. A collection of papers. – Academy of Sciences of the USSR, Vladivostok, 132 pp.

Ozerov, A. L., 1987b: A new species of the genus *Sepsis* (Diptera, Sepsidae) from the Near East. [Russian, with English summary.] – Vest. Zool. 1987 (6): 79-80, 3 figs.

Ozerov, A. L., 1989: Flies of the families Sepsidae and Piophilidae (Diptera) of the Zejskij State Nature Preserve. [In Russian.] – Ent. Obozr. 68 (4): 839-849, 2 figs. [English translation in 1990, Ent. Rev., Wash. 69 (4): 156-167.]

Ozerov, A. L., 1990: New data on the fauna and taxonomy of the African dipterans of the family Sepsidae (Diptera). [Russian, with English summary.] – Vest. Zool. 1990 (2): 15-20, 3 figs.

Ozerov, A. L., 1991a: The biology and morphology of the larvae of Palaearctic species of the genera *Meroplius* R.-D. and *Xenosepsis* Malloch (Diptera, Sepsidae). [Russian, with English summary.] – Nauch. Dokl. vyssh. Shk. Biol. Nauki 1991 (1): 49-55, 2 figs.

Ozerov, A. L., 1991b: Review of the subgenus *Platytoxopoda* (Diptera, Sepsidae: Toxopodinae). [Russian, with English summary.] – Byull. Mosk. Obshch. Ispyt. Prit. (Otdel Biol.) 96 (3): 48-53, 2 figs.

Ozerov, A. L., 1992a: On the taxonomy of flies of the family Sepsidae (Diptera). [Russian, with English summary.] – Byull. Mosk. Obshch. Ispyt. Prit. (Otdel Biol.) 97 (4): 44-47, 3 figs.

Ozerov, A. L., 1992b: New data on the neotropical sepsids (Diptera Sepsidae). – Russian ent. J. 1 (1): 81-86, 16 figs.

Ozerov, A. L., 1992c: New genus and three new species of sepsid flies (Diptera, Sepsidae) from Vietnam. [Russian, with English summary.] – Zool. Zh. 71 (3): 147-150, 3 figs.

Ozerov, A. L., 1993a: Six new species of the genus *Palaeosepsis* Duda (Diptera Sepsidae). – Russian ent. J. 2 (1): 63-71, 30 figs.

Ozerov, A. L., 1993b: A revision of the genus *Paratoxopoda* Duda (Diptera, Sepsidae). – Bull. Inst. r. Sci. nat. Belg. (Ent.) 63: 123-144, 80 figs.

Ozerov, A. L., 1993c: The results of the examination of the syntypes of *Sepsis insularis* Williston, 1896 and *Sepsis furcata* Melander et Spuler, 1917 (Diptera, Sepsidae). – Russian ent. J. 1 (2) [1992]: 75.

Ozerov, A. L., 1993d: Four new synonyms in the family Sepsidae (Diptera). [In Russian.] – Ent. Obozr. 72 (1): 248-249. [English translation in Ent. Rev., Wash. 72 (9): 99-101.]

Ozerov, A. L., 1994a: A review of the sepsid flies of Palaeosepsini, trib.n. (Diptera Sepsidae). – Russian ent. J. 2 (5-6) [1993]: 55-103, 176 figs.

Ozerov, A. L., 1994b: *Dicranosepsis planitarsis* sp. nov. - new species of Sepsidae (Diptera) from Vietnam. [Russian, with English summary.] – Byull. Mosk. Obshch. Ispyt. Prit. (Otdel Biol.) 99 (1): 113-114, 3 figs.

Ozerov, A. L., 1996: To the fauna and taxonomy of African Sepsidae (Diptera). – Russian ent. J. 4 (1-4) [1995]: 127-144, 55 figs.

Ozerov, A. L., 1997a: Some types of Sepsidae in the Berlin and Eberswalde Museums (Diptera). – Beitr. Ent. 47 (2): 477-487.

Ozerov, A. L., 1997b: A revision of the genus *Dicranosepsis* Duda, 1926 (Diptera, Sepsidae). – Russian ent. J. 5 (1-4) [1996]: 135-161, 111 figs.

Ozerov, A. L., 1997c: A new species of the genus *Nemopoda* Robineau-Desvoidy, 1930 [sic] (Diptera, Sepsidae) from the Far East of Russia. – Int. J. dipterol. Res. 8 (3): 159-162, 17 figs.

Ozerov, A. L., 1997d: Habitats of imagoes and larvae of *Ortalischema albitarse* (Zetterstedt) and *Susanomira caucasica* Pont (Diptera, Sepsidae). – Pp. 90 *in* Narchuk, E. P.: Diptera (Insecta) in ecosystems. – Zoological Institute, Russian Academy of Sciences, 159 pp.

Ozerov, A. L., 1998a: Some types of North American and African species of Sepsidae (Diptera). – Russian ent. J. 6 (1-2) [1997]: 85-88.

Ozerov, A. L., 1998b: The female abdomen and reproductive system of the Sepsidae. – Pp. 165-166 *in* Ismay, J. W.: Fourth International Congress of Dipterology, Abstracts Volume. – Oxford, 279 pp.

Ozerov, A. L., 1998c: Two new species of Sepsidae (Diptera). – Russian ent. J. 6 (3-4) [1997]: 83-85, 10 figs.

Ozerov, A. L., 1999a: A new species of the genus *Sepsis* Fallén (Diptera: Sepsidae) from North Ossetia. – Russian ent. J. 7 (1-2) [1998]: 57-58, 7 figs.

Ozerov, A. L., 1999b: Studies of Afrotropical Sepsidae (Diptera). I. A revision of the genus *Afrosepsis* Ozerov. – Int. J. dipterol. Res. 10 (1): 59-66, 5 figs.

Ozerov, A. L., 1999c: Studies of Afrotropical Sepsidae (Diptera). II. A revision of the genus *Meroplius* Rondani. – Int. J. dipterol. Res. 10 (2): 81-96, 10 figs.

Ozerov, A. L., 1999d: A review of the genus *Themira* Robineau-Desvoidy, 1830 (Diptera: Sepsidae) of the world, with a revision of the North American species. – Russian ent. J. 7 (3-4) [1998]: 169-208, 33 figs.

Ozerov, A. L., 1999e: 88. Fam. Sepsidae. [In Russian.] – Pp. 556-570, figs 342-351 *in* Ler, L. P.: Key to the insects of the Russian Far East. 6, Diptera and Siphonaptera, part 1. – Dal'nauka, Vladivostok, 665 pp., 393 figs, 1 map.

Ozerov, A. L., 1999f: Two species of Sepsidae (Diptera) described by J. C. H. de Meijere. – Russian ent. J. 8 (1): 51.

Ozerov, A. L., 2000a: The Sepsidae (Diptera) of Namibia, with descriptions of four new species. – Cimbebasia 16: 31-45, 36 figs.

Ozerov, A. L., 2000b: Studies of afrotropical Sepsidae (Diptera). III. A revision of the genus *Adriapontia* Ozerov, 1996. – Russian ent. J. 8 (4) [1999]: 269-278, 7 figs.

Ozerov, A. L., 2000c: 5.3.51. Sepsidae. – Pp. 120-121 *in* Ziegler, J. & Menzel, F.: Die historische Dipteren-Sammlung Carl Friedrich Ketel. Revision einer zwischen 1884 und 1903 angelegten Sammlung von Zweiflüglern (Diptera) aus Mecklenburg-Vorpommern. – Nova Supplta ent. 14: 266 pp., 42 figs, 3 tables.

Ozerov, A. L., 2000d: Studies of Afrotropical Sepsidae (Diptera). IV. Description of six new species and designation of lectotype for *Sepsis polychaeta* Duda, 1926. – Russian ent. J. 9 (1): 69-78, 6 figs.

Ozerov, A. L., 2000e: Revision of the Nearctic species of the genus *Sepsis* Fallén, 1810 (Diptera: Sepsidae). – Russian ent. J. 9 (2): 161-176, 7 figs.

Ozerov, A. L. & Meier, R., 1995: A key to the larvae of the Palaearctic genera of the Sepsidae (Diptera: Cyclorrhapha). – Annls Soc. ent. Fr. (n.s.) 31 (3): 259-283, 114 figs.

Pakalniškis, S., 1989: The list of the Lithuanian Diptera (to 1988 inclusive). [In Russian.] – Pp. 32-85 *in* Jonaitis, V.: New and rare for the Lithuanian SSR insect species. Records and descriptions of 1989. [In Lithuanian.] Institute of Zoology and Parasitology of the Academy of Sciences of the Lithuanian SSR, Lithuanian Entomological Society – Lithuanian Branch of All-Union Entomological Society, Vilnius, 88 pp.

Panzer, G. F. W., 1798: Favnae insectorvm germanicae initia oder Devtschlands Insecten 60: 1-24. – Felsecker, Nürnberg.

Papp, L., 1971a: Ecological and production biological data on the significance of flies breeding in cattle droppings. – Acta zool. hung. 17 (1-2): 91-105, 3 tables.

Papp, L., 1971b: Some faunistical and ecological data to the knowledge of flies developing in cattle droppings (Dipt.). [Hungarian, with English summary.] – Folia ent. hung. 24 (2): 427-441.

Papp, L., 1974a: Dipterological studies in a Hungarian horse farm (Diptera). – Folia ent. hung. 27 (1): 167-176, 7 figs, 3 tables.

Papp, L., 1974b: Dipterological studies in some Hungarian large-scale pig farms. – Acta agron. hung. 23 (1-2): 136-147, 3 figs, 2 tables.

Papp, L., 1975: Dipterological studies in some Hungarian and Afghan large-scale cattle farms. [Hungarian, with English summary.] – Folia ent. hung. 28 (1): 137-145, 2 figs, 3 tables.

Papp, L., 1976: Ecological and zoogeographical data on flies developing in excrement droppings (Diptera). – Acta zool. hung. 22 (1-2): 119-138, 7 tables.

Papp, L., 1985: Flies (Diptera) developing in sheep droppings in Hungary. – Acta zool. hung. 31 (4): 367-379, 4 tables.

Papp, L. & Garzó, P., 1985: Flies (Diptera) of pasturing cattle: some new data and new aspects. – Folia ent. hung. 46 (2): 153-168, 4 tables.

Parker, G. A., 1972a: Reproductive behaviour of *Sepsis cynipsea* (L.) (Diptera: Sepsidae). I. Preliminary analysis of the reproductive strategy and its associated behaviour patterns. – Behaviour 41 (1-2): 172-206, 26 figs, plates 10-12, 1 table.

Parker, G. A., 1972b. Reproductive behaviour of *Sepsis cynipsea* (L.) (Diptera: Sepsidae). II. The significance of the precopulatory passive phase and emigration. – Behaviour 41 (3-4): 241-250.

Payne, R. M., 1975: Insects at flowers of Elecampane (*Inula helenium* L.). – Entomologist's mon. Mag. 110 [1974]: 221.

Payne, R. M., 1979: Insects attracted to Alexanders (*Smyrnium olusatrum* L.). – Entomologist's mon. Mag. 113 [1977]: 233-234.

Perris, E., 1876: Nouvelles promenades entomologiques. [Part.] – Annls Soc. ent. Fr. (5) 6 (2): 177-244.

Persson, P. I., 1983: Flugor på Ölands Stora alvar. – Ent. Tidskr. 104 (3-4): 151-164, 1 fig., 1 table.

Peterson, A., 1960: Larvae of insects. Coleoptera, Diptera, Neuroptera, Siphonaptera, Mecoptera, Trichoptera. Part II. – A. Peterson, Columbus, Ohio, 416 pp., figs.

Petrova, B. K., 1968: Faunistic-ecological review of synantropical Diptera from the south of the Maritime Territory. [In Russian.] – Ent. Obozr. 47 (1): 95-105. [English translation in Ent. Rev., Wash. 47 (1): 54-58.]

Phillips, D. S. & Arthur, W., 1994: Observations on the distribution of seaweed fly larvae and other invertebrates within a wrack-bed. – Entomologist 113 (2): 154-163, 3 figs, 5 tables.

Pont, A. C., 1979: Sepsidae. Diptera Cyclorrhapha, Acalyptrata. – Handbk Ident. Br. Insects 10, 5 (c): 35 pp., 10 figs, 10 plates.

Pont, A. C., 1986a: Two additions to the list of British Sepsidae (Diptera). – Entomologist's mon. Mag. 122: 91-92.

Pont, A. C., 1986b: A revision of the Fanniidae and Muscidae described by J. W. Meigen (Insecta: Diptera). – Annln naturh. Mus. Wien (B) 87 [1983]: 197-253, 1 plate.

Pont, A. C., 1987a: 'The mysterious swarms of sepsid flies': an enigma solved? – J. nat. Hist. 21 (2): 305-317.

Pont, A. C., 1987b: Provisional atlas of the Sepsidae (Diptera) of the British Isles. – '1986', Biological Records Centre, Huntingdon, 33 pp.

Pont, A. C., 1987c: Two new genera of West Palaearctic Sepsidae (Diptera). – Entomologica scand. 18 (3): 265-272, 25 figs.

Pont, A. C., 1990: Notes on Finnish Sepsidae and Muscidae (Diptera). – Ent. fenn. 1 (3): 154.

Pont, A. C., 1995: The type-material of Diptera (Insecta) described by G. H. Verrall and J. E. Collin. – Oxford University Museum Publication 3: x + 223 pp., 8 plates. Clarendon Press, Oxford.

Pont, A. C. & Michelsen, V., 1982: The Muscoidea described by Moses Harris (Diptera: Fanniidae, Scathophagidae, Anthomyiidae, Muscidae). – Steenstrupia 8: 25-46.

Poppius, B. R., 1903: Blombiologiska iakttagelser. – Acta Soc. Fauna Flora fenn. 25 (1): 53 pp.

Poppius, B., Lundström, C. & Frey, R., 1916: Dipteren aus dem Sarekgebiet. – Naturw. Unters. Sarekgebirge 4 (Zoologie): 665-696, plate 10.

Porchinskiy, I. A., 1910: Recherches biologiques sur le *Stomoxys calcitrans* L. et biologie comparée des mouches coprophagues. [In Russian.] – Trudy Byuro Ent. 8 (8) [2]: 90 pp., 97 figs, 1 plate.

Pridantseva, E. A., 1967: The fauna of pasture flies (Diptera) connected with camels in South Tuva. [In Russian.] – Ent. Obozr. 46 (4): 827-834, 1 fig. [English translation in Ent. Rev., Wash. 46 (4): 491-494.]

Püchel, F., 1992: Schwingfliegen (Sepsidae) und Dungfliegen (Scathophagidae) (Diptera) aus Malaise-Fallen in Köln. – Decheniana Beih. 31: 417-430, 6 figs, 4 tables.

Randall, M., Coulson, J. C. & Butterfield, J., 1981: The distribution and biology of Sepsidae (Diptera) in upland regions of northern England. – Ecol. Ent. 6 (2): 183-190, 1 fig., 4 tables.

Rapp, W. F., 1946: The generic name *Pandora*. – Ann. Mag. nat. Hist. (11) 12 [1945]: 499-500.

Reed, H. B. J., 1958: A study of dog carcass communities in Tennessee, with special reference to the insects. – Am. Midl. Nat. 59 (1): 213-245, 4 tables, 3 figs.

Řehůřková, A. 1988: Synantropné dvojkrídlovce Diptera turistických stredísk západnej časti vysokých tatier. – Ent. Probl. 18: 211-232, 5 tables.

Remmert, H., 1955: Ökologische Untersuchungen über die Dipteren der Nord- und Ostsee. – Arch. Hydrobiol. 51: 1-53, 3 figs.

Reusch, T. & Blanckenhorn, W. U., 1998: Quantitative genetics of the dung fly *Sepsis cynipsea*: Cheverud's conjecture revisited. – Heredity 81: 111-119, 2 tables, 2 figs.

Ringdahl, O., 1931: Insektfaunan inom Abisko Nationalpark. III. Studier under ledning av Yngve Sjöstedt. 6. Flugor - Diptera brachycera. – K. svenska Vetensk Akad. Skr. Naturskydd. 18: 1-32.

Ringdahl, O., 1932: Anteckningar till det öländska alvarets dipterfauna. – Ent. Tidskr. 53 (2-3): 161-165.

Ringdahl, O., 1934: Anteckningar till svenska arter av fam. Sepsididae (Diptera). – Ent. Tidskr. 55 (1): 1-5, 13 figs.

Ringdahl, O., 1939: Diptera Brachycera i regio alpina. – Ent. Tidskr. 60 (1-2): 37-50.

Ringdahl, O., 1941: Bidrag till kännedomen om flugfaunan (Diptera Brachycera) på Hallands Väderö. – Ent. Tidskr. 62 (1-2): 1-23.

Ringdahl, O., 1947: Förteckning över flugor från Ölands alvar. – Ent. Tidskr. 68 (1-2): 21-28.

Ringdahl, O., 1948: Nya fynd av holometopa flugor. – Ent. Tidskr. 69 (1-2): 1-4.

Ringdahl, O., 1950: Dipterologiska anteckningar från sydsvenska mossar. – Ent. Tidskr. 71 (2): 111-119.

Ringdahl, O., 1951: Flugor från Lapplands, Jämtlands och Härjedalens fjälltrakter (Diptera Brachycera). – Opusc. ent. 16 (3): 113-186.

Ringdahl, O., 1952: Småplock ur Kullabergs flugfauna. – Fauna Flora, Upps. 47 (5-6): 234-244, 3 figs.

Ringdahl, O., 1954a: Nya fyndorter för norska Diptera. – Norsk ent. Tidsskr. 9 (1-2): 46-54.

Ringdahl, O., 1954b: Några dipterologiska anteckningar från Råå kärr och vassar. – Ent. Tidskr. 75 (2-4): 223-234.

Ringdahl, O., 1959a: Några tillägg till flugfaunan på Hallands Väderö. – Ent. Tidskr. 79 (3-4) [1958]: 120-122.

Ringdahl, O., 1959b: Flugor på strandängar i nordvästra Skåne. – Ent. Tidskr. 80 (1-2): 39-47.

Ringdahl, O., 1960: Flugfaunan på Kullaberg och Hallands Väderö. – Kullabergs Natur 2: 40 pp.

Robineau-Desvoidy, A. J. B., 1830: Essai sur les Myodaires. – Mém. prés. div. Sav. Acad. Sci. Inst. Fr. 2: 1-813.

Roháček, J. & Zuska, J., 1983: Diptera Acalyptrata rašelinišť severní Moravy (ČSSR). Část 3. Micropezidae, Psilidae, Sciomyzidae, Sepsidae. [Acalyptrate Diptera of peat-bogs in North Moravia.] – Čas. slezsk. Mus. Opave (A) 32: 1-22.

Rondani, C., 1856: Dipterologiae Italicae Prodromus. 1. Genera Italica ordinis Dipterorum ordinatim disposita et distincta et in familias et stirpes aggregata. – Stocchi, Parmae [= Parma], 226 pp.

Rondani, C., 1874: Species italicae ordinis Dipterorum (Muscaria Rndn.). Stirps XXI. – Tanipezinae Rndn. collectae et observatae. – Bull. Soc. ent. ital. 6 (3): 167-182.

Roper, P., 1993: Nocturnal invertebrates visiting flowers of Witch Hazels, *Hamamelis mollis pallida* and *Hamamelis japonica* cultivars, in January. – Entomologist's Rec. J. Var. 105 (7-8): 192-193.

Roser, C. L. F., 1840: Erster Nachtrag zu dem im Jahre 1834 bekannt gemachten Verzeichnisse in Württemberg vorkommender zweiflügliger Insekten. – Corresp Bl. k. württ. landw. Ver. (n.F.) 1 (1): 49-64.

Rühl, M., 1913: Liste neuerdings beschriebener oder gezogener Parasiten und ihrer Wirte. III. [Part.] – Societas Ent. 28 (19): 79-80.

Rühl, M., 1917: Liste neuerdings beschriebener oder gezogener Parasiten und ihre Wirte. VI. [Part.] – Societas Ent. 32 (12): 49-50.

Rühl, M., 1918: Liste neuerdings beschriebener oder gezogener Parasiten und ihre Wirte. VI. [Part.] – Societas Ent. 33 (7): 28.

Sabrosky, C. W., 1999: Family-group names in Diptera. An annotated catalog. – Myia 10: 1-360.

Schembri, S., Gatt, P. & Schembri, J., 1991: Recent records of flies from the Maltese Islands (Diptera). – Memorie Soc. ent. ital. 70 (1): 255-277.

Schrank, F. von P., 1803: Favna Boica. Durchgedachte Geschichte der in Baiern einheimischen und zahmen Thiere. Dritten und lezten [sic] Bandes erste Abtheilung. [3 (1)]. – Krüll, Landshut, viii + 272 pp.

Schulz, K., 1989: Vergleichende Untersuchung zur Biologie einiger kuhfladenbewohnender Arten der Gattung *Sepsis* (Diptera, Sepsidae). – Diplomarbeit des Fachbereichs Biologie, Freie Universität Berlin.

Schulz, K., 1999: The evolution of mating systems in black scavenger flies (Diptera: Sepsidae). – Dissertation, Department of Entomology, University of Arizona.

Schumann, H., 1961: Die Eier von Fliegen. – Mikrokosmos 50 (10): 297-300, 23 figs.

Schumann, H., 1962: Zur Morphologie einiger Larven der Familien Borboridae und Sepsidae (Diptera). – Mitt. zool. Mus. Berl. 38 (2): 415-450, 59 figs.

Schweiger, J., 1988: Beiträge zur Faunistik und Ökologie der Dipteren des Kuhfladens. Untersuchungen an zwei Berliner Standorten. – Diplomarbeit des Fachbereichs Biologie, Freie Universität Berlin.

Scott, H., 1920: Notes on (I) the parasitic staphylinid *Aleochara algarum* Fauvel, and its hosts, the phycodromid flies; (II) a case of supposed parasitism in the genus *Homalota*. – Entomologist's mon. Mag. 56 [= (3) 6]: 148-157, 2 figs.

Séguy, E., 1932: Contribution à l'étude des mouches phytophages de l'Europe occidentale. – Encycl. ent. (B II), Dipt. 6: 145-197, plates 2-6, figs 83-100, 103-123.

Séguy, E., 1934: F. Sepsidae. – Pp. 234-256, figs 270-354 in: Diptères (Brachycères) (Muscidae Acalypterae et Scatophagidae). Faune Fr. 28: 832 pp., 903 figs, 27 plates. Lechevalier, Paris.

Séguy, E., 1936: Voyage de MM. L. Chopard et A. Méquignon aux Açores (août-septembre 1930). X. Diptères. – Annls Soc. ent. Fr. 105 (1-2): 11-26, 2 figs.

Shtakel'berg, A. A. [Stackelberg], 1958: List of Diptera of the Leningrad region. III. Acalyptrata, part 1. [In Russian.] – Trudy zool. Inst. Leningr. 24: 103-246.

Shtakel'berg, A. A. [Stackelberg], 1970: 66. Family Sepsidae. – Pp. 175-180, figs 611-613 *in* Shtakel'berg, A. A. & Narchuk, E. P.: Keys to the insects of the European part of the USSR in five volumes (general editor G. Ya. Bey-Bienko). 5, flies, fleas, second part. – Nauka, Leningrad, 943 pp., 964 figs.

Siebke, H., 1853: Beretning om en i sommaren 1850 foretagen entomologisk reise i en deel af Gudbrandsdalen. – Nyt Mag. Naturvid. 7 (3): 253-306.

Siebke, H., 1866a: Entomologisk Reise i Romsdals amt i sommeren 1864. – Nyt Mag. Naturvid. 14 (4): 375-388.

Siebke, H., 1866b: Entomologiske Undersøgelser, foretagne i sommeren 1865. – Nyt Mag. Naturvid. 14 (4): 389-420.

Siebke, H., 1870: Beretning om en i sommeren 1869 foretagen entomologisk reise gjennem Ringerike, Hallingdal og Valders. – Nyt Mag. Naturvid. 17 (4): 246-314.

Siebke, H., 1873: Bidrag til Norges Insektfauna. Beretning om en i Østerdalen foretagen Reise i 1870. – Nyt Mag. Naturvid. 19 (1-2): 39-102.

Siebke, H., 1877: Enumeratio Insectorum Norvegicorum. Fasciculum IV, Catalogum Dipterorum continentem. – Broegger, Christianiae [= Oslo], xiv + 255 pp., 1 map.

Silén, F., 1906: Blombiologiska iakttagelser i södra Finland. – Meddn Soc. Fauna Flora fenn. 32: 120-134.

Silén, F., 1909: Blombiologiska iakttagelser i södra Finland, II. – Meddn Soc. Fauna Flora fenn. 35: 31-52.

Silva, V. C., 1990: Revision of the family Sepsidae of the Neotropical region. I. The genus *Meroplius* Rondani (Diptera, Schizophora). – Revta bras. Ent. 34 (4): 709-711, 9 figs.

Silva, V. C., 1992a: Levantamento preliminar de Sepsidae (Diptera, Schizophora) em Roraima, Projeto Maracá, com descrição de uma espécie nova. – Acta amazonica 21 [1991]: 369-374, 9 figs.

Silva, V. C., 1992b: Revision of the family Sepsidae of the Neotropical region. II. The genus *Meropliosepsis* Duda, 1926 (Diptera, Schizophora). – Revta bras. Ent. 36 (3): 549-552, 11 figs.

Silva, V. C., 1993: Revisão da familía Sepsidae na região Neotropical. III. Os gêneros *Palaeosepsis* Duda, 1926, *Archisepsis* gen. n. e *Microsepsis* gen. n.; chave para os gêneros neotropicais (Diptera, Schizophora). – Heringia (Ser. Zool.) 75: 117-170, 184 figs.

Silva, V. C., 1995: A new genus of Sepsidae (Diptera, Schizophora) from Nicaragua. – Studia dipterologica 2 (2): 203-206, 4 figs.

Smith, K. G. V., 1957: Some miscellaneous records of bred Diptera. – Entomologist's Rec. J. Var. 69 (10): 214-216.

Smith, K. G. V., 1975: The faunal succession of insects and other invertebrates on a dead fox. – Entomologist's Gaz. 26: 277-287, 1 table, plate 7.

Snowball, G. J., 1944: A consideration of the insect population associated with cow dung at Crawley, W. A. – Jl R. Soc. W. Aust. 28: 219-245, 2 tables 12 figs.

Soós, Á., 1945: Über die Sepsiden, Piophiliden und Drosophiliden des Karpatenbeckens. – Fragm. faun. hung. 8 (1-4): 18-23.

Soós, Á., 1958: Neubeschreibung der Zetterstedtschen Art *Themira gracilis* (Dipt., Sepsidae). – Acta zool. hung. 4 (1-2): 207-209, 5 figs.

Soós, Á., 1959: Diptera II. Torpikkelynélküli legyek I. Muscidae Acalyptratae I. 2. családsorozat: Schizophora - Homlokréses legyek. – Fauna hung. 48, XV (1): 88 + 3 pp., 37 figs.

Soós, Á., 1962: Sepsiden aus Ost- und Süd-Afrika (Diptera: Acalyptrata). (Ergebnisse der Forschungsreise Lindner 1958/59 - Nr. 13.) – Stuttg. Beitr. Naturk. 92: 1-5, 2 figs.

Soós, Á., 1964: Two new *Paratoxopoda* Duda species from East Africa (Diptera: Sepsidae). – Acta zool. hung. 10 (3-4): 451-456, 2 figs.

Soós, Á., 1972: Taxonomische und faunistische Untersuchungen über die mongolischen Sepsiden (Diptera). – Acta zool. hung. 18 (3-4): 353-370, 3 figs.

Soós, Á., 1981: Data to Muscidae Acalyptratae (Diptera) in the Hortobágy National Park, I. – Pp. 213-222 *in* Mahunka, S.: The Fauna of Hortobágy National Park. Volume 1. – Akadémiai Kiadó, Budapest, 415 pp.

Sowig, P., 1995: Habitatwahl und Lebensgeschichte bei dungbewohnenden Käfern der Gattung *Sphaeridium*: Wie werden die Imagines den ökologischen Ansprüchen ihrer Larven gerecht? (Coleoptera: Hydrophilidae). – Mitt. Dtsch. Ges. allg. angew. Ent. 10: 477-480, 2 figs.

Stæger, R. C., 1844: [A systematic review of the Danish species of the genus *Sepsis*, presented at the meeting of 12 March 1843.] – Pp. 22-36 *in* Schiødte, J. C.: Forhandlinger i det skandinaviske entomologiske Selskab. – Naturh. Tidsskr. (2) 1 (1): 16-70.

Steyskal, G., 1946: *Themira nigricornis* Meigen in North America, with a revised key to the Nearctic species of *Themira* (Diptera: Sepsidae). – Ent. News 57: 93-95.

Steyskal, G., 1965: Family Sepsidae. – Pp. 681-685 *in* Stone, A., Sabrosky, C. W., Wirth, W. W., Foote, R. H. & Coulson, J. R.: A Catalog of the Diptera of America North of Mexico. – Agric. Handbk U.S. Dep. Agric. 276: iv + 1696 pp.

Steyskal, G., 1987: 86. Sepsidae. – Pp. 945-950, 20 figs *in* McAlpine, J. F.: Manual of Nearctic Diptera. Volume 2. – Res. Branch Agric. Canada Monogr. 28: vi + 675-1332, figs.

Strand, E., 1913: Neue Beiträge zur Arthropodenfauna Norwegens nebst gelegentlichen Bemerkungen über deutsche Arten. XVIII. Weiteres über von mir gesammelte Diptera. – Nyt Mag. Naturvid. 51 (4): 323-329, 1 fig.

Strand, E., 1928: Miscellanea nomenclatorica zoologica et palaeontologica. I-II. – Arch. Naturgesch. 92 A 8 [1926]: 30-75.

Strobl, G., 1893: Neue österreichische Muscidae Acalypterae. [Part.] – Wien. ent. Ztg 12 (6): 225-231.

Strobl, G., 1894: Die Dipteren von Steiermark. II. Theil. – Mitt. naturw. Ver. Steierm. 30 [1893]: 1-152.

Suh, S. J. & Kwon, Y. J., 1994: Classification of the family Sepsidae from Korea (Diptera: Acalyptratae). – Insecta Koreana 11: 58-63, 2 figs.

Sukhacheva, E. I., 1963: The role of certain insect larva species in the epidemiology of ascariasis. [Russian, with English summary.] – Medskaya Parazit. 32 (5): 600-603, 1 fig., 2 tables.

Šulc, K., 1928: Biologický význam výzbroje předních samčích nohou u Sepsid (Muscidae). [Biologische Bedeutung der Bewaffnung der männlichen Vorderfüsse bei den Sepsiden (Muscidae).] [Czech, with German summary.] – Biol. Spisy vys. Šk. zvěrolék., Brno 7 (14): 181-194, 7 figs.

Šulc, K., 1929: Dosud neznámá osmeteria u Sepsid (Diptera). Bisher unbekannte Osmeterien bei den Sepsiden (Diptera). [Czech, with German summary.] – Biol. Listy 14 (5-6): 428-439, 12 figs.

Sychevskaya, V. I., 1957: Synanthropic flies in the region of Belovodsk (North Kirghisia). [In Russian.] – Ent. Obozr. 36 (1): 108-115, 3 tables.

Sychevskaya, V. I., 1966: On synanthropic flies of the Pamir. [Russian, with English summary.] – Zool. Zh. 45 (3): 390-399, 2 tables, 1 graph.

Sychevskaya, V. I., 1968: Synanthropic flies in the town of Osh. [In Russian.] – Pp. 49-56, 4 tables *in* Pek, L. V.: Entomological Investigations in Kirgiziya. – Akad. Nauk Kirgiz. SSR, Frunze, 128 pp.

Sychevskaya, V. I., 1972a: On the synanthropic flies of Yakutiya. [In Russian.] – Pp. 144-157, 2 tables *in* Ammosov, Yu. N.: Fauna and ecology of the insects of Yakutiya. – Akad. Nauk SSSR, Yakutsk, 180 pp.

Sychevskaya, V. I., 1972b: Synanthropic flies of Central Asia. [In Russian.] – Proc. 13th Int. Congr. Ent. 1968 3: 255-256.

Sychevskaya, V. I., 1972c: Synanthropic flies (Diptera) from the Lower Amu-Darya. [Russian, with English summary.] – Ent. Obozr. 41 (3): 534-553, 13 figs, 8 tables. [English translation in Ent. Rev., Wash. 41 (3): 320-331.]

Sychevskaya, V. I., 1981: The winter fauna of synanthropic flies in Uzbekistan. [In Russian.] – Pp. 87-91, 1 table *in* Kozlova, O. A.: Topical questions of medical parasitology. A collection of papers. – Tashkent, 96 pp.

Sychevskaya, V. I. & Vtorov, P. P., 1969: Synanthropic flies (Diptera) from mountain Kirghizia. [Russian, with English summary.] – Ent. Obozr. 48 (4): 816-830, 4 figs, 1 table. [English translation in Ent. Rev., Wash. 48 (4): 518-527.]

Tamarina, N. A., Devyatkin, A. L. & Khromova, L. A., 1982: Peculiarities of the fauna of pasture flies (Diptera) of the non-chernozem fauna of the European part of the USSR. [Russian, with English summary.] – Ent. Obozr. 61 (1): 84-90, 2 tables. [English translation in Ent. Rev., Wash. 61 (1): 73-80.]

Thompson, F. C., 1981: Nomenclature of the European species of *Neoascia* Williston (Diptera: Syrphidae). – Entomologica scand. 12: 470-478

Thompson, F. C. & Pont, A. C., 1994: Systematic database of *Musca* names (Diptera). A catalog of names associated with the genus-group name *Musca* Linnaeus, with information on their classification, distribution, and documentation. – Theses zoologicae, 20 [1993]: 219 (+ ii) pp.

Tiensuu, L., 1948: On the phenology of flower visiting by Diptera in Southern Finland. [Finnish, with English summary.] – Suom. hyönt. Aikak. 14, Suppl.: 212-218.

Timon-David, J., 1937: Recherches sur le peuplement des hautes montagnes. Diptères de la vallée de Chamonix et du massif du Mont Blanc. – Annls Fac. Sci. Marseille (2) 10 (1): 7-54, 14 figs.

Titchener, R. N., Newbold, J. W. & Wright, C. L., 1981: Flies associated with cattle in south west Scotland during the summer months. – Res. vet. Sci. 30: 109-113, 1 fig., 4 tables.

Tsacas, L., 1959: Contribution à l'étude des diptères du littoral marin de la région de Roscoff. (Travaux de la station biologique de Roscoff LIII.) – Archs Zool. exp. gén. 98, notes et revue 2: 62-92, 15 figs.

Tuomikoski, R., 1952: Über die Nahrung der Empididen-Imagines (Dipt.). – Suom. hyönt. Aikak. 18 (4): 170-181.

Ursu, A., 1965: Contribuţii la studiul taxionomic şi ecologic al muştelor sinantrope din ţara noastră. – Lucr. Şes. Ştiinţ. Insţ. Ig. Prot. Muncii 1965: 219.

Ursu, A., 1969a: Contributions à l'étude des diptères synanthropes de notre pays. [Romanian, with Russian and German summaries.] – Comunicări Zool., Bucuresti 7b: 315-332, 10 figs.

Ursu, A., 1969b: Contribuţii la cunoaşterea acarienilor, ectoparaziţi ai unor diptere sinantrope. – Ig. Microbiol. Epidem., Buc. 18 (9): 535-540, 1 table.

Ursu, A., 1973: Contribuţii la cunoaşterea dezvoltarii preimaginale la doua specii din familia Sepsidae (Diptera). – Communicări Zool., Bucuresti 1973: 78-84, 2 plates.

Ursu, A., 1984: Contribution à la connaissance de l'aréal et de la fréquence des sepsides (Diptera, Sepsidae) en Roumanie. – Trav. Mus. Hist. nat. 'Gr. Antipa' 25: 227-232, 1 fig.

Ursu, A. & Tudor, C., 1975: The parasitisation of the synanthropic Diptera by hymenopterous chalcidoids. – Trav. Mus. Hist. nat. 'Gr. Antipa' 16: 179-186, 1 fig., 2 tables.

Valiela, I., 1969: The arthropod fauna of bovine dung in central New York and sources on its natural history. – Jl N. Y. Ent. Soc. 77: 210-220, 2 tables.

Valiela, I., 1974: Composition, food webs and population limitation in dung arthropod communities during invasion and succession. – Am. Midl. Nat. 92(2): 370-385, 4 figs, 5 tables.

Verrall, G. H., 1886a: A hundred new British species of Diptera. [Part.] – Entomologist's mon. Mag. 22: 179-182, 1 fig.

Verrall, G. H., 1886b: A hundred new British species of Diptera. [Part.] – Entomologist's mon. Mag. 22: 230-234.

Verrall, G. H., 1909: British Flies. Vol. V. Stratiomyidae and succeeding families of the Diptera Brachycera of Great Britain. – Gurney and Jackson, London, [2+] 780 pp., 407 figs, frontispiece.

Veselkin, G. A., 1966: Flies (Diptera), satellites of domestic animals and man, in the southern part of the Tyumen region. [In Russian.] – Ent. Obozr. 45 (4): 779-792, 1 map, 1 table. [English translation in Ent. Rev., Wash. 45 (4): 439-447.]

Vibe, C., 1950: Some insects new to Greenland. – Ent. Meddr 25 (6): 419-420.

Villeneuve, J., 1920: Diptères inédits. – Bull. Soc. ent. Fr. 1919 (19): 352-355.

Wahlgren, E., 1915: Det öländska alvarets djurvärld. I. – Ark. Zool. 9 (19): 135 pp., 4 plates.

Wahlgren, E., 1917: Tvåvingar. Diptera. Andra underordningen Cyclorapha. Andra gruppen Schizophora. Fam. 5-12. 7. Fam. Tångflugor. Coelopidae. – Svensk

Insektfauna 11: 160-163. Almqvist & Wiksell, Uppsala.

Wahlgren, E., 1919: Tvåvingar. Diptera. Andra underordningen Cyclorapha. Andra gruppen Schizophora. Fam. 13-20. 17. Fam. Svängflugor. Sepsididae. – Svensk Insektfauna 11: 276-289, figs 133-135. Almqvist & Wiksell, Uppsala.

Wahlgren, E., 1920: De europeiska polaröarnas insektfauna, dess sammansättning och härkomst. – Ent. Tidskr. 41 (1): 1-23.

Walker, F., 1833: Observations on the British species of Sepsidae. – Entomol. Mag. 1 (3): 244-256.

Walker, F., 1871: List of Diptera collected in Egypt and Arabia, by J. K. Lord, Esq.; with descriptions of the species new to science. [Part.] – Entomologist 5: 339-346.

Ward, P. I., 1983: The effects of size on the mating behaviour of the dung fly *Sepsis cynipsea*. – Behav. Ecol. Sociobiol. 13: 75-80, 1 table, 2 figs.

Warncke, E., Terndrup, U., Michelsen, V. & Erhardt, A., 1993: Flower visitors to *Saxifraga hirculus* in Switzerland and Denmark, a comparative study. – Bot. helv. 103: 141-147, 1 fig., 1 table.

Weinberg, M., 1982: Dipteran type specimens recently introduced in the collections of the 'Grigore Antipa' Natural History Museum. – Trav. Mus. Hist. nat. 'Gr. Antipa' 24: 291-296.

Werner, D., 1997: Die Dipterenfauna verschiedener Mülldeponien und Kompostierungsanlagen in der Umgebung von Berlin unter besonderer Berücksichtigung ihrer Ökologie und Bionomie. – Studia dipterologica Supplement 1: 176 pp., 75 figs, 31 tables.

Westwood, J. O., 1840: Order XIII. Diptera Aristotle. (Antliata Fabricius. Halteriptera Clairv.). – Pp. 125-154 *in:* An Introduction to the modern classification of insects; founded on the natural habitats and corresponding organisation of the different families. Synopsis of the genera of British insects. – Longman, Orme, Brown, Green, and Longmans, London, 158 pp.

Wharton, R. A. & Moon, R. D., 1979: Puparia of cyclorrhaphous Diptera from bovine dung in open pasture and rangeland in the transition zone of Western North America. – Ann. ent. Soc. Am. 72: 80-89, 18 figs.

Wheeler, M. R., 1969: A note on Dr. O. Duda's collecting area at Habelschwerdt, Germany. – Entomologist 102 (9): 205-206, 1 map.

Wiedemann, C. R. W., 1830: Aussereuropäische zweiflügelige Insekten. Zweiter Theil. [2.] – Schulz, Hamm, xii + 684 + xi pp., 5 plates.

Wilcke, H. C. D., 1789: CII. Politia naturae quam praeside D. D. Car. Linnaeo proposuit H. Christ. Daniel. Wilcke, Stockholmiensis. Upsaliae 1760. Mart. 29. – Amoenitates Academicae, Ed. 2, 6: 17-39.

Woodley, N. E. & Hilburn, D. J., 1994: The Diptera of Bermuda. – Contr. Am. ent. Inst. 28 (2): 64 pp.

Wulp, F. M. Van der, 1864: Iets over de in Nederland waargenomen Sepsinen. – Tijdschr. Ent. 7 (4): 129-143, plate 8 with 14 figs.

Wulp, F. M. Van der, 1871a: Dipterologische Aanteekeningen. N°. 3. [Part] – Tijdschr. Ent. 14 (4): 186-204.

Wulp, F. M. Van der, 1871b: Dipterologische Aanteekeningen. N°. 3. [Concl.] – Tijdschr. Ent. 14 (5): 205-210, plate 8 with 11 figs.

Zenk, D., 1990: Die Rektaldrüse der Sepsidae (Diptera): Chemie, Morphologie und biologische Bedeutung. – Diplomarbeit der Fakultät Biologie, Universität Bayreuth.

Zerbe, F. 1993: Innerartliche Größenvariabilität und Paarungsverhalten bei *Sepsis punctum* (Fabricius, 1794) [Diptera, Sepsidae]. – Diplomarbeit der Fakultät Biologie der Julius-Maximilians-Universität Würzburg.

Zerbe, F. 1995: Innerartliche Größenvariabilität und Paarungsverhalten der Schwingfliege *Sepsis punctum* (Diptera, Sepsidae). – Sber. Ges. naturf. Freunde Berl. (N. F.) 34: 117-130, 1 table, 4 figs.

Zetterstedt, J. W., 1837: Conspectus familiarum, generum et specierum Dipterorum, in Fauna Insectorum Lapponica descriptorum. – Isis, Jena 21 (1): 28-67.

Zetterstedt, J. W., [1838]: Sectio tertia. Diptera. – Pp. 477-868 *in* Insecta Lapponica. – '1840', Leopold Voss, Lipsiae [= Leipzig], vi + 1140 pp.

Zetterstedt, J. W., 1847: Diptera Scandinaviae disposita et descripta. Tomus sextus. [6]. – Ex officina lundbergiana, Lundae [= Lund]: 2163-2580.

Zetterstedt, J. W., 1849: Diptera Scandinaviae disposita et descripta. Tomus octavus seu supplementum, continens conspectum synopticum familiarum, generum et specierum, addenda, corrigenda et emendanda tomis septem prioribus. [8]. – Ex officina lundbergiana, Lundae [= Lund]: 2935-3366.

Zetterstedt, J. W., 1852: Diptera Scandinaviae disposita et descripta. Tomus undecimus. [11]. – Ex officina lundbergiana, Lundae [= Lund]: i-xii, 4091-4545.

Zetterstedt, J. W., 1855: Diptera Scandinaviae disposita et descripta. Tomus duodecimus seu Supplementum Tertium, continens addenda, corrigenda & emendanda tomis undecim prioribus. [12]. – Ex officina lundbergiana, Lundae [= Lund]: i-xx, 4547-4942.

Zetterstedt, J. W., 1860: Diptera Scandinaviae disposita et descripta. Tomus decimus qvartus seu ultimus, continens addenda, corrigenda & emendanda tomis prioribus, una cum indice alphabetico novarum specierum hujus & praecedentis tomi, atque generico omnium tomorum. [14]. – Ex officina lundbergiana, Lundae [= Lund]: 6191-6609.

Zimsen, E., 1954: The insect types of C. R. W. Wiedemann in the Zoological Museum in Copenhagen. – Spol. zool. Mus. Haun. 14: 43 pp.

Zimsen, E., 1964: The type material of I. C. Fabricius. – Munksgaard, Copenhagen, 656 pp., 2 plates.

Zuska, J., 1960: Beitrag zur Kenntnis der Systematik, Faunistik und Ökologie der tschechoslowakischen Arten der Familie Sepsidae (Diptera). – Čas. čsl. Spol. ent. 57 (1): 21-30.

Zuska, J., 1965: Notes on the Palaearctic species of the genus *Nemopoda* Robineau-Desvoidy (Diptera, Sepsididae). – Acta ent. bohemoslovaca 62 (4): 308-313, 16 figs.

Zuska, J., 1967: Diptera from Nepal. Sepsidae. – J. Nat. Hist. 2 (1): 131-136, 2 tables.

Zuska, J., 1968: A revision of the genus *Australosepsis* (Diptera, Sepsidae). – Acta ent. bohemoslovaca 65 (6): 470-476, 9 figs.

Zuska, J., 1970a: A revision of the genus *Parapalaeosepsis* (Diptera, Sepsidae). – Acta ent. bohemoslovaca 67 (1): 50-61, 34 figs.

Zuska, J., 1970b: Zoogeographic aspects of the European fauna of the family Sepsidae (Diptera). – Polskie Pismo ent. 40 (3): 605-610, 1 map, 2 tables.

Zuska, J., 1971: Three new synonyms in the European Sepsidae (Diptera). – Acta ent. bohemoslovaca 68 (1): 58-59.

Zuska, J., 1972: Revision of Oriental and Australasian species of *Meroplius* and *Xenosepsis* (Diptera, Sepsidae). – Acta ent. bohemoslovaca 69 (1): 60-68, 20 figs.

Zuska, J., 1974: Asian *Themira* (Diptera: Sepsidae): descriptions of two new species and distributional notes. – Proc. ent. Soc. Wash. 76 (2): 190-197, 15 figs.

Zuska, J., 1977a: Re-description of the genus *Nemopoda* and a revision of the Ethiopian species (Diptera, Sepsidae). – Acta ent. bohemoslovaca 74 (1): 35-40, 8 figs.

Zuska, J., 1977b: Family Sepsidae. – Pp. 174-181 *in* Delfinado, M. D. & Hardy, D. E.: A catalog of the Diptera of the Oriental region. Volume III. Suborder Cyclorrhapha (excluding division Aschiza). – University Press of Hawaii, Honolulu, x + 854 pp.

Zuska, J., 1980: Family Sepsidae. – Pp. 591-596 *in* Crosskey, R. W.: Catalogue of the Diptera of the Afrotropical region. – British Museum (Natural History), London, 1437 pp., 1 map.

Zuska, J., 1989: 88. Family Sepsidae. – Pp. 572-574 *in* Evenhuis, N.L.: Catalog of the Diptera of the Australasian and Oceanian regions. – Special Publications of the Bernice Pauahi Bishop Museum 86: 1155 pp., 1 plate, 2 maps. Bishop Museum Press, Honolulu, & E.J. Brill, Leiden.

Zuska, J. & Barták, M., 1997: Sepsidae. – Pp. 74-75 *in* Chvála, M.: Check list of Diptera (Insecta) of the Czech and Slovak Republics. – 'Karolinum', Charles University Press, Prague, 130 pp.

Zuska, J. & Colless, D. H., 1984: Australian Sepsidae (Diptera). – J. Aust. ent. Soc. 23 (1): 59-67, 14 figs.

Zuska, J. & Laštovka, P., 1969: Species-composition of the dipterous fauna in various types of food-processing plants in Czechoslovakia. – Acta ent. bohemoslovaca 66 (4): 201-221, 4 tables.

Zuska, J. & Pont, A. C., 1984: Family Sepsidae. – Pp. 154-167 *in* Soós, Á. & Papp, L.: Catalogue of Palaearctic Diptera. Volume 9. Micropezidae - Agromyzidae. – Akadémiai Kiadó, Budapest, 460 pp.

Zvereva, E. L., 1982: Hen manure as a habitat for coprobiont flies. [Russian, with English summary.] – Ent. Obozr. 61 (3): 485-490, 3 tables.

Taxonomic index

Synonyms are in *italics*. The main entry for each taxon is in **bold**.

FAUNA ENTOMOLOGICA SCANDINAVICA

EDITED BY

N.P. KRISTENSEN and V. MICHELSEN

1. ROZKOŠNÝ, R. *The Stratiomyioidea (Diptera) of Fennoscandia and Denmark.* 1973. ISBN 87 87 49100 1
2. FIBIGER, M; KRISTENSEN, N.P. *The Sesiidae (Lepidoptera) of Fennoscandia and Denmark.* 1974. ISBN 87 87 49102 8
3. CHVÁLA, M. *The Tachydromiinae (Dipt. Empididae) of Fennoscandia and Denmark.* 1975. ISBN 87 87 49104 4
4. LOMHOLDT, O. *The Sphecidae (Hymenoptera) of Fennoscandia and Denmark.* 2nd ed. 1984. ISBN 87 87 49106 0
5. SPENCER, K.A. *The Agromyzidae (Diptera) of Fennoscandia and Denmark.* 1976. ISBN 87 87 49108 7
6. TRAUGOTT-OLSEN, E.; NIELSEN, E.S. *The Elachistidae (Lepidoptera) of Fennoscandia and Denmark.* 1977. ISBN 87 87 49114 1

7/1. OSSIANNILSSON, F. *The Auchenorrhyncha (Homoptera) of Fennoscandia and Denmark.* 1. Introduction, infraorder Fulgoromorpha. 1978. ISBN 87 87 49124 9

7/2. OSSIANNILSSON, F. *The Auchenorrhyncha (Homoptera) of Fennoscandia and Denmark.* 2. The families Cicadidae, Cercopidae, Membracidae and Cicadellidae (excl. Deltocephalinae). 1981. ISBN 87 87 49136 2

7/3. OSSIANNILSSON, F. *The Auchenorrhyncha (Homoptera) of Fennoscandia and Denmark.* 3. The family Cicadellidae: Deltocephalinae, catalogue, literature and index. 1983. 87 87 49113 3

8. COLLINGWOOD, C. *The Formicidae (Hymenoptera) of Fennoscandia and Denmark.* 1979. ISBN 87 87 49128 1
9. HEIE, O.E. *The Aphidoidea (Hemiptera) of Fennoscandia and Denmark.* I. General part. The families Mindaridae, Hormaphididae, Thelaxidae, Anoeciidae, and Pemphigidae. 1980. 87 87 49134 6
10. BILÝ, S. *The Buprestidae (Coleoptera) of Fennoscandia and Denmark.* 1982. ISBN 87 87 49142 7
11. HEIE, O.E. *The Aphidoidea (Hemiptera) of Fennoscandia and Denmark.* II. The family Drepanosiphidae. 1982. ISBN 87 87 49144 3
12. CHVÁLA, M. *The Empidoidea (Diptera) of Fennoscandia and Denmark.* II. General part. The families Hybotidae, Atelestidae and Microphoridae. 1983. ISBN 87 87 49107 9
13. BENGTSSON, B. *The Scythrididae (Lepidoptera) of Northern Europe.* 1984. ISBN 90 04 07312 4
14. ROZKOŠNÝ, R. *The Sciomyzidae (Diptera) of Fennoscandia and Denmark.* 1984. ISBN 90 04 07592 5

15/1. LINDROTH, C.H. *The Carabidae (Coleoptera) of Fennoscandia and Denmark.* 1. 1985. ISBN 90 04 07727 8

15/2. LINDROTH, C.H. *The Carabidae (Coleoptera) of Fennoscandia and Denmark.* 2. With an Appendix on the Family Rhysodidae. 1986. ISBN 90 04 08182 8

16. HOLST, K.TH. *The Saltatoria (Bush-crickets, Crickets and Grass-hoppers) of Northern Europe.* 1986. ISBN 90 04 07860 6
17. HEIE, O.E. *The Aphidoidea (Hemiptera) of Fennoscandia and Denmark.* III. Family Aphididae: Subfamily Pterocommatinae and Tribe Aphidini of Subfamily Aphidinae. 1986. ISBN 90 04 08088 0
18. HANSEN, M. *The Hydrophiloidea (Coleoptera) of Fennoscandia and Denmark.* 1987. ISBN 90 04 08183 6
19. PAPE, TH. *The Sarcophagidae (Diptera) of Fennoscandia and Denmark.* 1987. ISBN 90 04 08184 4
20. HOLMEN, M. *The aquatic Adephaga (Coleoptera) of Fennoscandia and Denmark.* I. Gyrinidae, Haliplidae, Hygrobiidae and Noteridae. 1987. ISBN 90 04 08185 2
21. LILLEHAMMER, A. *Stoneflies (Plecoptera) of Fennoscandia and Denmark.* 1988. ISBN 90 04 08695 1

22. BILÝ, S.; MEHL, O. *Longhorn beetles (Coleoptera, Cerambycidae) of Fennoscandia and Denmark.* 1989. ISBN 90 04 08697 8
23. JOHANSSON, R.; NIELSEN, E.S.; NIEUKERKEN, E.J. VAN; GUSTAFSSON, B. *The Nepticulidae and Opostegidae (Lepidoptera) of North and West Europe.* 1991. ISBN 90 04 08698 6
24. ROGNES, K. *Blowflies (Diptera, Calliphoridae) of Fennoscandia and Denmark.* 1991. ISBN 90 04 09304 4
25. HEIE, O.E. *The Aphidoidea (Hemiptera) of Fennoscandia and Denmark.* IV. Family Aphididae: Part 1 of Tribe Macrosiphini of Subfamily Aphidinae. 1991. ISBN 90 04 09514 4
26. OSSIANNILSSON, F. *The Psylloidea (Homoptera) of Fennoscandia and Denmark.* 1992. ISBN 90 04 09610 8
27. LUFF, M.L. *The Carabidae (Coleoptera) larvae of Fennoscandia and Denmark.* 1993. ISBN 90 04 09836 4
28. HEIE, O.E. *The Aphidoidea (Hemiptera) of Fennoscandia and Denmark.* V. Family Aphididae: Part 2 of tribe Macrosiphini of Subfamily Aphidinae. 1993. ISBN 90 04 09899 22
29. CHVÁLA, M. *The Empidoidea (Diptera) of Fennoscandia and Denmark.* III. Genus *Empis*. 1994. ISBN 90 04 09663 9
30. OLMI, M. *The Dryinidae and Embolemidae (Hymenoptera: Chrysidoidea) of Fennoscandia and Denmark.* 1994. ISBN 90 04 10224 8
31. HEIE, O.E. *The Aphidoidea (Hemiptera) of Fennoscandia and Denmark.* VI. Family Aphididae: Part 3 of tribe Macrosiphini of subfamily Aphidinae, and family Lachnidae. 1995. ISBN 90 04 10354 6
32. NILSSON, A.N.; HOLMEN, M. *The aquatic Adephaga (Coleoptera) of Fennoscandia and Denmark.* II. Dytiscidae. 1995. ISBN 90 04 10456 9
33. ANDERSEN, S. *The Siphonini (Diptera: Tachinidae) of Europe.* 1996. ISBN 90 04 10731 2
34. GØNGET, H. *The Brentidae (Coleoptera) of Northern Europe.* 1997. ISBN 90 04 10847 5
35. FJELLBERG, A. *The Collembola of Fennoscandia and Denmark.* Part 1: Poduromorpha. 1998. ISBN 90 04 11241 3
36. CHANDLER, P.J. *The Flat-footed Flies (Diptera: Opetiidae and Platypezidae) of Europe.* 2001. ISBN 90 04 12023 8
37. PONT, A.C.; MEIER, R., *The Sepsidae (Diptera) of Europe.* 2002. ISBN 90 04 12477 2